Micellization, Solubilization, and Microemulsions

Volume 2

Micellization, Solubilization, and Microemulsions

Volume 2

Edited by

K.L. Mittal

IBM Corporation
East Fishkill Facility
Hopewell Junction, New York

Plenum Press · New York and London

Library of Congress Cataloging in Publication Data

Symposium on Micellization, Solubilization, and Microemulsions, Albany, 1976.
  Micellization, solubilization, and microemulsions.
  Symposium held at the Seventh Northeast regional meeting of the American Chemical
Society.

  Includes indexes.
  1. Micelles—Congresses. 2. Solubilization—Congresses. 3. Emulsions—Congresses. I. Mittal,
K. L., 1945-   II. Title. III. Title: Microemulsions.
QD549.S976 1976   541'.3451   77-1126

ISBN-13: 978-1-4613-4159-8   e-ISBN-13: 978-1-4613-4157-4
DOI: 10.1007/978-1-4613-4157-4

Proceedings of the second half of the International Symposium on Micellization,
Solubilization, and Microemulsions held at the Seventh Northeast Regional Meeting of the
American Chemical Society in Albany, New York, August 8—11, 1976

© 1977 Plenum Press, New York
Softcover reprint of the hardcover 1st edition 1977
A Division of Plenum Publishing Corporation
227 West 17th Street, New York, N.Y. 10011
All rights reserved

PREFACE

This volume and its companion Volume 1 chronicle the proceedings of the International Symposium on Micellization, Solubilization, and Microemulsions held under the auspices of the Northeast Regional Meeting of the American Chemical Society in Albany, New York, August 8-11, 1976. The technical program consisted of 48 papers by 91 authors from twelve countries. The program was divided into six sessions, and Dr. Hartley delivered the Concluding Remarks. Subsequently, six more papers were contributed for the proceedings volumes with the result that these volumes contain 51 papers (three papers are not included for certain reasons) and are divided into seven parts. The first three parts are embodied in Volume 1 and the remaining four parts constitute Volume 2; each part is followed by a Discussion Section. Dr. Hartley's Concluding Remarks are included in both volumes.

When the idea of arranging a symposium on micelles was broached to me, I accepted it without an iota of hesitation. I had two options: either to make it a one- or two-sessions symposium, or bring together the global community actively carrying out research in this area. I opted for the latter as the subject was very timely and I knew that it would elicit very enthusiastic response. In order to broaden the scope of the symposium, I suggested that the theme of the symposium should be Micellization, Solubilization and Microemulsions.

Two salient features of this symposium should be mentioned: (i) a truly international symposium of this magnitude rarely occurs at a Regional Meeting, and (ii) I do not know whether there was ever a symposium of this quality, magnitude, and breadth of coverage on this topic.

Micelles are colloidal species in solution that are produced by aggregation of a relatively large number (from 20 up to thousands) of small amphiphilic molecules or ions. Micellar systems are usually described as association colloids. The fundamental characteristics of micelle-forming monomers is their amphiphilicity – the presence of both polar and nonpolar portions in the same molecule. The great variety of possible monomers produces micelles of widely differing surface composition and interior. The applications of amphiphilic substances ranges from A to Z (anesthesiology to zoology) and their micelle formation has important ramifications. Micelle formation occurs by cooperative association of monomers over a narrow concentration range known as the critical micellization concentration (c.m.c.).

Their special structural characteristics and their ability to solubilize otherwise insoluble substances render micelles both important and useful. Among other things, micelles act as good model systems and are excellent catalysts for a host of reactions. The many and varied applications of micelles and microemulsions are summarized in the opening chapter of these proceedings volumes.

The earlier research activity in the area of micelles was primarily carried out by colloid chemists but a glance at the literature on this topic attests that presently researchers from many disciplines are actively pursuing research in this area. The energy crisis has given an impetus for increased work on application of microemulsions and micellar solutions for tertiary oil recovery. Also, permanent storage of energy through light-driven redox reactions is shown to be feasible using micellar surfactant systems. It should be added that the availability of sophisticated instrumentation has been a boon in micellar research.

These proceedings volumes contain a comprehensive coverage of both theoretical and practical aspects of micellization, solubilization, and microemulsions. These volumes bring together the latest theoretical and experimental research activities being carried out by scientists from various disciplines, bring out clearly the interdisciplinary and multidisciplinary aspects and importance of the subject symposium. The topics covered include: history, applications, and prospects of micelles; thermodynamics and kinetics of micellization, application of fast reaction kinetics; theoretical developments in understanding monomer-micelle equilibria and stepwise aggregation; stepwise aggregation and the concept of c.m.c.; micelles of ionic and nonionic surfactants in aqueous and nonaqueous media; micelles as model systems; micelles and oil recovery; mixed micelles; application of spectroscopic techniques to understand mechanisms of reactions and interactions in micellar media; micellar catalysis of a variety of reactions; solubilization of polar and nonpolar substances; formation and structure of microemulsions and reactions in microemulsion media.

These volumes – a product of the efforts of more than one hundred scientists representing many countries – attest to the brisk research activity taking place in the realm of micelles and microemulsions, and all signals indicate that this tempo will be continued. As we probe more into this fascinating area of micelles and microemulsions, more new vistas will emerge.

Acknowledgments. First of all I am thankful to Dr. G. S. Hartley for his presence at the Symposium. His presentation, "Micelles: Retrospect and Prospect" was the highlight of the program. Thanks are due to the officials of the Northeast Regional Meeting for sponsoring the event, to IBM Corporation for permitting me to organize the symposium and to edit these volumes. Thanks are due to our secretary, Ms. Julie Hrib, for helping with the correspondence typing. I am thankful to my wife, Usha, for sacrificing many hours which rightfully belonged to her, and to my daughter, Anita, and son, Rajesh, for being very nice kids and letting their Daddy work at home. The reviewers should be thanked for their many valuable comments on the manuscripts. The enthusiasm and cooperation of the contributors, particularly those from overseas, is sincerely acknowledged. The Petroleum Research Fund of the American Chemical Society should be thanked for partial travel assistance to three speakers.

K. L. Mittal

IBM Corporation
East Fishkill Facility
Hopewell Junction, New York   12533

CONTENTS OF VOLUME 2

PART VII. GENERAL PAPERS

CONTENTS OF VOLUME 1

PART III.   MICELLES IN NONAQUEOUS MEDIA

Part IV

Reactions in Micelles and Micellar Catalysis

in Aqueous Media

THE KINETIC THEORY AND THE MECHANISMS OF MICELLAR EFFECTS ON CHEMICAL REACTIONS

K. Martinek, A. K. Yatsimirski, A. V. Levashov
and I. V. Berezin

Lomonosov State University of Moscow
U.S.S.R.

A few years ago we suggested, as an explanation of micellar effects on chemical reactions, a comprehensive kinetic theory which takes into consideration a partition of the reagents between the bulk and micellar "phases", the simultaneous course of the reaction in the two phases and the shift of the apparent ionization constant of one of the reagents under the action of the surface micelle charge. In terms of this theory, from "surfactant concentration versus overall rate" profiles one can obtain partition coefficients of reagents between the bulk and micellar phases and a true rate constant of the reaction going in a micellar medium. That the kinetic equations are true is confirmed by the fact that the partition coefficients obtained in this way are in conformity with the values obtained by other methods (gel filtration, solubilization, spectrophotometric titration, etc.).

INTRODUCTION

The number of papers dealing with the kinetics of chemical reactions in the presence of surfactants has greatly increased since 1960 (see exhaustive review[1]). However, it is only recently that a clear-cut physico-chemical concept has gradually been formed, in terms of which the role of different factors in the micellar effects (i.e. a heightening of the reagents' concentration in the micellar "phase", their mutual orientation in the surface layer of the micelle, the influence of the micellar medium, etc.) can be formulated.[2] It is natural that such a concept should be based on the present day knowledge about the physico-chemical properties of surfactant solutions (see excellent review[3], for example). On the other hand, the study of the mechanisms of the micellar effects in chemical reactions may furnish new information about micelles.

THE KINETIC THEORY

The most interesting events in the history of the development of the kinetic theory of micellar catalysis are discussed by Romsted.[3] At present the most successful is the pseudo-phase treatment of micellar effects on the reaction rates elaborated in Moscow State University.[2,4,5] The effect of surfactants on the kinetics of an nth order reaction was analyzed in Reference 6. Let us consider, by way of example, the second order reaction.[4-9]

The Second Order Reactions

Considering the kinetics of the chemical interaction A + B → products, we suppose that (i) the solution consists of two "phases" (i.e. a bulk phase and a micellar phase)[10] and that (ii) a definite distribution of the reagents between the two phases does exist:[11]

$$(A)_{\text{in bulk phase}} \underset{\longleftarrow}{\overset{P_A}{\rightleftharpoons}} (A)_{\text{in micelles}}$$

$$(B)_{\text{in bulk phase}} \underset{\longleftarrow}{\overset{P_B}{\rightleftharpoons}} (B)_{\text{in micelles}} \tag{1}$$

with the partition coefficients being expressed as

$$P_A = [A]_m/[A]_b$$

$$P_B = [B]_m/[B]_b \tag{2}$$

Here and below indices m and b refer to the micellar and the bulk phases, respectively. If the reaction proceeds in the two phases:

$$(A + B)_{\text{in bulk phase}} \xrightarrow{k_b} \text{products}$$

$$(A + B)_{\text{in micelles}} \xrightarrow{k_m} \text{products} \tag{3}$$

the overall reaction rate averaged in relation to the volume of the whole system, may be expressed via reaction rates in the micellar (v_m) and the bulk (v_b) phases in the following way:

$$v = k_{app} [A]_t [B]_t = v_m CV + v_b (1 - CV) =$$
$$= k_m [A]_m [B]_m CV + k_b [A]_b [B]_b (1 - CV) \tag{4}$$

where C is the total surfactant concentration from which CMC is subtracted, V is the molar volume of the surfactant. The relationship between the total concentrations of the reagents, $[A]_t$ and $[B]_t$, and their true concentrations in the respective phases is described by Equations (2) and the material balance equations:

$$[A]_t = [A]_m CV + [A]_b (1 - CV)$$
$$[B]_t = [B]_m CV + [B]_b (1 - CV) \tag{5}$$

Equation (4) is based on the assumption (iii) that the reagent does not affect the properties of the micelles and, what is most important, does not shift CMC. This means that the equation holds only at sufficiently low concentrations of the reacting substances-- in fact they should be considerably lower than that of the surfactant. These conditions not only minimize the effect of the reagents on the micelle formation but also ensure the correctness of relations (2), which seems to be true for dilute solutions only.

If one makes the assumption (iv) that the exchange of molecules between the phases occurs very rapidly, i.e. the chemical reaction (3) does not alter the partition equilibrium (1), then from Equation (4) the rate constant can be represented as follows:

$$k_{app} = \frac{k_m P_A P_B CV + k_b (1-CV)}{\{1 + (P -1)CV\}\{1 + (P -1)CV\}} \tag{6}$$

In the case of diluted surfactant solutions (assumption (v)), when the volume fraction of the micellar phase is small (CV << 1) and if both reagents strongly bind with the micelles, i.e. P_A and P_B >> 1 (assumption (vi)), Equation (6) may be simplified as

$$k_{app} = \frac{(k_m/V) K_A K_B C + k_b}{(1 + K_A C)(1 + K_B C)} \tag{7}$$

where the binding constants will be expressed in the following way:

$$K_A = (P_A - 1)V$$
$$K_B = (P_B - 1)V$$

(8)

We are dealing here with a simplest pseudo-phase model that presupposes even distribution of the reagents over the entire volume of the micelle. It goes without saying that in the general case this model may be made more comprehensive. Firstly, if the reagent is an ionic or polar compound, the micellar phase will be represented only by the surface layer of the micelle, which could be assumed to have a certain width and hence, volume. Equation (6) will then be valid but the micellar volume, V, will be "effective."[6,12] Secondly, the molecules of the same kind may be oriented in the micelle in a different way. Consequently, kinetic parameter k_m is actually an averaged one for all the sorption states of the reagent's molecule.

Equations (6) and (7) allow the character of the micellar effects to be analyzed both in relation to the reaction rate constants in the two phases and in relation to the binding constants of the reagents. It is obvious that if one knows the values of P_A and P_B (from an independent experiment), and makes reasonable assumptions about the k_m/k_b ratio, one may in principle predict the character of the dependence of k_{app} on the concentration of the surfactant. Figure 1 shows some peculiar cases of relationship (6).

However, the main advantage of the kinetic theory put forward by us lies elsewhere: (1) analyzing the experimentally obtained profile of "k_{app} vs. C", one may find both the binding constants and the true rate constant of the reaction proceeding in the micellar phase (see below); (2) separate determination of the values characterizing the binding ability and the reactivity of the reagents allows one to judge about the mechanism of the micellar effect.

Let us analyze in more detail a system where in the presence of micelles a considerable acceleration of the reaction occurs, i.e. the rate of the reaction in the bulk phase is neglected (in terms of Equation (7) $k_m K_A K_B C/V \gg k_b$). In this particular case the "k_{app} vs. C" profile has a maximum (see curves in Figure 1) and the maximal acceleration observed at the optimal concentration of the surfactant, $C_{opt} = 1/\sqrt{K_A K_B}$, is equal to

$$\left(\frac{k_{app}}{k_b}\right)_{max} = \frac{k_m}{k_b} \cdot \frac{K_A K_B}{V(\sqrt{K_A} + \sqrt{K_B})^2}$$

(9)

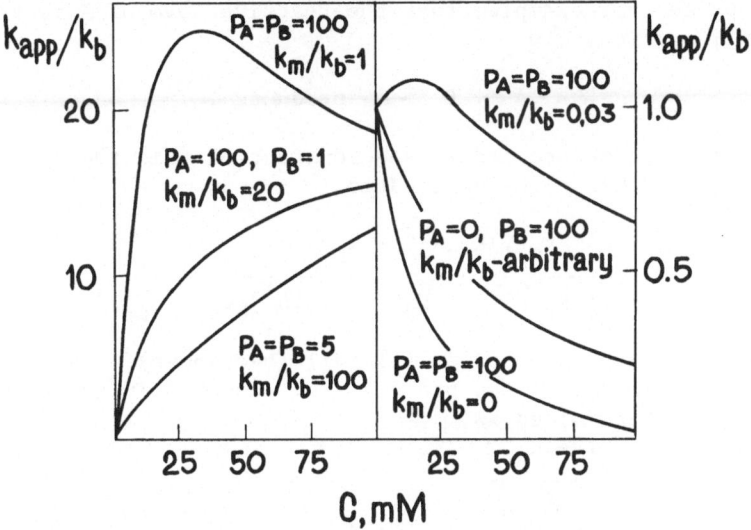

Figure 1.   The theoretical plots of apparent second order rate constant (k_{app}) versus surfactant concentration (C), drawn in accordance with Equation (6) at various values of P_A, P_B and k_m/k_b (with
the use of V = 0.3 1/mole[13]).

The ratio of k_m/k_b characterizes the change in the reactivity when
the reagents are transferred from water to the micelle and, consequently, reflects the specific effect of the micelles on the transition state of the reaction (in particular, the value of k_m depends
on the micellar medium and on the orientation of the reagent's
molecules sorbed by the micelle).   The second factor on the right
hand side of Equation (9) (which includes the binding constants) is
indicative of the "trivial" effect of the acceleration of the reaction due to reagents being concentrated in the micellar phase.

     The second order reaction with ionogenic reagent.   The reaction
may be additionally accelerated (or decelerated) because of the shift
in the apparent value of pK_a of the ionogenic reagent in the presence
of micelles.[5,7]   Considering the kinetics of the chemical interaction
of reagents A and B, one of which has an ionogenic group:

$$A \rightleftharpoons A^- + H^+ \qquad\qquad (10)$$

let us suppose, for example, that the reactive form of A is represented by its anion, i.e.

$$A^- + B \longrightarrow \text{products} . \tag{11}$$

Then the apparent second order rate constant of this reaction will be described by the following equation:

$$k_{app} = \frac{(k_m/V)K_{A^-}K_B C + k_b}{(1 + K_{A^-}C)(1 + K_B C)(1 + [H^+]_b/K_{a,app})} \tag{12}$$

which is valid with assumptions (i)-(vi) (see above). The apparent dissociation constant of the process (10) observed in the presence of micelles is determined as

$$K_{a,app} = K_a \frac{1 + K_{A^-}C}{1 + K_A C} \tag{13}$$

Here K_a corresponds to the value measured in the absence of micelles, and the binding constant for the anion looks like

$$K_{A^-} = (P_{A^-} - 1)V \tag{14}$$

where

$$P_{A^-} = [A^-]_m/[A^-]_b .$$

Let us analyze Equation (12) in the conditions when in the presence of micelles there occurs a considerable acceleration of the reaction (i.e., $k_m K_{A^-} K_B C/V \gg k_b$). At $[H^+]_b \gg K_a$ (when the reactive ionic form is present in a small amount) the maximal acceleration observed at $C_{opt} = 1/\sqrt{K_A K_B}$ is equal to

$$\left(\frac{k_{app}}{k_{app, C=0}}\right)_{max} = \frac{k_m}{k_b} \cdot \frac{K_A K_B}{V(\sqrt{K_A} + \sqrt{K_B})^2} \cdot \frac{K_{A^-}}{K_A} \tag{15}$$

The first two factors in the right hand part of the Equation (15) are the same as in Equation (9) and the last factor (equal to K_{A^-}/K_A) is actually the extreme value of $K_{a,app}$ shift, as follows from Equation (13) (with $C \to \infty$). If the anion binds with micelles better than the electroneutral form of the reagent (i.e. $K_{A^-}/K_A > 1$, which is the case with cationic micelles), the shift of $pK_{a,app}$ will lead to the acceleration of the reaction, and vice versa.

As is obvious, the effect of the $pK_{a,app}$ shift is, in fact, also due to the concentration factor because it also reflects the preferential heightening of the concentration of one ionic form over the other in the micellar phase.

Second order reactions with one of the reagents being a micelle former. In terms of the pseudo-phase treatment, the micellar effect on the kinetics of the reaction involving a micelle former was also discussed.[14,15]

APPLICATION OF THEORETICAL EQUATIONS TO EXPERIMENTAL DATA

Estimation of the Binding Constants of the Reagents Basing on the Kinetic Data

The analysis of the dependence of the first order reaction rate on the surfactant concentration is simple enough.[2] Therefore we analyze here in some detail the second order reactions.[2,7,8]

To be able to find from the experimental data (obtained as the "k_{app} vs. C" profile) the binding constants, K_A and K_B, and also the value of k_m/V (proportional to the true rate constant in the micellar phase), Equation (7) is transformed as follows:

$$\frac{C}{k_{app} - k_b} = \alpha + \beta\, C\, \frac{k_{app}}{k_{app} - k_b} + \gamma C^2 \frac{k_{app}}{k_{app} - k_b} \tag{16}$$

where

$$\alpha = V/k_m\, K_A\, K_B$$
$$\beta = \alpha(K_A + K_B) \tag{17}$$
$$\gamma = \alpha\, K_A\, K_B$$

The value of α can be found as the ordinate intercept of the curve plotted in coordinates of Equation (16), $C/(k_{app} - k_b)$ vs. $C\, k_{app}/(k_{app} - k_b)$. Then the results of the experiment may be analyzed in terms of linear Equation (18), which follows from (16):

$$\frac{C/k_{app} - \alpha}{C} \left(1 - \frac{k_b}{k_{app}}\right) = \beta + \gamma C \tag{18}$$

Evidently β and γ may be found as the ordinate intercept and the slope of the corresponding straight line plotted in coordinates $(C/k_{app} - \alpha)(1 - k_b/k_{app})/C$ vs. C. By way of example in Figure 2 is given a graphical analysis of the kinetic data obtained from the system where in the presence of micelles a marked acceleration of the reaction takes place (i.e. $k_{app} \gg k_b$). Finally, knowing

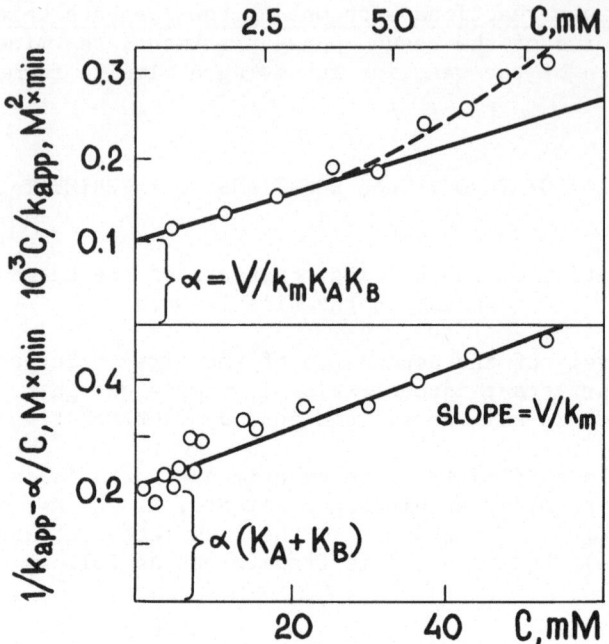

Figure 2. Graphical determination of binding and rate constants in accordance with Equations (16) and (18) (from the experimental data[8]).

α, β and γ one may, with the help of Equations (17), calculate the values of K_A, K_B and k_m/V.

Comparison of the Values of the Binding Constants Found by Different Methods

That the kinetic equations proposed are correct is supported by the fact that the binding constants found by different methods agree satisfactorily. The effect of cationic (CTAB) and anionic (SDS) surfactants in substitution reactions of benzimidazole (to be more exact, acylation with p-nitrophenyl heptanoate[7] and arylation with 2,4-dinitrofluorobenzene[8,16]) has been studied in the most comprehensive fashion. The values of the binding constants of these reagents with micelles are presented in Table I.

Table I. Binding Constants (1/mole) Characterizing the Incorporation of the Reagents into the CTAB and SDS micelles.[16]

	REAGENT		
Surfactant	Benzimidazole	p-Nitrophenyl heptanoate	2,4-Dinitrofluorobenzene
CTAB	33^a 37^b 40^c	3600^d 3800^c	27^d 21^c
SDS	28^a 30^b 30^c	1500^d 2000^c	18^e 23^c

[a]From the dependence of the difference spectrum of benzimidazole on the concentration of the surfactant.

[b]From the dependence of the apparent value of pK_a of benzimidazole on the concentration of the surfactant.

[c]From the kinetic data analyzed in terms of Equation (7).

[d]From the dependence of the solubility of the reagent on the concentration of the surfactant.

[e]Found by gel filtration.[11]

Determination of the Reaction Rate Constant in the Micellar Phase

For a first order reaction, the value of the true rate constant in the micellar phase may be easily found experimentally. In this case the dependence of the apparent rate constant, k_{app}, upon the concentration of the surfactant tends to achieve the limiting value, k_m, in accordance with the hyperbolic law.[6]

With high order reaction, this value is more difficult to find. For example, to determine the true rate constant of the bimolecular reaction in the micellar phase from the effective value of k_m/V, measured experimentally (in terms of Equations (16)-(18), see Figure 2), one should know the molar volume of the surfactant. (This independent parameter is also indispensable for calculating partition coefficients, P_A and P_B, from the experimentally obtained binding constants, K_A and K_B; see Equation (8).) This is rather difficult as the molar volume, which is a part of Equations (4)-(9), (12) and (14) is in fact effective and indicative of only a part of the surfactant molecule that is localized in the region of the micelle in which the reagents are actually solubilized and a chemical reaction occurs. In other words, effective parameter V is to be found with the help of some other independent method. This, however, is hardly possible at present, as the localization of molecules and ions sorbed by the micelle is not very well known. Nevertheless, to make a rough estimation of the value of V, one should bear in mind that all the reactions studied involve polar molecules sorbed, in all probability, in the charged and hydrated surface layer of the micelle. Therefore one may assume[17] that the layer in which reactions usually occur on micelles include the Stern layer and 3-4 methylene groups of the surfactant molecule. In the case of sodium dodecyl sulphate, the volume of such a layer is equal to about 3/4 of the total volume of the micelle.[2] For other surfactants, too, the difference between the total and the effective volume does not seem to be great. This means that the total molar volume of the surfactant may be assumed as being its effective volume V. This value is easy to calculate[2,11] basing on the molecular weight of the surfactant and the density of the micelles, which, by the way, is largely the same for different surfactants and equals 0.9-1.1 g/cm^3; see Reference 18.

SPECIFICITIES OF THE KINETICS OF THE REACTIONS INVOLVING AN IONIC REAGENT

There are no objections of the principal nature to the applicability of the above described kinetic theory to the analysis of micellar effects in the reactions involving an ionic reagent. Ions, like organic molecules, become distributed between the micellar and bulk (aqueous) phases; consequently, all the micellar effects may be explained in terms of the above described factors of the heightening of the reagents' concentration in the micellar phase and change in the inherent reactivity of the substances. However, ionmolecular reactions in the presence of surfactants possess certain peculiarities associated with the fact that distribution of ions, unlike that of nonpolar substances, very strongly depends on the state of the surface layer of the micelles. This manifests itself, first and foremost, in the fact that the reactions involving an ionic reagent are inhibited by added salt[19,20] or by an excessive concentration of the surfactant.[6] The reason for this is that an increase in the counterion concentration leads to a decrease in the sorption ability of micelles towards an oppositely charged reagent.[7,16,21] As a result, application of kinetic equations of type (6) or (12) for experimental data analysis (in the form of k_{app} vs. C profiles) is hindered by the fact that the value of the binding constant for the ionic reagent is not constant, but is a function of the concentration of the surfactant (or of that of the added salt). Let us review briefly the quantitative aspects of the problem.

The partition coefficient of ion I may be represented in the following form[22] (for univalent ions):

$$P_I = P_{I,o} \, e^{-\psi e/kT} \tag{19}$$

where ψ is the surface potential of the micelle, $P_{I,o}$ is the factor reflecting the contribution of nonelectrostatic (for example, hydrophobic) forces to ion-micelle interactions. It is obvious that the effects analyzed are associated with the changes in the value of ψ which determines the contribution of the electrostatic ion-micelle interaction. With high values of $\psi \gg kT/e$ (usually dealt with in colloid chemistry of surfactants) the following approximation is valid:[22]

$$\psi = \frac{kT}{e} \ln \frac{2000\pi\sigma^2}{DNkTC_i} = \frac{kT}{e} \ln \frac{A}{C_i} \tag{20}$$

where σ is the surface density of the micelle charge, D is the dielectric permeability, C_i is the total concentration of uni-univalent electrolyte in the aqueous phase equal to

$$C_i = CMC + C_s \tag{21}$$

where C_s is the concentration of the added salt. Having substituted (20) into (19), we shall have

$$P_I = P_{I,o} \, A/C_i \, . \tag{22}$$

Equation (22) predicts a hyperbolic decrease in coefficient P_I and hence in the value of k_{app} (which is proportional to it) as the concentration of an added electrolyte grows. This agrees qualitatively with experimental data,[19,20] for example. Unfortunately, Equation (20) and, consequently, (22), is inferred from a purely electrostatic model, which neglects the nature of the ions, i.e. their ability to get involved into additional nonelectrostatic interactions. Therefore Equation (22) does not reflect the strong dependence of the inhibitory effect on the nature of the salt.[7] The second drawback of this equation is that it does not give the dependence of P_I on the concentration of the surfactant; this is also due to the fact that the model based on Equation (20) is too simple. However, more realistic models would be mathematically

cumbersome.[23]   Therefore the dependence of P_I on the surfactant
concentration is usually accounted for empirically.   Two such
approaches were made use of in the study of acylation of benzimidazole anion by p-nitrophenyl carboxylates, which involves cationic
micelles.

The first approach.   In the presence of a sufficiently high
concentration of an added salt (for example, more than 0.1 M KNO_3),
the change in the concentration of the surfactant hardly produces
any effect on the binding of micelles with an ionic reagent.   In
other words, in the reaction of type (11) the value of K_{A^-} is
constant; therefore analysis of the experimental K_{app} vs. C profiles may be made with the help of Equation (12) only, without additional assumptions.   Moreover, Equation (12) may be simplified to
(7), if to take into account that, with a high concentration of
counterions, the electrostatic surface potential of the micelle is
fully "quenched" and hence an anion of benzimidazole binds with a
cationic micelle not stronger than an electroneutral form, see
Reference 7.   As a result, the experimental data could be analyzed
with the help of the method shown in Figure 2.

The second approach.   In a general case, the dependence of
the binding constant of the ionic reagent on the concentration of
the surfactant may be found in an independent experiment.   For
example, in Reference 16 use was made for this purpose of the
dependence of the apparent value of pK_a of an ionogenic reagent
(benzimidazole) on the concentration of CTAB.   The values of
$pK_{a,app}$ were determined by spectrophotometric titration (Figure
3A) and the $pK_{a,app}$ vs. C profile found (Figure 3B) was analyzed
in terms of Equation (13).   The effective value of the binding
constant calculated in such a way for anionic reagent (K_{A^-}) is
represented as a function of the concentration of the surfactant
in Figure 3C.

The third approach employed in Reference 12 to analyze acid
hydrolysis of methyl orthobenzoate in the presence of anionic
micelles was essentially that of using the values of ψ obtained
with different concentrations of SDS in an independent experiment,
and Equation (19).

A semiempirical approach has been recently put forward by
Romsted,[3] who proceeded from the concept (based on experimental
data of Mukerjee, Mysels, Stigter and many other scientists) that
the total concentration of small hydrophilic counterions in the
surface layer of micelles, equal β/V, is constant (β is the degree
of binding of counterions that is close to unity):

$$[X]_m + [I]_m \;=\; \beta/V \tag{23}$$

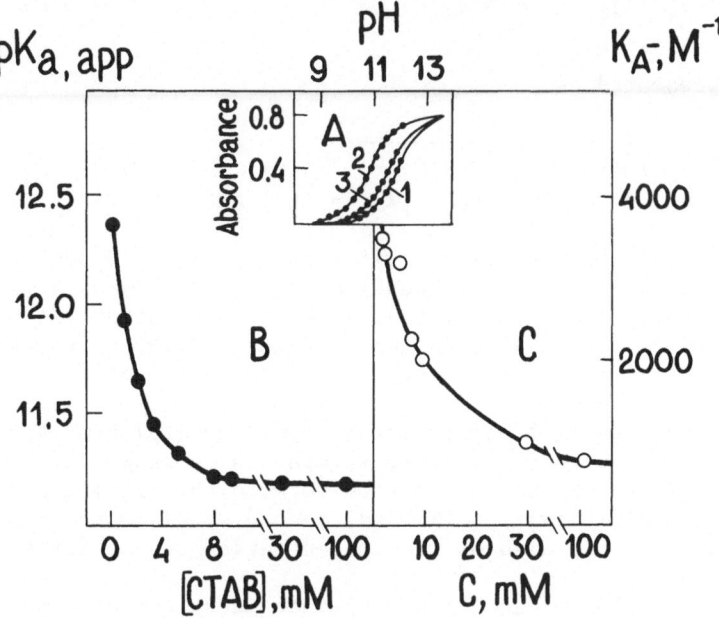

Figure 3. (A) Determination of the apparent value of pK_a of benzimidazole by spectrophotometric titration (283 nm): 1 - no
surfactant, 2 - 10 mM CTAB, 3 - 10 mM CTAB and 0.12 M KNO$_3$.
(B) Dependence of the apparent value of pK_a of benzimidazole on
the concentration of CTAB. (C) Dependence of constant K_{A^-}, characterizing the binding of benzimidazole anion with the micelle, on
the concentration of CTAB.[16]

In accordance with this, the distribution of two species of ions,
reactive (I) and nonreactive (X), between the bulk and the micellar
phases has been described by Romsted as follows:

$$K = [I]_b [X]_m / [I]_m [X]_b \qquad (24)$$

We suggest that equation (24) be analyzed on the condition
that $[I]_t \ll [X]_t$, which agrees with assumption (iii) (see above).
This allows Equation (23) to be simplified as $[X]_m \simeq \beta/V$. In this
case the partition coefficient of the ionic reagent, P_I, equal to

$$P_I = \frac{[I]_m}{[I]_b} = \frac{1}{K}\frac{[X]_m}{[X]_b} \tag{25}$$

may be presented as

$$P_I = \frac{\beta}{KV(C_s + CMC + (1-\beta)C)} \tag{26}$$

if to take into account that $[X]_t = C_s + CMC + C = [X]_m CV + [X]_b$.

Our Equation (26) agrees with the hyperbolic dependence P_I vs. C predicted by theoretical Equation (22). This is associated with the fact that Equation (25) rearranged from Equation (24) suggested by Romsted is not an alternative to Equation (19) as the ratio of $[X]_m/[X]_b$ in its turn is determined by the value of ψ.

The advantage of semiempirical equation (26) over theoretical equation (22) is that, firstly, it contains a parameter reflecting the specificity of the interaction between the micelle and a non-reactive (inhibitory) ion (K) and, secondly, gives the dependence not only on C_s but also on the concentration of the surfactant.

Substitution of Equation (26) into Equations of type (6) or (12) or (13) allows one to obtain a theoretical dependence of the value of the apparent rate (k_{app}) or equilibrium ($K_{a,app}$) constants on the concentration of a surfactant just for the reaction involving an ionic reagent.

THE MECHANISMS OF THE MICELLAR EFFECTS ON THE RATE AND THE EQUILIBRIUM OF CHEMICAL REACTIONS

The factors responsible for the micellar effects on chemical reactions may be divided in two types according to their physico-chemical nature: (1) the change in the reactivity of the substances transferred from water (or, in a general case, from the bulk phase) to the micellar phase; and (2) a heightening in the reagents' concentration in the micellar phase; see Equation (9). The effects of the first type are characterized by the rate constant ratio, k_m/k_b, and may be due to the influence of the microenvironment (including the electrostatic interaction of the transition state of the reaction with the surface charge of the micelle) and to the mutual orientation of the reagents in the micelle. The contribution of the concentrating of the reagents to the micellar effects is largely determined by the effectiveness of the ionic and hydrophobic interactions between the molecules of the reagents and the micelles. The effect of the apparent pK_a shift of the ionogenic reagent belongs actually to the second type, as it reflects preferential concentrating of one ionic form over the other; see Equation (15). Let us analyze these factors one by one.

   The true reactivity in the micellar phase.   The study of the
first order reactions furnishes most reliable information about the
value of the reaction rate constants in the micellar phase.   Here
k_m may be found without making any assumptions about the molar
volume of the surfactant (see above).   Most interesting data on this
matter are discussed in Reference 2.   Let us consider the second
order reactions.

   The micellar medium produces a strong effect on the rate of
acylation of imidazole derivatives by p-nitrophenyl heptanoate.[24,25]
The experimental results are presented in terms of the Brønsted
relationship (see Figure 4).   For electroneutral nucleophiles, the
true reaction rate constant in the micelles is by two orders of
magnitude lower than in water; this seems to be due to the fact that
the low dielectric permeability and the weak solvating ability of
the micellar medium produce an unfavorable effect on the formation
of the polar transition state of the reaction (resembling in structure a tetrahedral complex).   On the contrary, the micellar medium

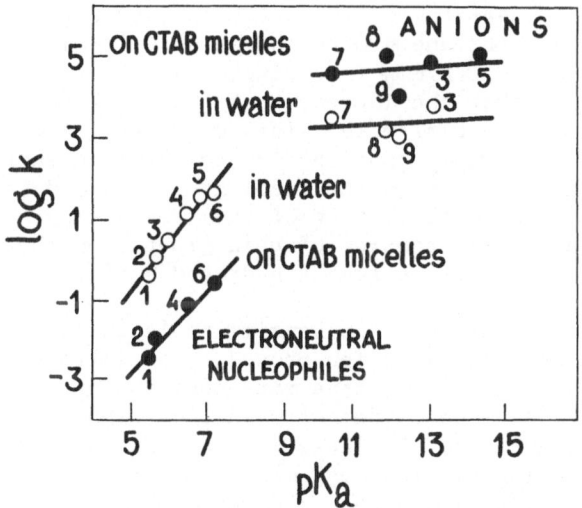

Figure 4.   True second order rate constants (1/mole min^{-1}) for
acylation of a series of imidazole derivatives by p-nitrophenyl-
heptanoate versus pK$_a$ values for the nucleophiles.[24,25]   Compounds:
(1) N-methylbenzimidazole, (2) N-phenylimidazole, (3) 4(5)-phenyl-
imidazole, (4) N-benzylimidazole, (5) N-benzoyl-L-histidine, (6)
N-heptylimidazole, (7) 5(6)-nitrobenzimidazole, (8) 4(5)-bromo-
imidazole, (9) benzimidazole.

produces a favorable effect on the reaction of imidazole anions.
On their being transferred from an aqueous to a micellar medium, they
increase their reactivity by more than by one order of magnitude
(see Figure 4). This evidently happens because, as a result of
sorption, the anion on the micelle becomes (at least in part) desol-
vated.

The orientation of the reagents' molecules sorbed on the
micelle. It was found for acylation of m-bromo-benzaldoxime by p-
nitrophenyl carboxylates[5] that the second order rate constant in
the micellar phase (k_m) is approximately equal to that for water
(k_b) but only in the case of p-nitrophenyl acetate. Increase in
the length of the aliphatic hydrocarbon chain in the acyl moiety
of the ester causes a decrease in the k_m, whereas k_b remains un-
changed. This seems to be due to the fact that, on being sorbed
on the micelle, the molecules of the reagents localize in different
parts of the micelle, i.e. the anion of the oxime - on the surface
of the micelle and the ester group - inside it. Such localization
of the reagents should prevent them from interacting. In other
words, formation of the transition state of the reaction involves
a thermodynamically unfavorable transfer of the ester molecule
(or its acyl fragment) from the poorly hydrated depth of the micelle
to its surface. And indeed a linear correlation was found to
exist between the logarithm of the rate constant ratio, k_m/k_b, and
the number of the methylene groups in the side fragment of the ester
molecule, $H(CH_2)_n CO(O)C_6H_4NO_2$ (see Figure 5). From the slope of
curve 1 it follows that the increment of the free energy of activation is equal to 0.4 kcal/mole. This means that each methylene
group makes an unfavorable contribution to the free energy of
activation of the reaction in the micellar phase, and the value of
this contribution is equal to that of the free energy of the thermodynamically unfavorable transfer of this group from the hydrophobic
core of the micelle to its strongly hydrated surface layer.

In the case of acylation of benzimidazole anion by p-nitrophenyl carboxylates the situation proved to be different. In the
first place, the value of the true rate constant of the reaction
in CTAB micelles is by one order of magnitude higher than in the
aqueous phase (Figure 4). Besides, increase in the hydrophobicity
of hydrocarbon residue in the acyl moiety of the ester molecule does
not affect the ratio between the true rate constants of acylation
in the micellar and aqueous phases (Figure 5). These results were
accounted for by the fact that the reaction involving benzimidazole
anion proceeds not in the surface layer but in the more hydrophobic
region of the micelle, probably, in the site of localization of the
acyl moiety of the ester.[25]

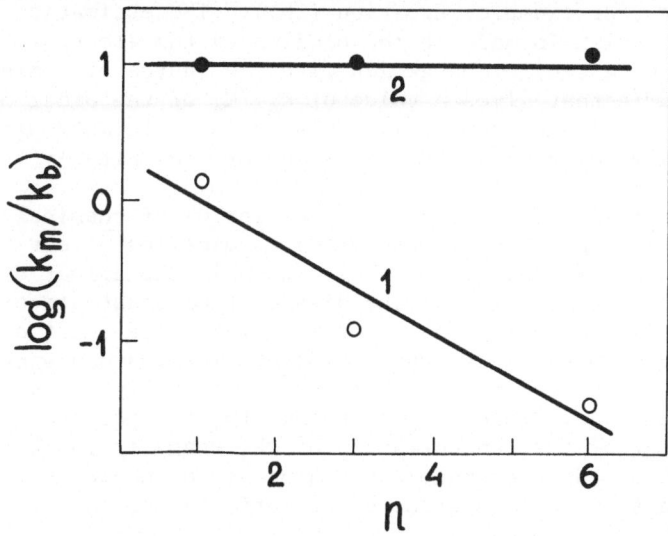

Figure 5. Logarithmic ratios of the second order rate constants
for acylation of m-bromobenzaldoxime (curve 1) and benzimidazole
(curve 2) anions by p-nitrophenylcarboxylates in micelles (k_m) and
in water (k_b) as a function of the number (n) of methylene groups
in a molecule of the acylating agent, $H(CH_2)_n CO(O)C_6H_4NO_2$. For
details see References 5, 7 and 25.

A heightening of the reagents' concentration in the micellar
phase. The binding of the reagents' molecules with micelles (and
hence a heightening of their concentration in the micellar phase)
is mostly due to ionic (electrostatic) and hydrophobic interactions.
With binding constants of the order of 10 to 10^3 1/mole (see Reference 2 and the table, for example) the contribution of the concentration factor to the second order reaction rate should give a 10-
to 10^3-fold acceleration; this follows from theoretical Equation
(9). For higher order reactions, the effect of a concentration
heightening may be still greater.[6] It is noteworthy that for many
chemical reactions, the only source of micellar catalysis is
heightening of the reagents' concentration in micelles.[2,5,6,8,9]

The shift of the apparent pK_a value of an ionogenic reagent
under the action of the surface charge of the micelle also makes a
considerable contribution to the kinetic micellar effects. The
extreme values of this effect are determined by the ratio of the
binding constants of the ionic and electroneutral forms of the

reagent (see, for example, Equation (15)). The estimation of the contribution which is made to the binding by the electrostatic interaction of a single charge ion with the charge of a micelle of ionogenic surfactant gives a value of K_A^-/K_A of the order of 10 to 100 (see review in Reference 2). This is in conformity with the experimentally observed pK_a shift by one or two units.[1,26]

The effect of micelles on the equilibrium of chemical reactions. In addition to producing kinetic effects, micelles affect the equilibrium of chemical reactions. Well known is the effect of ionogenic surfactants micelles on the dissociation constants of acids and bases[26] (see Equation (13), for example). The shift of the non-ionic equilibrium under the action of micelles was studied with the help of the reaction of benzaldehyde and aniline, which yields benzylideneaniline.[6] One of the reasons for the equilibrium shift is heightening of the concentration of the reagents in the micelles. This means that the phenomenon is analogous to acceleration of chemical reactions in the presence of surfactants.

In the light of the above facts, the term "micellar catalysis" may be regarded as being somewhat incorrect, for a "true" catalyst, as follows from the definition of Reference 27, must not shift the equilibrium to a considerable degree (this discrepancy was first noticed in References 2 and 6). However, this term, much used in the literature, is very handy indeed, because, firstly, the surfactant is not utilized in the reaction, and secondly, it produces a marked effect on the reaction rate if added in low concentrations (0.001 - 0.01 M). The total contribution of various mechanisms (heightening of the reagents' concentration in the micelles, apparent pK_a shift, effect of the micellar microenvironment, etc.) sometimes results in a 10^5 and more acceleration of chemical reactions.[7,15]

REFERENCES

1. J. H. Fendler and E. J. Fendler, "Catalysis in Micellar and Macromolecular Systems", Academic Press, New York, 1975.
2. I. V. Berezin, K. Martinek, and A. K. Yatsimirski, Russ. Chem. Revs. (Usp. Khim.) 42, 787 (1973).
3. L. Romsted, Ph.D. Thesis, Indiana University, Bloomington, 1975.
4. I. V. Berezin, K. Martinek, and A. K. Yatsimirski, Dokl. Akad. Nauk SSSR 194, 840 (1970).
5. A. K. Yatsimirski, K. Martinek, and I. V. Berezin, Tetrahedron 27, 2855 (1971).
6. K. Martinek, A. K. Yatsimirski, A. P. Osipov, and I. V. Berezin, Tetrahedron 29, 963 (1973).
7. K. Martinek, A. P. Osipov, A. K. Yatsimirski, and I. V. Berezin, Tetrahedron 31, 709 (1975).

8. A. K. Yatsimirski, Z. A. Streltsova, K. Martinek, and I. V. Berezin, Kinetics and Catalysis (Russ.), <u>15</u>, 354 (1974).
9. A. P. Osipov, K. Martinek, A. K. Yatsimirski, and I. V. Berezin, Izv. Akad. Nauk SSSR (ser. khim.) No. 9, 1984 (1974).
10. K. Shinoda in "Proc. IVth Intern. Cong. Sur. Act. Sub.", Vol. 2, p. 527, Gordon and Breach, New York, 1967.
11. D. G. Herries, W. Bishop, and F. M. Richards, J. Phys. Chem. <u>68</u>, 1842 (1964).
12. K. Shirahama, Bull. Chem. Soc. Japan <u>48</u>, 2673 (1975).
13. J. M. Corkill, J. F. Goodman, and T. Walker, Trans. Faraday Soc. <u>63</u>, 768 (1967).
14. K. Martinek, A. V. Levashov, and I. V. Berezin, Tetrahedron Letters No. 15, 1275 (1975).
15. A. V. Levashov, K. Martinek, and I. V. Berezin, Bioorg. Khim. (Russ.) <u>2</u>, 98 (1976).
16. A. K. Yatsimirski, A. P. Osipov, K. Martinek, and I. V. Berezin, Kolloid. Zhurn. (Russ.) <u>37</u>, 526 (1975).
17. E. H. Cordes and R. B. Dunlap, Accounts Chem. Res. <u>2</u>, 329 (1969).
18. P. Mukerjee, J. Phys. Chem. <u>66</u>, 1733 (1962).
19. L. R. Romsted and E. H. Cordes, J. Amer. Chem. Soc. <u>90</u>, 4404 (1968).
20. C. A. Bunton and L. R. Robinson, J. Org. Chem. <u>34</u>, 773 (1969).
21. W. K. Matthews, J. W. Larsen, and M. J. Pikal, Tetrahedron Letters 513 (1972).
22. K. Shinoda, T. Nakagawa, B. I. Tamamushi, and T. Isemura, "Colloidal Surfactants", Academic Press, New York, 1963.
23. H. Morawetz, "Macromolecules in Solution", Interscience Pub., New York, 1963.
24. K. Martinek, A. P. Osipov, A. K. Yatsimirski, and I. V. Berezin, Tetrahedron Letters, No. 15, 1729 (1975).
25. K. Martinek, A. P. Osipov, A. K. Yatsimirski, and I. V. Berezin, Bioorg. Khim. (Russ.) <u>1</u>, 469 (1975).
26. P. Mukerjee and K. Banerjee, J. Phys. Chem. <u>68</u>, 3567 (1964).
27. P. G. Ashmore, "Catalysis and Inhibition of Chemical Reaction", Butterworths, London, 1963.

APPENDIX

Romsted[3] analyzes Equations (23) and (24), in combination with the material balance equations

$$[I]_t = [I]_m CV + [I]_b (1 - CV) \quad \text{and}$$

$$[X]_t = [X]_m CV + [X]_b (1 - CV) ,$$

with two simplifications: namely,

$$[I]_t \gg [I]_m CV \tag{27}$$

and $K \leq 1$. This means that he deals only with the systems where all the ionic reagent is in the aqueous phase. This may be true only in two cases, i.e., firstly, with a high concentration of an added electrolyte when almost all the molecules of the ionic reagent have been forced out of the micelle by nonreactive counterions. This case is in conformity with Approach 1 (see text), i.e. change in the concentration of the surfactant does not affect the binding of the reactive ion. Hence the kinetic data (obtained for the constant concentration of an added salt) may be analyzed directly in terms of Equation (6) or (7).

The second case is when the concentration of the ionic reagent is high in comparison with that of the surfactant, when the quantity of the bound reagent is low compared to the free one, even in the absence of the added electrolyte. Then, however, the value of $[I]_m$ inevitably becomes so high that practically the whole of the micellar surface is occupied by the reactive ions. Indeed, having substituted $[I]_t$ from Equation [A-17] (given in Reference 3 and described as $[I]_m = \beta[I]_t/V([I]_t + [X]_t K)$) into inequality $[I]_t \gtrsim 10 \, [I]_m CV$, which follows from (27), we have

$$[I]_m \geq \frac{1}{V} (\beta - 0.1 \frac{K[X]_t}{C}) \tag{28}$$

or, in the first approximation, $[I]_m \geq \beta/V$, if to take into consideration that $K \leq 1$ and $[X]_t$ is of the order of C (in the absence of an added salt). As β/V is the total concentration of all counterions in the surface layer, inequality (28) means that in Romsted's conditions (27) the micelle is fully saturated with the reagent's ions.

Consequently, in these conditions, Romsted neglects the requirement for "diluted solution of the reagent in the micellar phase" (see assumption (iii) in the text) and it should therefore be expected that the partition coefficient of the second (nonionic) reagent and the rate constant in the micellar phase (k_m) must depend on the concentration of the reactive ion (as is the case in the "real solution", where the coefficient of activity is to be taken into account).

A GENERAL KINETIC THEORY OF RATE ENHANCEMENTS FOR REACTIONS BETWEEN ORGANIC SUBSTRATES AND HYDROPHILIC IONS IN MICELLAR SYSTEMS

Laurence S. Romsted

Department of Chemistry, University of California

Santa Barbara, California 93106

Kinetic equations for first, second, and third order reactions between organic molecules and hydrophilic ions in micellar solutions are derived. The approach depends upon the validity of two new assumptions. First, in accord with other recent treatments, rate enhancements are assumed to arise primarily from the large increase in the relative concentrations of the organic and hydrophilic substrates upon incorporation into the micellar phase. Second, the Stern layer of the micelle is assumed to be saturated with respect to its counterion.

The predictions of the equations are qualitatively in accord with the reaction rate versus surfactant concentration profiles for first, second, and third order reactions under a variety of experimental conditions. They also account for the effect of added salt on the rate of second order reactions. For selected cases, the equations are used to calculate the micellar rate constant and the ion exchange constants. The results are in accord with the assumptions of the theory. Finally, the theory predicts several new and as yet unobserved experimental results.

INTRODUCTION

A substantial body of data has established beyond reasonable
doubt that the partitioning of substrates into micelles in aqueous
solutions and not the effect of micelle formation on the aqueous
phase produces the observed catalysis or inhibition when the surfactant concentration exceeds critical micelle concentration (cmc).[1-3]
Many of the factors contributing to the observed changes have been
isolated, but much of the detail is not yet clearly understood.
General trends for the effect of increasing surfactant or salt concentration on the rate dimensions of first, second, and in some
cases third order reactions have been found. There are, however,
exceptions both apparent and real; and the situation is complicated
by a lack of adequate structural models for the micelle under many
experimental conditions. Consequently, a general theory of reaction
kinetics in micellar systems has yet to be firmly established.

Early work in micellar catalyzed reactions was pursued with
the hope that micelles might function as simple models for enzymes.[4,5]
This approach, which directed both the design of experiments and the
interpretation of results, produced a number of initial successes.
Much of the data at that time could be understood in terms of the
micelle's potential ability to alter the relative energies of the
ground states and transition states of reactions through a combination of hydrophobic and electrostatic interactions between the
micelle and organic substrates. The ground state for reactive
hydrophilic ions was assumed to be the aqueous phase. Like enzymes,
micelles were not supposed to significantly alter the distribution
of hydrophilic ions in solution.

However, this approach, and the rate equations developed from
it, cannot encompass the radically different rate dimensions for
first, second, and third order reactions as a function of surfactant concentration, nor the different effects produced by increasing
the hydrophilic counterion concentration on first and second order
reactions.[1,6] Finally, conceptualizing micelles as simple enzymes
masks a crucial difference between micellar and enzymic solutions.

In studies of enzyme catalyzed reactions, substrate concentrations are usually orders of magnitude larger than the enzyme concentration. Thus even when the enzyme is saturated with substrate,
the enzyme's presence does not significantly alter the relative
concentrations of substrates in the aqueous phase. In micellar
solutions, however, the concentration of at least one of the substrates and the micelles are often within the same order of magnitude. This difference is crucial for second and higher order
reactions. If two or more substrates partition strongly in favor
of the micellar phase, then the small total volume of the first
formed micelles can produce a large relative increase in concentration of substrates on the microscopic level. This concentration

effect (and the concomitant loss of translational and possibly
rotational entropy) is sufficient to produce a dramatic increase
in the observed reaction rate.

Recognition of this fact has produced several new kinetic
models over the past few years.[7,8] The most successful of these is
found in the work of Berezin and coworkers.[8,9] They derived rate
expressions for the effect of increasing micelle concentration on
the observed rate of reaction for second and higher order reactions
between two or more polar and/or charged organic molecules. The
rate constant in the micellar phase calculated from the equations
was usually less than or equal to the rate constant for the reaction
in water. Their results suggested that catalysis by micelles does
not require, for example, electrostatic stabilization of the transition state relative to the ground state.

While reactions between organic substrates and hydrophilic
ions generally exhibit reaction rate/surfactant concentration profiles similar to those between organic substrates, there are numerous exceptions. And none of the models successfully accounts for
all the observed changes in reaction rate produced by added salts.
The greatest problem, however, is that while the binding of organic
substrates can be described by a simple partition function, a large
quantity of experimental evidence indicates that this approach cannot be used for the binding of hydrophilic ions.[6]

What is needed to solve these problems is a sufficiently precise model for the distribution of hydrophilic ions in micellar
solutions, one that accounts for both the effect of increasing
ionic strength and ion size. In 1964, Stigter published such a
model to calculate the specific adsorption potential of counterions
to the Stern layer of the micelle.[10]

The purpose of this paper is to develop and illustrate a hope-
fully general kinetic model for second and higher order reactions
between organic substrates and hydrophilic ions in micellar solutions. The model is, in essence, a judicious combination of the
kinetic analysis developed by Berezin et al. and Stigter's model
for the distribution of small ions.

DISTRIBUTION OF HYDROPHILIC IONS IN MICELLAR SOLUTIONS

The applicability of this kinetic treatment depends primarily
upon the validity of one new assumption: The Stern layer of the
micellar phase is saturated with respect to its hydrophilic counterions. At constant temperature, surfactant chain length and head
group structure, for ions that undergo primarily electrostatic
interactions with the micellar surface, and in the absence of

hydrophobic counterions and nonelectrolyte additives, the degree of
ionization (and the Stern layer concentration of counterions) is
essentially independent of the surfactant concentration and the
ionic strength. A corollary to this assumption is that in micellar
solutions containing mixtures of two or more hydrophilic counterions, the relative concentration of counterions in the Stern layer
will depend only on their individual specific adsorption potentials
and thus their mole fraction ratios in solution. The evidence for
this assumption is basically circumstantial, relying primarily on
the presence or absence of trends rather than precise quantitative
results. First, the concept of a Stern layer saturated with
counterions is in accord with but not required by the micelle model
developed by Stigter (see Figure 1). To overcome the difficulties
caused by the smooth surfaced Gouy-Chapman model for the electrical
double layer, Stigter developed the concept of the rough surfaced
micelle first proposed in 1955.[11] For a spherical micelle containing
n surfactant monomers, $(1-\alpha)n$ fully hydrated counterions are allowed
to penetrate between the fully hydrated head groups in the Stern
layer. The remaining αn counterions are distributed in the Gouy-
Chapman layer. The micelle interior is assumed to be "liquid-like"
hydrocarbon. Calculated values of the specific adsorption potential
of counterions to sodium dodecylsulfate (SDS) and dodecylammonium
chloride micelles showed no specific trends with increasing ionic

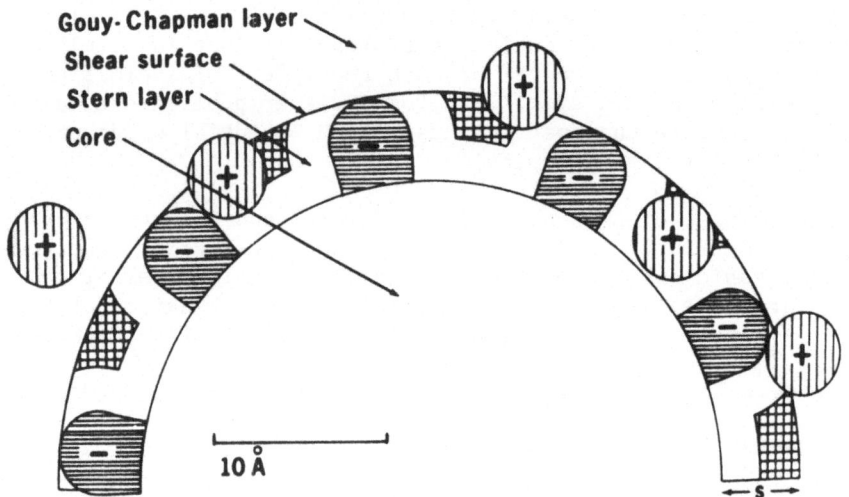

Figure 1. Partial cross section of a sodium dodecylsulfate
micelle. Crosshatched area in the Stern layer is available to
the centers of the hydrated sodium ions.[10]

strength (counterion concentration). Also, numerical differences
between the Stern potential and the zeta potential (assumed to be
a measure of the potential at the shear surface (see Figure 1)) did
not change much with increasing counterion concentration. Consequently, even though the Stern potential decreases continuously
with increasing ionic strength, the potential drop across the Stern
layer remains constant. If reactions take place in the Stern layer
as is generally assumed, then the results of Stigter's model directly
contradict an assumption of the enzyme model--that the parallel
decrease in reaction rate and surface potential caused by increasing counterion concentration is due to a decrease in the electrostatic stabilization of the transition state or increased stabilization of the ground state.[4] A more reasonable conclusion from
Stigter's model would be that both the organic substrate and the
counterions in the Stern layer experience an invariant potential
with increasing hydrophilic counterion concentration.

     Second, because the degree of ionization, α, is a measure of
the distribution of counterions between the micellar and aqueous
phases, changes in α should reflect changes in the Stern layer concentration of counterions. A variety of experimental techniques
have been used to measure α, including light scattering,[12,13], conductance,[14] electromotive force measurements,[15] and ion selective
electrodes.[16] While numerical agreement between the various methods
for the same surfactant is seldom very good, several consistent
trends have appeared.[6,17] Generally, α increases with increasing
temperature, nonelectrolyte concentration, surfactant head group
size, and the hydrated radius of the hydrophilic counterion (following a Hofmeister series) and decreases with increasing surfactant chain length. However, no consistent trends have been found
for α with increasing surfactant and hydrophilic counterion concentration. Instead, depending upon the method used and the
surfactant studied, α may increase, decrease or remain constant.
The extent of the change is usually small and within the range of
α values found for most surfactants, $\alpha = 0.1$ to 0.3.[6] The exceptions can generally be attributed to the presence of a second
binding force such as charge transfer or hydrophobic interactions
between the micelle and its counterions in addition to the coulombic
interaction.[18-20]

     Finally, even if the Stern layer concentration of counterions
(and α) does change somewhat with increasing surfactant or salt
concentration, the change may not make a significant contribution
to the observed change in reaction rate. Using Stigter's molecular
dimensions for SDS and an aggregation number of 62,[21] the Stern layer
concentration of sodium ions in moles of sodium ions per liter of
Stern layer volume is estimated to vary between 4.5 $\underline{M}$ ($\alpha=0.1$) and 3.5$\underline{M}$
($\alpha = 0.3$).[6] The difference between these two sodium ion concentra-
tions is very small, especially when compared to the concentration

changes of 10 to 100 fold in added counterions used during experiments to measure the change in α as a function of counterion concentration.

In summary, micelles can increase dramatically the relative concentrations of both the organic substrate and the hydrophilic ion, and since the Stern layer can accommodate only a limited number of ions, nonreactive counterions exert their inhibiting effect on reaction rates of second and higher order reactions primarily by exchanging with reactive counterions in the Stern layer and not by decreasing the surface potential of the micelle.

DERIVATION OF THE KINETIC EQUATION

The derivation of the second order rate constant for a reaction between an organic substrate and a hydrophilic ion will follow the method of Berezin et al.[8] Micelles are considered to be a separate but uniformly distributed pseudophase.[22] This simplifying assumption makes the final kinetic expression independent of changes in micelle size and shape.

An organic substrate, A, in a micellar solution is assumed to partition between the aqueous and micellar phases according to a simple distribution function:

$$P_a = \frac{A_m}{A_w} \tag{1}$$

Here and subsequently, m and w denote the micellar and aqueous phases. P_a is the partition constant. All concentrations are expressed in molarity. The value of P_a will be determined by the magnitude of the hydrophobic and electrostatic interactions between the substrate and the micellar phase.

The distribution function for the second substrate, the reactive hydrophilic ion, I, must also contain terms for the distribution of the nonreactive counterion, X, whose concentration is equal to the total surfactant concentration, C_t, plus the concentration of added salt ($X_t = C_t + MX$). The concept that the Stern layer is saturated with respect to its counterions is expressed by

$$I_m + X_m = \beta S$$

where β is the degree of counterion binding to the Stern layer ($\beta = 1 - \alpha$), and S is the molar density of the micellar phase expressed in moles of surfactant per liter of micellar phase. This assumption is reasonable primarily because the nonreactive counterion concentration is usually much larger than the reactive ion

concentration so that differences in their binding constants will
not be important, but also because differences in their binding
constants are not very large (i.e. the variation in α is small).

The two counterions are assumed to exchange rapidly between
the micellar and aqueous phases:

$$I_m + X_w \rightleftarrows I_w + X_m$$

and their distribution can be expressed by the ion exchange constant, K, where

$$K = \frac{I_w X_m}{I_m X_w} \tag{3}$$

One important consequence of this approach is that even when the
nonreactive counterion concentration is much greater than the reactive ion concentration ($X_t \gg I_t$), the concentration of the reactive
ion in the micellar phase can be greater than its concentration in
the aqueous phase ($I_m \gg I_w$).

Finally, with the reasonable assumption that all species diffuse between phases more rapidly than the rate of reaction in either
phase, the velocity of the reaction averaged over the whole solution
volume is

$$v_{ave} = k_2 A_t I_t = k_m A_m I_m CV + k_w A_w I_w (1-CV) \tag{4}$$

The overall second order rate constant and the rate constants in
each phase are respectively, k_2, k_w, and k_m; A_t and I_t are the total
concentrations of the reactants; C is the surfactant concentration
in micellar form, $C = C_t - cmc$; and V is the molar volume of the
surfactant ($V = 1/S$).

Equations (1) through (4) are combined with the appropriate
materials balance expressions for A, I, and X, simplified, and
rearranged to give the rate of the reaction in terms of the overall
second order rate constant:

$$k_2 = \frac{k_m \beta S K_a (C_t-cmc)}{[K_a(C_t-cmc) + 1][I_t + X_t K]} + \frac{k_w}{[K_a(C_t-cmc) + 1]} \tag{5}$$

When $P_a \gg 1$ as is generally the case, $K_a = P_a V$. The major simplifying assumption used in the derivation, $I_t \gg I_m CV$, limits the
applicability of Equation (5) to low micelle concentrations. The
complete derivation of Equation (5) has been published elsewhere.[6]

If the assumptions used in this derivation are correct, then any difference in value between k_w and k_m represents the medium effect of the micellar phase on the reaction compared to the aqueous phase. The difference in free energy of activation for the reaction in each phase will appear in the value of these rate constants free of concentration effects.

As will be shown below, Equation (5) meets the most important criteria for predicting the effect of increasing surfactant concentration on the observed second order rate constant. Using values for the constants that mimic typical experimental conditions, the Equation predicts a maximum in the reaction rate/surfactant concentration profile.

QUALITATIVE COMPARISONS BETWEEN THEORY AND EXPERIMENT

To illustrate the qualitative predictive power of the kinetic equations a FORTRAN program was written to produce relative rate constants ($k_{rel} = k_2/k_w$) as a function of two independent variables.[6] In each of these examples, the rate constants in the micellar and aqueous phases are assumed to be equal ($k_m = k_w = 1$). This assumption gives full play to the concentration effect. Values for the constants used to produce the curves are specified in the figure legends. The families of curves produced in these plots will be compared with published results for changes in equivalent experimental variables.

Figure 2 is a family of curves produced to illustrate the effect of increasing substrate hydrophobicity, which is mathematically equivalent to increasing the binding constant, K_a, as the surfactant concentration increases. The general shapes of the curves in Figure 2 can be interpreted as the summation of two opposing effects. Once the surfactant concentration has exceeded the cmc, the relative concentrations of the organic substrate and the hydrophilic ion increase rapidly in the Stern layer of the micellar phase. The larger the binding constant, K_a, the greater the concentration increase, the faster the rate increase, and the greater the rate attainable at a lower surfactant concentration. The rate maximum also shifts to lower surfactant concentration with increasing K_a because the concentration effect is opposed by a continuous decrease in the Stern layer concentration of the reactive counterion, I_m. The nonreactive counterion concentration increases continuously ($X_t = C_t$), while the reactive counterion concentration is constant. Consequently, because there is a limited number of binding "sites" available, the ratio I_m/X_m decreases continuously. This effect will predominate at higher surfactant concentration producing the observed maximum followed by the gradual decrease in the observed rate of reaction.

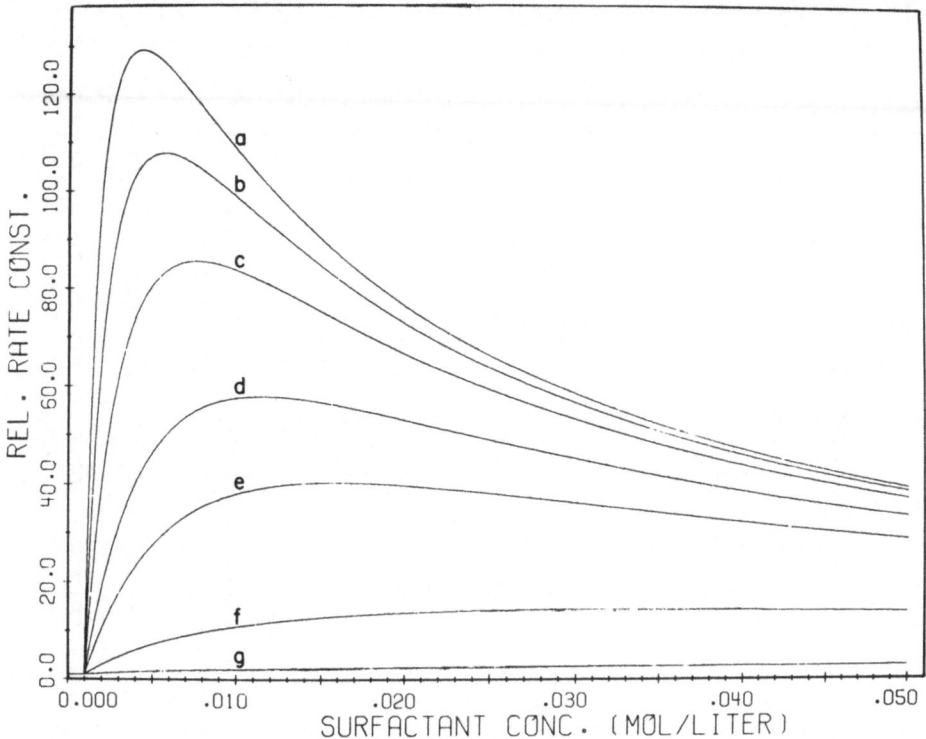

Figure 2. Computer generated plots of the change in the relative
rate constant for a second order reaction, k_2, as a function of the
surfactant concentration, C_t. The substrate binding constant, K_a,
is the second independent variable. K_a = (a) 1000, (b) 500,
(c) 250, (d) 100, (e) 50, (f) 10, (g) 1; cmc = 0.001; k_m = k_w = 1;
β = 0.8; S = 6.0; K = 1.0; I_t = 0.01; X_t = C_t.

Figure 3 shows the observed rate constants for the basic
hydrolysis of p-nitrophenylacetate (PNPA), -hexanoate (PNPH), and
-laurate (PNPL) as a function of surfactant concentration for a
series n-alkyltrimethylammonium bromide surfactants (n = 10, 12,
14, 16, 18).[23] Carbonate buffer was used to control the pH, so
that k_{rel} is proportional to k_2 and k_{obs}. For those surfactants
that produce significant catalysis, the general shape of the curves
is quite similar to those in Figure 2. In TDTAB and HDTAB, for
example, the rate maximum clearly increases and shifts to lower
surfactant concentrations with increasing substrate hydrophobicity.

The differences between the plots in Figures 2 and 3 can be
rationalized by assuming that k_w > k_m (PNPA) > k_m (PNPH) ≃ k_m(PNPL).
This assumption correlates the following facts: (a) The differences

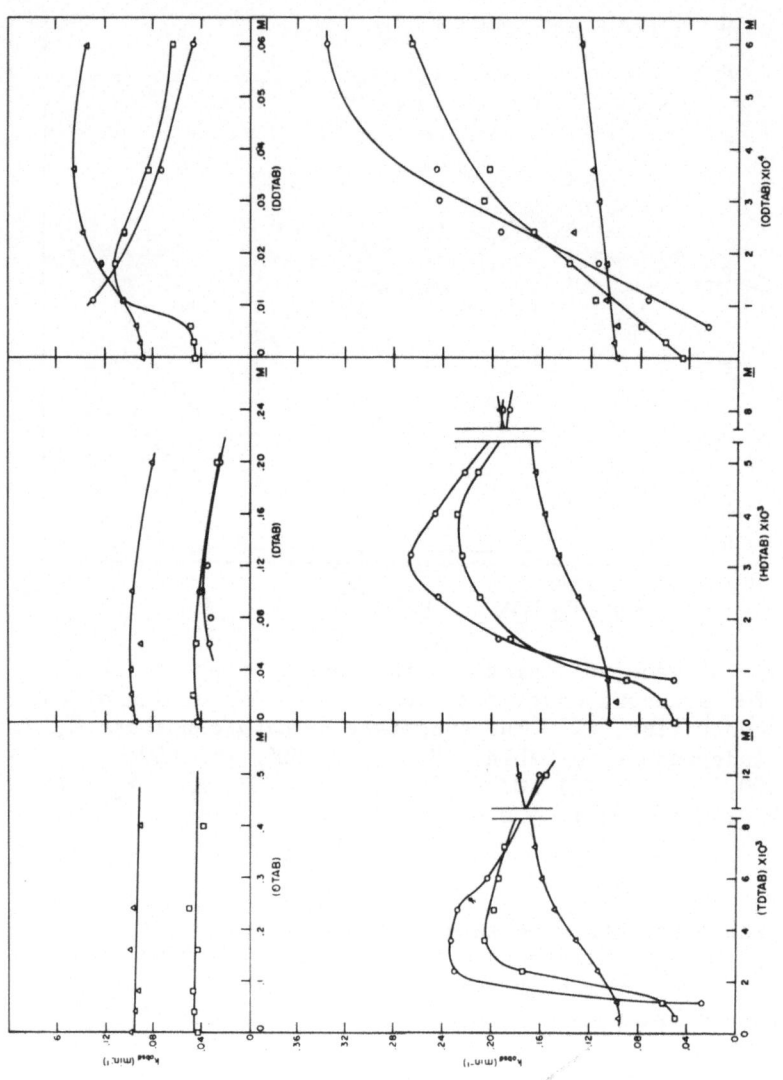

Figure 3. First order rate constants for the basic hydrolysis of p-nitrophenyl acetate (△), p-nitrophenyl hexanoate (□), and p-nitrophenyl laurate (○) at 25° and pH 10.07 plotted as a function of the concentration of a series of n-alkyltrimethylammonium bromides: From left to right starting at the top, the n-alkyl groups are octyl, decyl, dodecyl, tetradecyl, hexadecyl, and octadecyl.[23]

in maximum attainable rates are modest compared to those in Figure
2 even though the binding constants (as measured independently by
gel filtration) are very different, 1.6×10^4 $\underline{M}^{-1}$ for PNPH and
33 $\underline{M}^{-1}$ for PNPH and 33 $\underline{M}^{-1}$ for PNPA. (b) At high surfactant concentrations the rate constants for PNPL and PNPH are below that of
PNPA for all surfactants except ODTAB. (c) The order for maximum
rate enhancement with substrate hydrophobicity is reversed in OTAB,
DTAB and DDTAB. Also, because their cmc's are quite high, the
I_m/X_m ratio is already small at the cmc, so that little if any
catalysis can be produced by the concentration effect. The
results for PNPL are slightly different because the substrate is
insoluble in water and no catalysis can be observed until the
micelle concentration is sufficient to solubilize it. The curves
for the reactions in ODTAB are incomplete because the surfactant
precipitates at high concentrations.

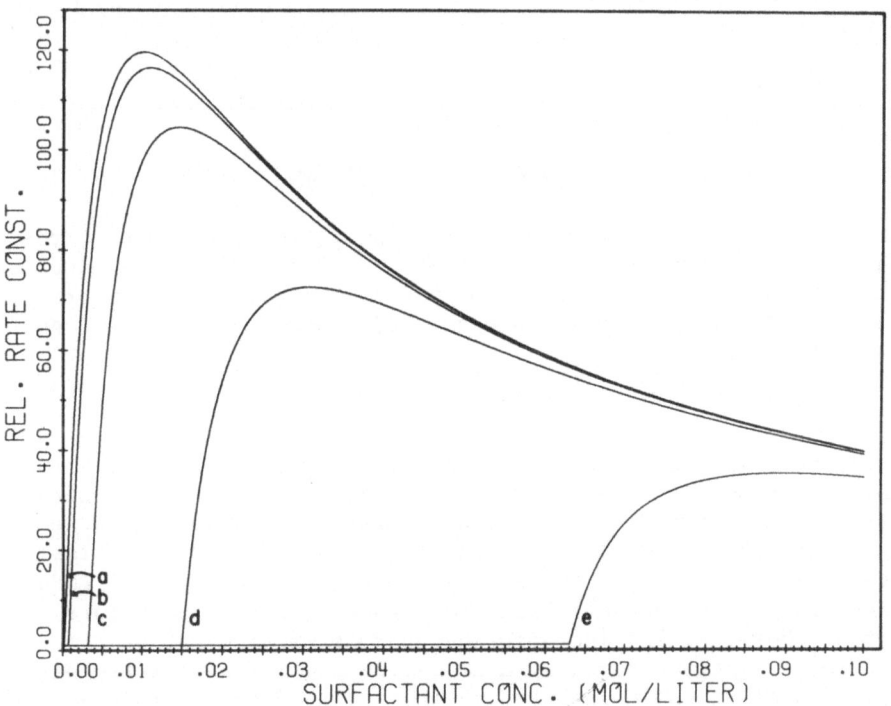

Figure 4. Computer generated plots of the change in the relative
rate constant for a second order reaction, k_2, as a function of the
surfactant concentration, C_t. The cmc is the second independent
variable. cmc = (a) 0.00016, (b) 0.00071, (c) 0.0032, (d) 0.015,
(e) 0.063; $k_w = k_m = 1$; $K_a = 100$; $\beta = 0.8$; $S = 6.0$; $K = 1.0$; $I_t =$
0.01; $X_t = C_t$.

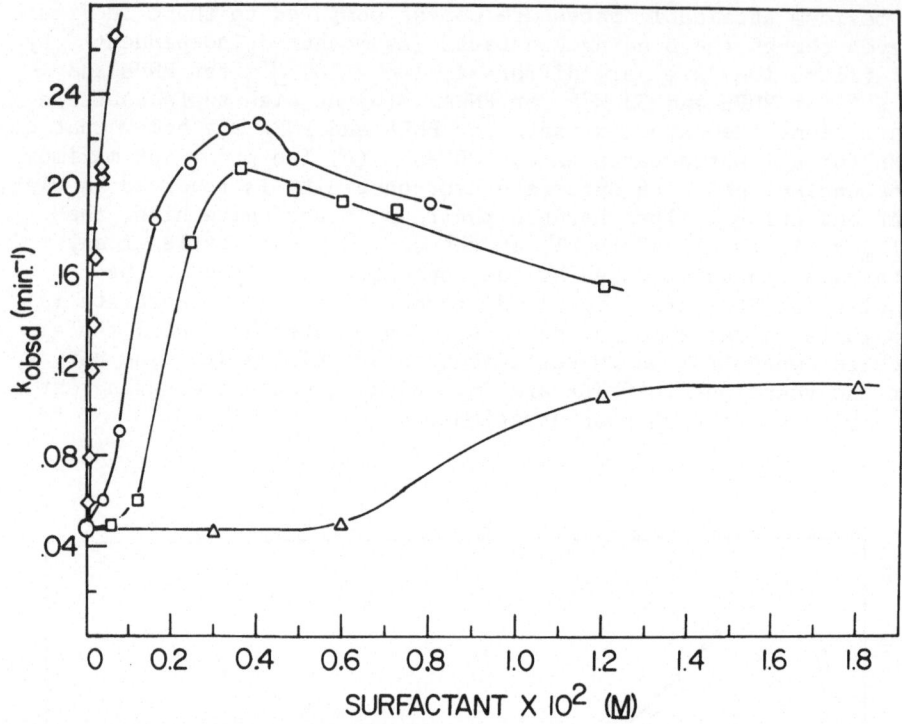

Figure 5. First order rate constants, k_{obsd}, for the basic hydrolysis of p-nitrophenylhexanoate at 25° and pH of 10.07 as a function
of the concentration of a series of n-alkyltrimethylammonium bromides: n-dodecyl- ($\triangle$), n-tetradecyl- ($\square$), n-hexadecyl- (0), and
n-octadecyl ($\diamondsuit$).

    Figure 4 is a computer plot and Figure 5 is replotted experi-
mental data from Figure 2 for PNPH which show the effect of increasing surfactant chain length (or its mathematical equivalent a decreasing cmc) on reaction rate/surfactant concentration profiles.
The general shapes of the curves for the hydrolysis data for PNPH
in Figure 5 look remarkably like the computer plots. Even more
important, however, the increase in the rate maximum can be
accounted for without invoking a change in the micellar rate constant. A smaller value for the cmc means that micelles form at
lower surfactant concentrations. At lower surfactant concentrations, the I_m/X_m ratio in the Stern layer is higher, making the
concentration effect more important and the potential maximum rate
increase higher.

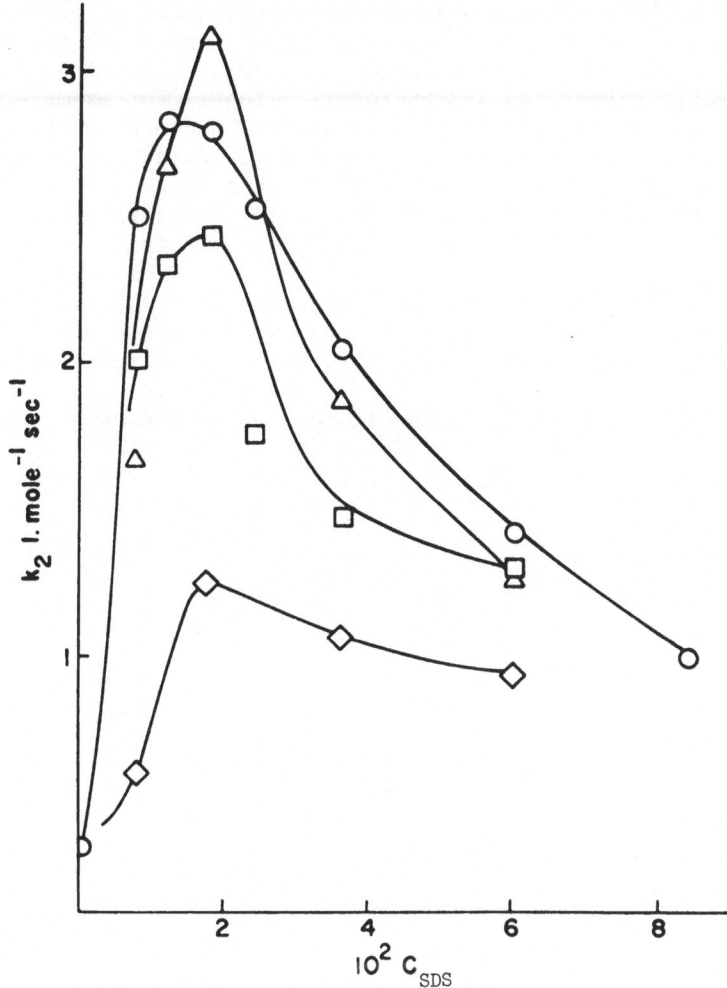

Figure 6. Variation of $k_2 = k_\psi/C_{H^+}$ with the concentration of SDS in dilute HCl at 25.0° for the acid catalyzed hydrolysis of p-nitrobenzaldehyde diethyl acetal: ($\triangle$) 10^{-3} M HCl; (O) 3.16×10^{-3} M HCl; ($\square$) 10^{-2} M HCl; ($\diamond$) 3×10^{-2} M HCl.[27]

The analysis outlined above is supported by the experimental results from other micelle catalyzed reactions which showed similar trends for both increasing substrate hydrophobicity and increasing surfactant chain length.[24-26]

Another significant observation first discussed clearly by Berezin et al. is that even when the surfactant concentration is

high enough to bind all the organic substrate to the micelle, the
experimental second order rate constant is an "apparent" and not a
true rate constant.[8] This conclusion also holds for reactions
between organic substrates and hydrophilic ions. For example, Bunton
and Wolfe measured the effect of increasing SDS concentration on the
second order rate constant for the specific acid catalyzed hydrolysis of p-nitrophenylbenzaldehyde diethylacetal at several different
acid concentrations (Figure 6).[27] The second order rate constant
decreases steadily with increasing acid concentration.

Equation (5) fits this result because k_2 is predicted to be an
inverse function of the reactive ion concentration (in this case,
H^+). A computer plot illustrating the effect of increasing the
concentration of the reactive counterion is shown in Figure 7. Like
Figure 6, the family of curves in Figure 7 shows the obvious maximum

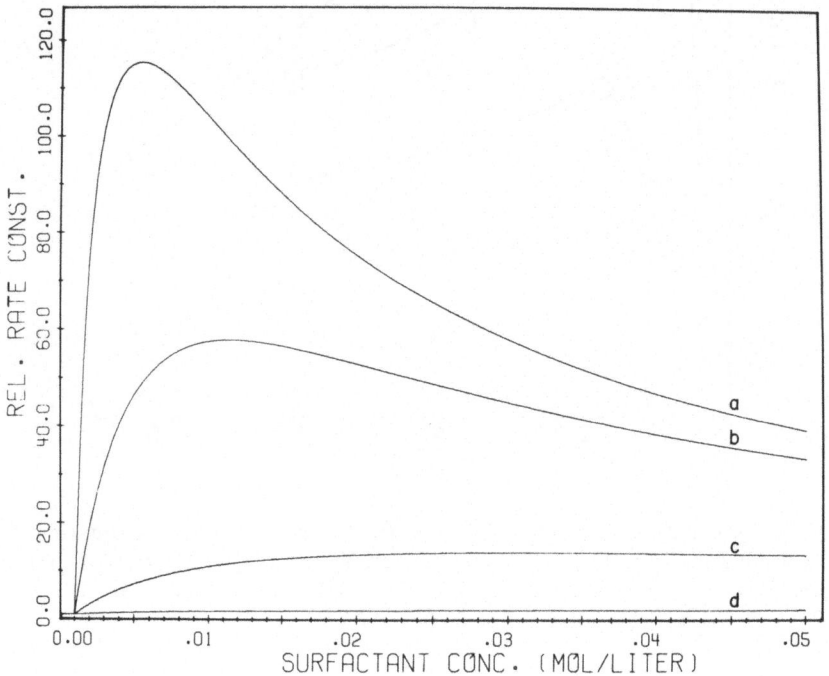

Figure 7. Computer generated plots of the change in the relative
rate constant for a second order reaction, k_2, as a function of the
surfactant concentration, C_t. The concentration of the reactive
counterion, I_t, is the second independent variable. I_t = (a)
0.001, (b) 0.01, (c) 0.1, (d) 1.0; cmc = 0.001; k_w = k_m = 1;
K_a = 100; β = 0.8; S = 6.0; K = 1.0; X_t = C_t.

in the profile approaching an almost level plateau as the concentration of the reactive ion is increased.

When the second order rate constant is calculated in terms of
the hydrogen ion activity (a_{H^+}) as measured by the hydrogen electrode instead of the total hydrogen ion concentration, the maximum
in the catalysis curves disappears.[27] This result, in accord with
the authors' interpretation, indicates that for each incremental
increase in micelle concentration there is an incremental uptake of
hydrogen ion from the aqueous phase by the micellar phase.

Consequently, once all the organic substrate has been bound,
the observed second order rate remains constant ($k_2 = k_{obs}/a_{H^+} =$
constant). The adsorption of significant quantities of reactive
counterion by the micellar phase is in accord with the model
developed here, but not with the enzyme model which assumes that
the ground state for the hydrophilic ion is in the aqueous phase.
Additional examples are available in the literature which show that
the observed second order rate constant is an apparent rather than
a true rate constant, with its value being an inverse function of
the concentration of the reactive counterion.[28,29]

Finally, Equation (5) predicts that if the ratio of the reactive to nonreactive counterion concentrations (I_t/X_t) is held
constant while the surfactant concentration is allowed to increase,
then an plateau instead of a rate maximum will be observed in the
catalysis curves. This result has been observed several times; for
the base catalyzed hydrolysis of PNPH in tetradecyltrimethylammonium
chloride,[23] the acid catalyzed hydrolysis of methyl ortho-benzoate
in SDS,[30] and the addition of cyanide ion to N-alkyl-3-carbamoylpy-
ridinium ions in TDTAB.[6]

PREDICTIONS, FIRST AND THIRD ORDER REACTIONS

There is an interesting, previously unanticipated, and as yet
unobserved result predicted by Equation (5) that provides a power-
ful test of the kinetic theory. When a micellar system contains
only reactive counterions, and is free of nonreactive counterions
($X_t = 0$), then the reactive ion concentration will be equivalent to
the surfactant concentration ($I_t = C_t$). Potential surfactant systems would be n-alkylsulfonic acids or long-chained quaternary
ammonium hydroxides. Under these conditions, Equation (5) simplifies considerably, and when written in terms of the first order
rate constant ($k_2 = k_1 C_t$) reduces to

$$k_1 = \frac{k_m \beta S K_a (C_t - cmc) + k_w C_t}{[K_a (C_t - cmc) + 1]} \tag{6}$$

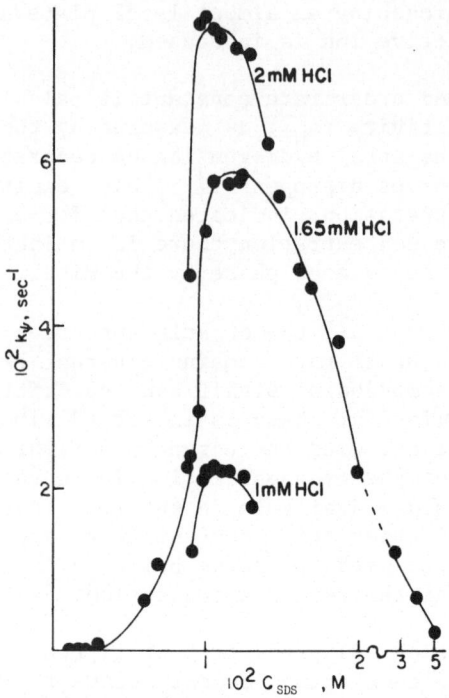

Figure 8. First order rate constants, k_ψ, for the acid catalyzed benzidine rearrangement against the SDS concentration at 25° in 1 mM, 1.65 mM, and 2 mM HCl.[31]

This Michaelis-Menten type of relation predicts a plateau instead of a rate maximum for a second order reaction at high surfactant concentrations.

The same analysis used to derive Equation (5) can be applied to first order reactions of organic substrates. The result is

$$k_1 = \frac{k_m K_a (C_t - cmc) + k_w}{[K_a (C_t - cmc) + 1]} \qquad (7)$$

This expression is formally identical to the one derived according to the assumptions of the enzyme model[1] and, of course, is exactly like the one derived by Berezin et al.[8] It predicts the often observed plateau at high surfactant concentrations following the complete adsorption of the substrate into the micellar phase. Equation (7) predicts that first order reactions will be zero order

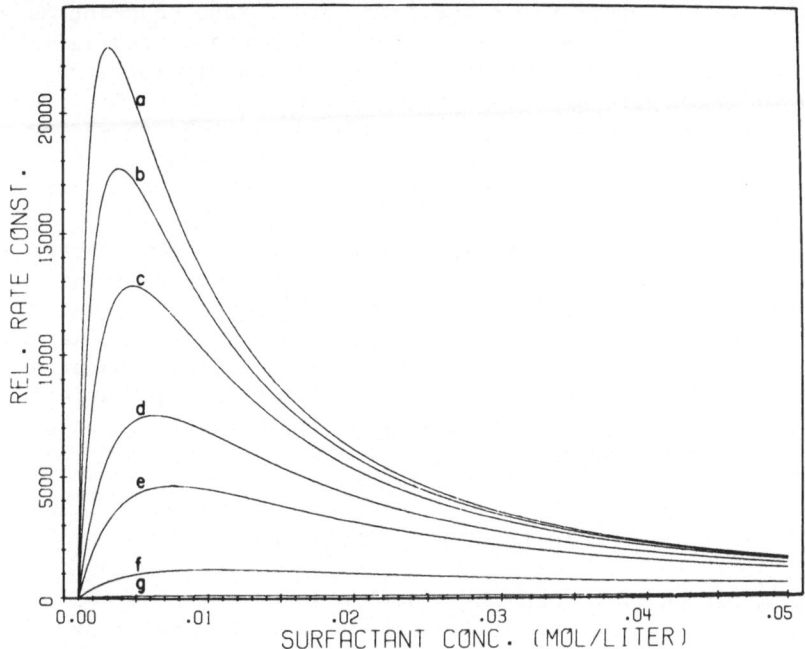

Figure 9. Computer generated plots of the change in the relative rate constant for a third order reaction k_3, as a function of the surfactant concentration, C_t. The substrate binding constant, K_a, is the second independent variable. K_a = (a) 1000, (b) 500, (c) 250, (d) 100, (e) 50, (f) 10, (g) 1.0; cmc = 0.001; $k_w = k_m = 1$; $\beta = 0.8$; S = 3.0; K = 1.0; $I_t = 0.01$; $X_t = C_t$.

with respect to concentration changes and types of hydrophilic ions present in solution. This prediction is not completely in accord with experiment, since both small rate decreases and increases have been observed with added hydrophilic counterions. What is important here is that the extent of the change is much smaller than that observed for the effect of added salt on second order reactions.

Bunton and Rubin recently measured the effect of increasing SDS concentration on the specific acid catalyzed benzidine rearrangement (Figure 8).[31] This is a third order reaction overall, first order in benzidine and second order in hydrogen ion. Equation (8) is derived by applying the theoretical approach developed here to third order reactions of this type. Figure 9 is a family of curves

$$k_3 = \frac{k_m K_a (C_t - cmc) \beta^2 S^2}{[K_a(C_t - cmc) + 1][I_t + X_t K]^2} + \frac{k_w}{[K_a(C_t - cmc) + 1][1 - (C_t - cmc)V]} \tag{8}$$

generated by the computer for the third order rate constant as a
function of increasing surfactant concentration and increasing substrate binding constant. Both plots show extremely large and rapid
changes in rate up to and away from the rate maximum. The predicted and observed rate increases are much larger than those for
second order reactions, with the changes in the shape of the profile
taking place over a much narrower range of surfactant concentration.

Equation (8) predicts three additional and as yet unobserved
results: (a) a second order dependence of the rate constant on
the concentration of both the reactive and nonreactive counterion
concentrations; (b) a plateau in the value of the first order rate
constant when $X_t = 0$ and all the organic substrate is bound; and
(c) a plateau in the third order rate constant at high surfactant
concentrations when the reactive/nonreactive counterion concentration ratio is held constant.

QUANTITATIVE CORRELATIONS

The addition of hydrophilic ions inhibits the observed rate
of reaction of a number of second order reactions, often by an
order of magnitude or more. The concentration of added electrolyte
required to produce these effects is quite small, enough to produce
barely detectable changes in the rate of the same reaction in water.
This effect has often been ascribed to a progressive neutralization
of the charge of the Stern layer, and a concomitant decrease in the
extent of the electrostatic stabilization of the transition state
relative to the ground state. Here it is assumed that this effect
is negligible, that the potential drop and concentration of the
counterions are both constant in the Stern layer, and that changes
in the rate of reaction depend only on the reactive ion concentration in the Stern layer, which in turn depends upon the relative
concentration of reactive and unreactive counterions in solution
and the size of the ion exchange constant, K.

If these assumptions are correct, then at surfactant concentrations well above the cmc and sufficient to incorporate all the
organic substrate, Equation (5) can be simplified and rearranged
into an expression that predicts a linear relation between the
reciprocal of the second order rate constant and the concentration
of added counterion, Equation (9). A similar expression has been
derived in terms of

$$\frac{1}{k_2} = \frac{I_t}{k_m \beta S} + \frac{K}{k_m \beta S} X_t \tag{9}$$

the enzyme model, but based on fundamentally different assumptions.[1]
The presence of a constant amount of other types of counterions will

simply add additional constant terms to Equation (9). If values
for I_t, β, and S are known or assumed, values for both the micellar
rate constant, k_m, and the ion exchange constant, K, can be calculated.

A number of salt effect studies have been published allowing
sample calculations to be made. However, the reactions were usually run in the presence of buffer added to control the pH, creating
several difficulties in interpretation. If the buffer controls
the concentration of the reactive counterion, it is impossible to
assign a value to I_t, and at least one form of the buffer may interact with the micelle altering its properties in an uncontrolled and
unknown way.

Table I contains the micellar rate constant, k_m, the ion exchange constant, K, and the rates of ion exchange constants,
K_X/K_F, calculated according to Equation (9) for the effects of
added counterion on two different reactions. Baumrucker et al.
measured the effect of increasing concentration of F^-, Cl^-, Br^- and
NO_3^- ions on the rate of addition of cyanide ion to N-dodecyl-3-
carbamoylpyridimium bromide in n-tetradecyltrimethylammonium
bromide (TTAB).[32] The reciprocal plots of $1/k_2$ versus the concentration of added salt were linear for F^-, Cl^-, and Br^- ions, while
the plot for NO_3^- was decidedly nonlinear at high concentrations.

Each incremental increase in surfactant concentrations results
in an equivalent increase in counterion concentration, so that
Equation (9) also predicts a linear relationship between the
reciprocal of the second order rate constant and increasing surfactant concentrations--once all the organic substrate is incorporated
into the micellar phase. Table I also contains the values of k_m
and K calculated from the data of Baumrucker et al. for the same
reaction as above measured as a function of increasing TTAB concentration (up to 0.8 $\underline{M}$) under nearly identical conditions.[32]

The same calculations were also carried out for the effect of
the same set of counterions on the basic hydrolysis of PNPH in n-
tetradecyltrimethyl ammonium chloride (TTACl).[23] Because buffer
was used to control the pH, values for k_m and K could not be estimated. However, the ratio K_X/K_F (X = Br^-, NO_3^-, Cl^-) could still
be determined and the results are shown in Table I. As before,
the reciprocal plots for F^-, Cl^-, and Br^- were essentially linear,
while the one for NO_3^- was slightly curved.

The results of these calculations are important for several
reasons. First, when the calculated values of k_m are compared with
the second order rate constant in water for the addition of cyanide
ion to the water soluble substrate N-propyl-3-carbamoylpyridinium
iodide (see Table I), $k_m < k_w$. Preliminary calculations on the

Table I. Calculated Values for the Micellar Rate Constant, k_m, the Ion Exchange Constant, K, and the Ion Exchange Constant Ratio, K_X/K_F

Counterion	Br(TTAB)[a]	Br	NO_3	Cl	F
k_m[b] $M^{-1}min^{-1}$	0.24	0.30	0.29	0.28	0.27
k_w[c] $M^{-1}min^{-1}$		(0.84)			
K[b]		0.019	0.037	0.024	0.019	0.0026
K_X/K_F[b]		14.5	9.4	7.6	1.0
K_X/K_F[d]		18.5	25.1	5.9	1.0

a. Addition of cyanide ion (0.004 $\underline{M}$) to n-dodecyl-3-carbamoylpyridinium bromine in TTAB at 25°, with 0.001 M hydroxide added to insure that all cyanide is present as the anion.[32] b. The same reaction as in (a), but with increasing counterion concentration (to 0.5 $\underline{M}$) in 0.02 $\underline{M}$ TTAB and 0.001 $\underline{M}$ cyanide ion at 30° with the pH maintained at 10.4 by 0.01 $\underline{M}$ triethylamine-ammonium buffer.[32] The value of k_m was calculated from $k_m\beta S$ using the assumed values: β=0.8 and S=3.0 moles per liter of micellar phase. c. Second order rate constant for the addition of cyanide ion to N-propyl-3-carbomoylpyridinium iodide measured at 25° with an ionic strength of 0.5.[33] d. Base catalyzed hydrolysis of PNPH in 0.009 $\underline{M}$ TTACl and 0.02 $\underline{M}$ trimethylamine-ammonium chloride buffer at 25°. Counterion concentration was increased up to 0.2 $\underline{M}$.[23]

other micelle catalyzed reactions indicate that the observed rate enhancements for reactions between hydrophilic ions and organic substrates might also be determined primarily by the concentration effect.[6] Second, the assumptions used to derive Equation (9) are tentatively confirmed since the values of k_m and K, determined from added bromide ion and increasing TTAB, are quite close. Third, the absolute values of K for the cyanide addition reaction are all less than one, confirming the reasonable suspicion that cyanide ion binds more tightly to micelles than any of the added counterions. Fourth, for the halide ions, the values for the K_X/K_F ratios for the two reactions are quite close and independent of the type of reaction used to measure them, providing additional support for the assumption of a Stern layer saturated with counterions. A complete discussion of these calculations is published elsewhere.[6]

CONCLUSION

If successful, the model developed here will enhance the power of reaction kinetics as a probe of micelle structure. The micellar rate constant, k_m, should be a measure of the effect of micelles on the energetics of a reaction free from concentration effects. And if the value of the ion exchange constant, K, proves to be independent of the type of reaction studied, then it is a new measure of the relative binding power of counterions to the micelle Stern layer.

ACKNOWLEDGMENTS

This work would not have been possible without the continued support of Jean and Eugene over the last several years. I am also indebted to the American Chemical Society for permission to reproduce Figures 1, 3, and 6; to D. Stigter for the use of Figure 1; to C. A. Bunton and B. Wolfe for the use of Figure 6; and to C. A. Bunton and R. J. Rubin and Pergamon Press for permission to reproduce Figure 8.

REFERENCES

1. C. A. Bunton, Prog. Solid State Chem., **8**, 239 (1973).
2. J. H. Fendler, E. H. Fendler, "Catalysis in Micellar and Macromolecular Systems", Academic Press, New York, 1975.
3. E. H. Cordes, Editor, "Reaction Kinetics in Micelles", Plenum Press, New York, 1973.
4. E. H. Cordes, R. B. Dunlap, Accts. Chem. Res., **2**, 329 (1969).
5. E. H. Cordes, C. Gitler, Prog. Biorg. Chem., **2**, 1 (1973).
6. L. Romsted, Thesis, Indiana University, 1975.
7. K. Shirahama, Bull. Chem. Soc. Jap., **48**, 2673 (1975).
8. I. V. Berezin, K. Martinek, A. K. Yatsimirskii, Russ. Chem. Revs., **42**, 787 and references (1973).
9. K. Martinek, A. P. Osipov, A. K. Yatsimirskii, V. A. Dadali, I. V. Berezin, Tetrahedron Letters, 1279 and references (1975).
10. D. Stigter, J. Phys. Chem., **68**, 3603 (1964).
11. D. Stigter, K. J. Mysels, J. Phys. Chem., **59**, 45 (1955).
12. E. W. Anaker, in "Cationic Surfactants", E. Jungerman, Editor, p. 203, Marcel Dekker, New York, 1970.
13. D. Stigter, J. Phys. Chem., **70**, 1323 (1966).
14. K. C. Evans, J. Chem. Soc., Pt. **1**, 579 (1956).
15. C. Botré, V. L. Crescenzi, A. Mele, J. Phys. Chem., **63**, 297 (1969).
16. J. W. Larsen, L. G. Magid, J. Amer. Chem. Soc., **96**, 5774 (1969).
17. G. C. Kresheck, in "Water: A Comprehensive Treatise", F. Franks, Editor, Vol. 4, p. 95, Plenum Press, New York, 1975.

18. A. Ray, P. Mukerjee, J. Phys. Chem., 70, 2138 (1966).
19. P. Mukerjee, A. Ray, J. Phys. Chem., 70, 2150 (1966).
20. K. D. Heckmann, R. F. Woodbridge, in "Proc. IVth Intern. Cong.
     Sur. Act. Sub.", Vol. 2, p. 519, Gordon and Breach, New York,
     1967.
21. K. J. Mysels, L.H. Princen, J. Phys. Chem., 63, 1696 (1959).
22. K. Shinoda, in "Proc. IVth Intern. Cong. Sur. Act. Sub.",
     Vol. 2, p. 527, Gordon and Breach, New York, 1967.
23. L. R. Romsted, E. H. Cordes, J. Amer. Chem. Soc., 90, 4404
     (1968).
24. J. Albrizzio, J. Archila, T. Rudolfo, E. H. Cordes, J. Org.
     Chem., 37, 871 (1972).
25. J. L. Kurz, J. Phys. Chem., 66, 2239 (1962).
26. H. Nogami, Y. Kanakubo, Chem. Pharm. Bull. Tokyo, 11, 943
     (1963).
27. C. A. Bunton, B. Wolfe, J. Amer. Chem. Soc., 95, 3742 (1973).
28. C. A. Bunton, L. G. Ionescu, J. Amer. Chem. Soc., 95, 2912
     (1973).
29. C. A. Bunton, B. Wolfe, J. Amer. Chem. Soc., 96, 7747 (1974).
30. R. B. Dunlap, E. H. Cordes, J. Amer. Chem. Soc., 90, 4395
     (1968).
31. C. A. Bunton, R. J. Rubin, Tetrahedron Letters, 55 (1975).
32. J. Baumrucker, M. Calzadilla, M. Centeno, G. Lehrman, M.
     Urdaneta, P. Linquist, D. Dunham, M. Price, B. Sears, E. H.
     Cordes, J. Amer. Chem. Soc., 94, 8164 (1972).
33. R. N. Linquist, E. H. Cordes, J. Amer. Chem. Soc., 90, 1269
     (1968).

LASER PHOTOLYSIS STUDIES OF PHOTO REDOX PROCESSES IN

MICELLAR SOLUTION

M. Grätzel

Hahn-Meitner-Institut für Kernforschung Berlin GmbH

Berlin, West Germany

This paper deals with electron transfer reactions involving excited states of probe molecules which are solubilized in aqueous micellar solutions. Three different processes are discussed:

(i) Photoejection of electrons from P* inside the micelle into the aqueous phase

(ii) Transfer of electrons from triplet states P^T to acceptors located outside the micelle

(iii) Transfer of electrons from donors outside the micelle to P^T.

The kinetics of these reactions are examined by kinetic spectroscopy technique. It is shown that both the charge of the micelles as well as the free energy change associated with the redox reactions control the rate of electron transfer. Subsequent dark reactions of the radical ions produced were also investigated. These can be directed by suitable choice of the micellar system to make possible conversion of light into chemical energy.

INTRODUCTION

Light initiated redox reactions may be characterized by a common thermodynamic feature: If both the electron donor (D) and acceptor (A) participating in the electron transfer reaction are present in their ground states the process is endoergic and hence cannot occur spontaneously

$$A + D \rightarrow A^- + D^+ \qquad\qquad \Delta G > 0 \qquad\qquad (1)$$

If, on the other hand, one of the reactants is excited by light to a higher electronic state the free energy change associated with a photo redox reaction becomes negative:

$$A + D^* \rightarrow A^- + D^+ \qquad\qquad\qquad\qquad (2a)$$
$$\Delta G < 0$$
$$A^* + D \rightarrow A^- + D^+ \qquad\qquad\qquad\qquad (2b)$$

This inversion of sign of ΔG is explained by a shift in the standard redox potential of the A/A^- or D^+/D redox couple occurring upon light excitation

$$E_0(A^*/A^-) \simeq E_0(A/A^-) + \Delta E^* \qquad \Delta E^*: \text{ excitation energy} \qquad (3a)$$

$$E_0(D^+/D^*) \simeq E_0(D^+/D) - \Delta E^* \qquad\qquad\qquad\qquad (3b)$$

Equation (3) expresses the fact that the excited state of a molecule is a better electron acceptor as well as a better electron donor when compared to the ground state.

In a photo redox reaction the light can be regarded as an electron pump which drives the reaction against a positive gradient of free energy. Hence in such a process light energy is converted into chemical energy. This effect has potential importance since it may be used for the exploitation of solar energy to produce a chemical fuel. A major difficulty in achieving this goal is the reversible character of photo redox processes. In homogeneous solutions the forward electron transfer described by Equation (2) is in general followed by a rapid back reaction, i.e.

$$A^- + D^+ \rightarrow A + D \qquad\qquad\qquad\qquad (3)$$

As a result a cycle of reactions is performed leading to thermal degradation of the light energy absorbed by the system. This behavior is illustrated schematically in Figure 1.

Conversion of light energy into a storable chemical fuel can only be achieved if provisions are made to prevent the back reaction (3). In this respect aqueous micellar solutions offer distinct advantages over the homogeneous solutions, as will be shown in the present paper. It is possible by suitable choice of the micellar

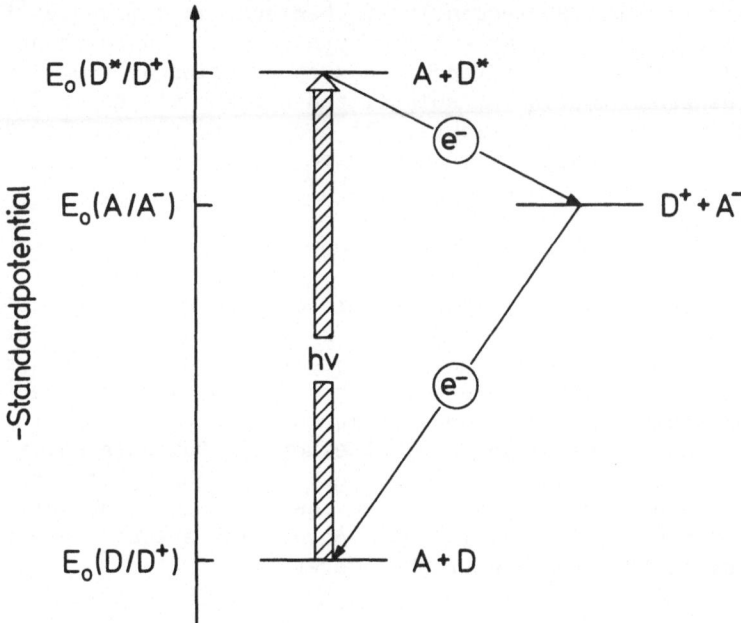

Figure 1. Energy diagram of a light induced redox reaction.

system to direct a light induced redox reaction as well as subsequent dark reactions in a way desired to produce chemical fuel from light.

EXPERIMENTAL

Materials

Sodium lauryl sulfate (Merck, "for tenside investigations") and dodecyltrimethylammonium chloride (Eastman) were purified by recrystallization from alcohol-ether mixtures. N,N,N',N'-tetramethylbenzidine and duroquinone (Aldrich) were recrystallized from ethanol.

Sample Preparation

Tetramethylbenzidine (TMB) was solubilized in micelles by injecting an aliquot of a stock solution in benzene into the aqueous

micellar solution. Subsequently the benzene was removed by flushing argon through the sample for several hours. Duroquinone is
readily dissolved when stirred in a soap solution at room temperature. The concentration of solubilized species was determined by
spectrophotometric measurements. If necessary, the samples were
deoxygenated by bubbling with highly purified argon.

Apparatus

Laser photolysis experiments were carried out using a Q-
switched Korad K1QP ruby laser. The 347.1 nm pulse had a duration
of 15 ns and maximum energy of 100 mjoules as measured by a bolo-
meter. Transient species were detected by fast kinetic spectroscopy and fast DC conductance techniques. A detailed description
of this setup has been published elsewhere.[1] Suitable cutoff filters
were placed in the analyzing light beam to prevent photolysis.
Solutions were flowed through a 1 cm^2 quartz cell to obviate interference from product accumulation. Absorption spectra were recorded
with a Unicam SP .70DC spectrophotometer.

RESULTS AND DISCUSSION

a)   Photo Redox Reactions in Which the Excited Species
     Acts as an Electron Donor

We first consider a situation in which the electron acceptor
acts also as the solvent for the photoactive species. In this case
the photo redox reaction comprises ejection of an electron from
the excited donor into the liquid whereby cation radicals and electrons are produced. If water is used as the solvent

$$D* + H_2O \rightarrow D^+ + e^-_{aq} \tag{4}$$

where the symbol e^-_{aq} stands for hydrated electrons.
Reaction (4) has attracted attention as a possible pathway for
photochemical production of hydrogen from water since hydrated
electrons produced in the photoionization event are known to undergo the diffusion controlled reaction

$$e^-_{aq} + e^-_{aq} \rightarrow H_2 + 2\ OH^- \tag{5}$$

The two major problems which heretofore have prevented successful
application of such a system are: (a) the majority of organic
sensitizers with ionization potentials low enough to permit electron ejection by visible or near u.v. light are sparsely soluble or
insoluble in water; (b) rapid recombination of e^-_{aq}/D^+ ion pairs
leads to annihilation of electrons before hydrogen can be formed.

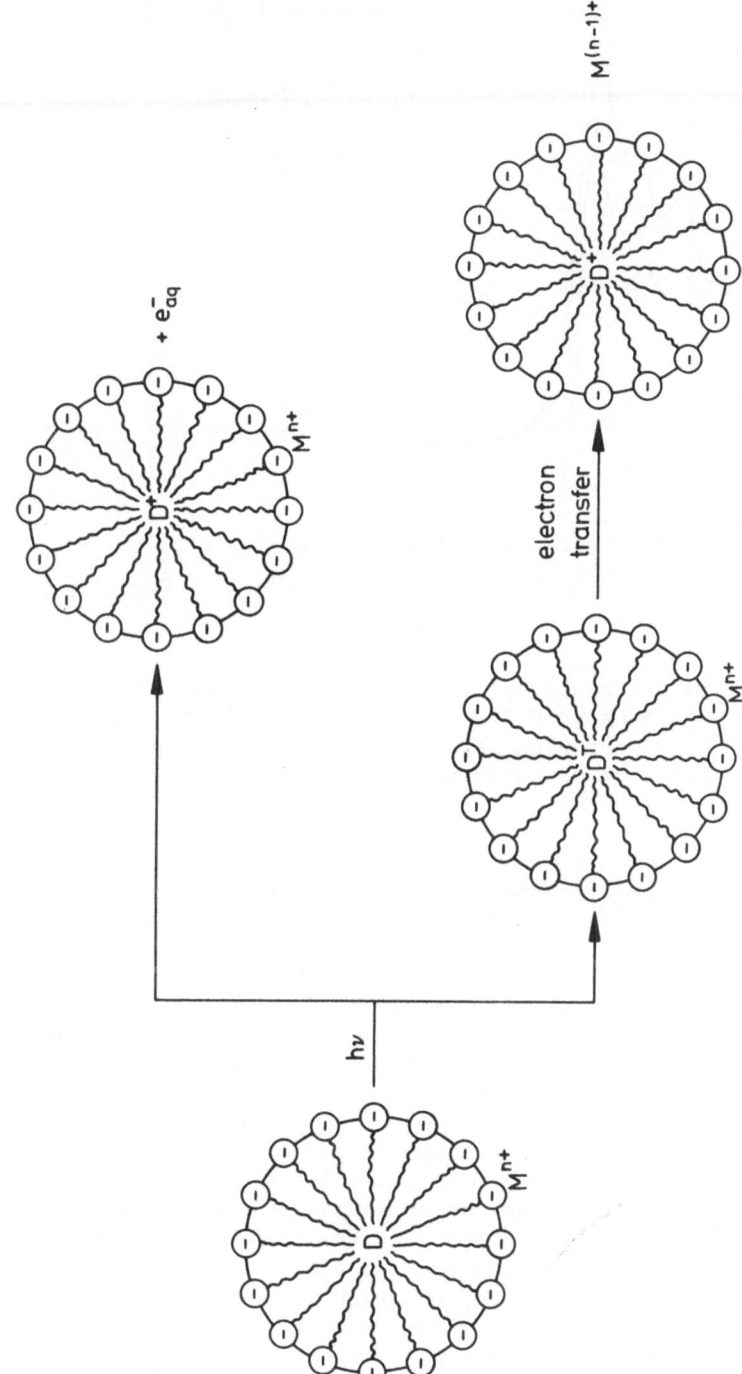

Figure 2. Schematic representation of photoionization and electron transfer processes in solutions of surfactant micelles containing a solubilized photoactive probe D. The electron acceptor is M^{n+} located in the Stern layer of the micelle and the electron is transferred through the Stern layer from the triplet (D^T).

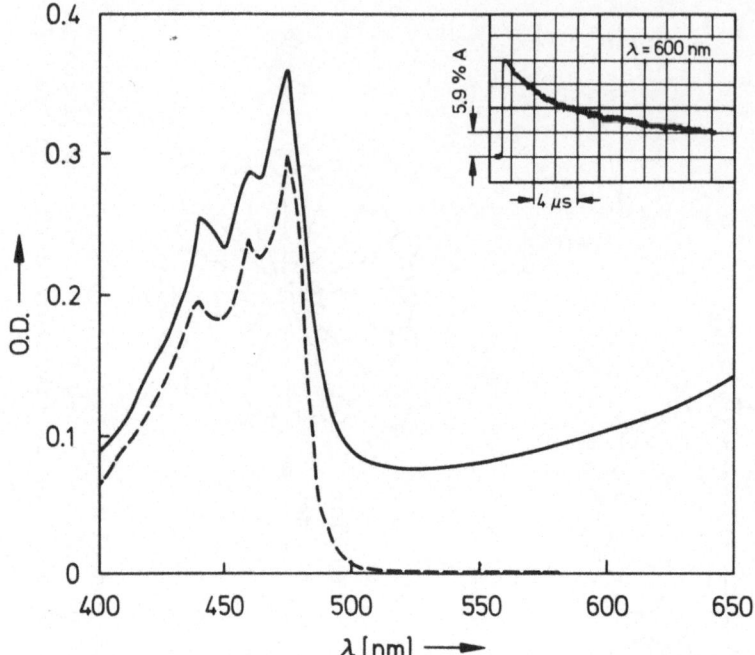

Figure 3. Spectra of transients obtained in the laser photolysis of 5 x 10⁻⁵M TMB in aqueous 0.1 M sodium lauryl sulfate: (–) end of pulse, (---) 10 ms after pulse. Insert: oscilloscope trace showing kinetic behavior of solvated electrons.

These obstacles may be overcome by usage of anionic micellar solutions. A schematic illustration of a photoionization event in such a system is given in Figure 2.

The sensitizer D is solubilized in the hydrophobic interior of the micellar aggregate whose negative surface charge prevents reentry of the hydrated electron and recombination with the parent D^+. These phenomena will now be illustrated by laser photolysis results obtained from solutions of TMB in sodium lauryl sulfate micelles.[2]

Transitory spectra obtained from the 347.1 nm laser photolysis of 5 x 10⁻⁵ M TMB in 0.1 M NaLS micellar solution are presented in Figure 3. Considering first the end of the pulse spectrum one notices significant differences between this and the transitory spectra observed in methanolic or cyclohexane solution. A strong band appears in the long wavelength region which is absent in the latter solvents and which is removed by adding electron scavengers to the solution. It is attributed to the hydrated electron (e^-_{aq}). The three peaks at shorter wavelengths, on the other hand, are readily identified with the TMB^+ spectrum. Apparently, TMB^+ and hydrated electrons are the main species present after the laser pulse:

$$TMB \quad \overset{h\nu}{\rightarrow} \quad TMB^+ + e^-_{aq} \qquad\qquad \text{anionic micelles} \qquad\qquad (6)$$

This contrasts to the results obtained in methanolic or cyclohexane solution where TMB triplets were formed predominantly by the 347.1 nm light flash:

$$TMB \quad \overset{h\nu}{\rightarrow} \quad TMB^T \qquad\qquad\qquad \text{homogeneous solution} \qquad\qquad (7)$$

The relative efficiency of these two competing channels in the TMB photolysis can be expressed quantitatively by the ratio of cation radical and triplet concentrations

$$r = [TMB^+]/[TMB^T]$$

     This parameter was found to be 0.17 in methanol while in NaLS micellar solution a value $r = 6$ was obtained implying that photo-ionization of TMB is strongly promoted by anionic micelles. Similar enhancements have been found with other photoactive donors such as phenothiazine.[3] These micellar effects can be accounted for by invoking a tunneling mechanism for the transfer of an electron from the micellar into the aqueous phase. The energetics of such a pro-cess are depicted in Figure 4. Here the redox potential of the D/D^+ couple in the micelle is expressed on an energy scale where the energy of the electron in vacuum is set equal to zero. $E_o(D^+/D)$ expressed on such a scale is equal to the Fermi energy of this redox system ε_F[4] which is given by the sum of the ionization energy (IP_g) of the donor in the gas phase and the electronic polarization energy $(E_{pol}(D^+))$ of the cation D^+. It is apparent from Figure 4 that the energy of the laser light is not sufficient to produce free electrons in the micelle whose energy is given by the V_o level. By absorp-tion of laser light the $\varepsilon_F(D^+/D)$ value is shifted to a value which lies by an amount of ΔE below V_o. However, this Fermi level of the excited donor is close to the value of $\varepsilon_F(aq/e^-_{aq})$, the Fermi energy of hydrated electrons in water, which allows for electron tunneling from D* across the micellar Stern layer to aqueous traps surrounding the micelle. This process is facilitated further by the negative potential difference between the micellar interior and bulk water which also inhibits recombination of geminate ion pairs. The latter factors which promote ionization are absent in alcoholic solution. Even if the initial photoejection occurred here with the same proba-bility as in NaLS micelles, ion pairs could recombine, a reaction which is a probable source of the high triplet yield found for TMB in methanol.

     It is illustrative to examine the kinetic behavior of the two transients, TMB^+ and e^-_{aq} produced by the laser pulse in more detail. The oscillogram inserted in Figure 3 shows the time dependency of

<u>Electron tunneling picture of photoionization processes.</u>

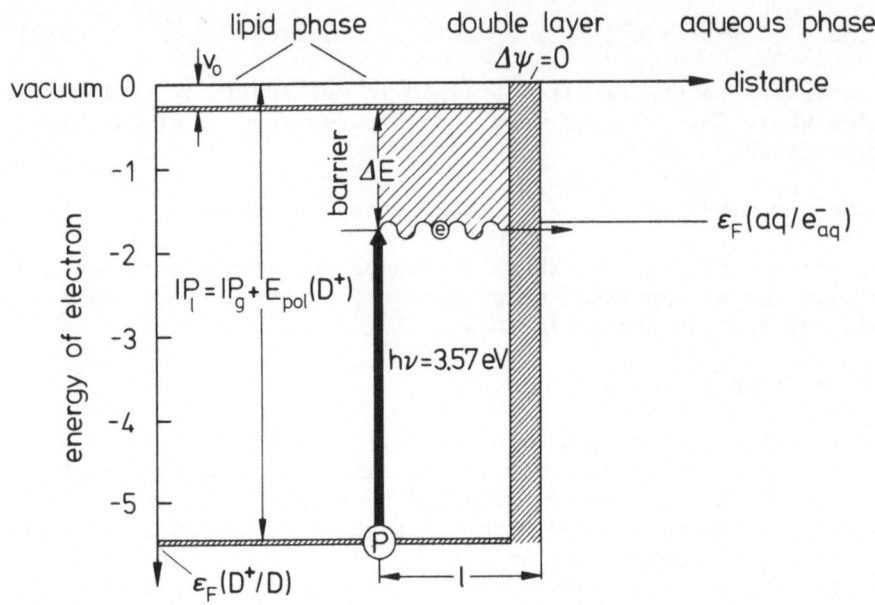

Figure 4. Tunneling model for the photoejection of an electron
from a micelle (lipid phase) into water. The potential difference
between the phases is assumed to be zero.

the absorption at 600 nm, where the hydrated electron contributes
more than 90% to the total optical density change. The photoejection of electrons from micelles into the water and their subsequent
hydration is a very rapid process which is reflected by the immediate
increase of optical density at 600 nm occurring simultaneously with
the laser pulse. Subsequently the absorption signal decays via a
second order kinetics with a first halflife time of 6 μs. As has
been pointed out above, recombination of e_{aq}^- with the parent cation
inside the micelle, i.e. the reverse of photoreaction (5), is prevented by the negative micellar surface potential. The most likely
reaction of e_{aq}^- in the absence of other scavengers is therefore
conversion into hydrogen via reaction (5) for which a rate constant
of 6×10^9 $M^{-1}s^{-1}$ has been reported.[5] From this value and e_{aq}^- concentrations of 10^{-5} to 2×10^{-5} M produced by the laser pulse,
halflife times of 6-12 μs are predicted in agreement with the
experimental observation.
     One point of particular interest and potential practical impor-
tance is whether photoionization processes in micellar systems can

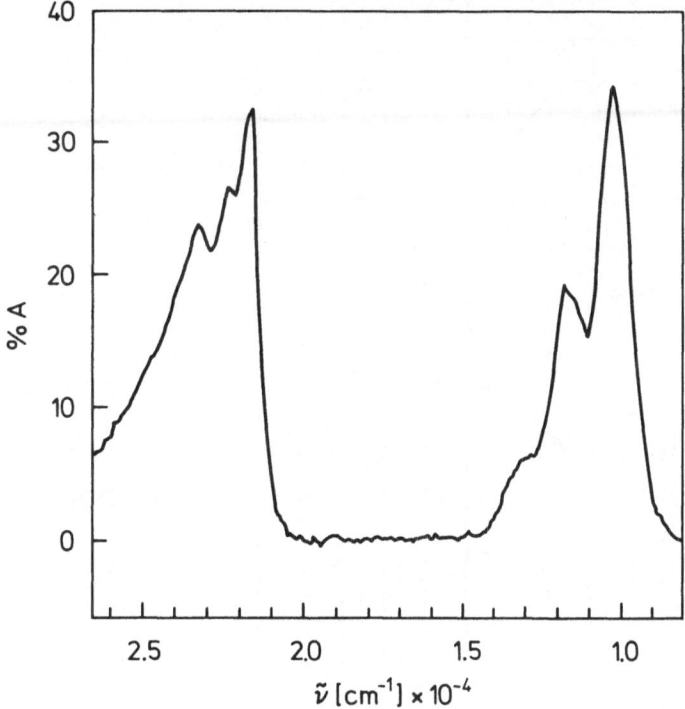

Figure 5. Spectrum of TMB$^+$ cation radical after formation by a ~5 min exposure to sunlight of a deaerated 5 x 10^{-5} M TMB, 0.1 M sodium lauryl sulfate aqueous micellar solution.

be applied to photochemical utilization of solar light energy. In a first exploration of such possibilities we have checked the effect of sunlight on a solution of TMB in NaLS which was of identical composition to the one used in the laser experiments. After thorough deaeration the solution contained in a Pyrex flask was placed in the sunlight. A few minutes' irradiation was sufficient to produce an intense greenish-yellow color which corresponds to an absorption spectrum shown in Figure 5. This spectrum is identical with the TMB$^+$ spectrum. In particular, the absorption around 475 nm matches exactly with the transitory optical density in Figure 3 observed 10 ms after the laser pulse. These observations indicate that photoionization of TMB is readily achieved by sunlight which has a much lower intensity than the laser flash. The monophotonic mechanism of hydrated electron production is thus confirmed. Useful ways of exploiting the strong reducing power of hydrated electrons are conceivable. For example, e_{aq}^- is readily convertible into hydrogen or may be employed to reduce carbon dioxide. However, a second redox system for restoration of the dye from its oxidized form has

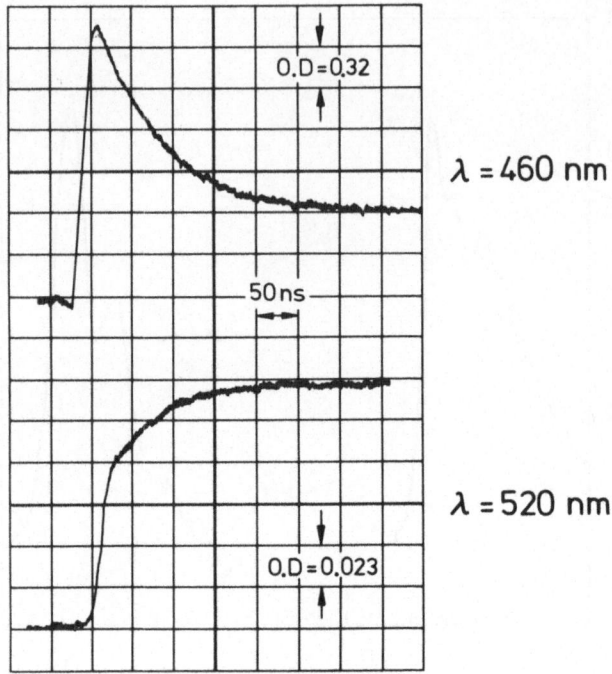

Figure 6. Oscilloscope traces showing PTH^T decay (upper trace) and the PTH^+ formation (lower trace) during the laser photolysis of 5×10^{-5} M PTH in MeOH containing 3×10^{-3} M $Eu(NO_3)_3$.

to be found to make feasible practical application of the photo-ionization process. In addition, it would be desirable to explore photoionizable dyes with absorptions extending further in the visible than those of phenothiazine or TMB in order to assure better utili-zation of the light present in the solar spectrum.

Monophotonic and biphotonic[6,7] ionization processes provide good examples for successful prevention of undesirable back reac-tions by micellar systems. In the following section we shall examine micellar effects on another kind of photo redox reaction of type (2a). This time a metal ion M^{n+} rather than the solvent acts as an electron acceptor and triplet state phenothiazine (PTH^T) represents the excited donor molecule:

$$PTH^T + M^{n+} \rightarrow PTH^+ + M^{(n-1)+} \tag{8}$$

An illustration of this reaction is given in Figure 6, which shows oscillograms obtained from the laser photolysis of pheno-thiazine in methanol containing 3×10^{-3} M $Eu(NO_3)_3$. The optical density at 460 and 520 nm increases during the laser pulse reflec-ting the formation of triplets. The optical signal at 460 nm where PTH^T has a maximum absorption decreases rapidly after a few hundred nanoseconds until a plateau is attained. Conversely, at 520 nm, the wavelength of the PTH^+ absorption maximum, the OD increases further after the pulse until a final level is reached. The kine-tics of the 520-nm growth match those of the 460-nm decay indica-ting the formation of PTH^+ during the reaction of Eu^{3+} with PTH^T. Apparently, the quenching process involves electron transfer from phenothiazine triplets to europium ions. This conclusion is sub-stantiated by examining the transitory spectrum after completion of the triplet decay, Figure 7. The spectrum is identical with

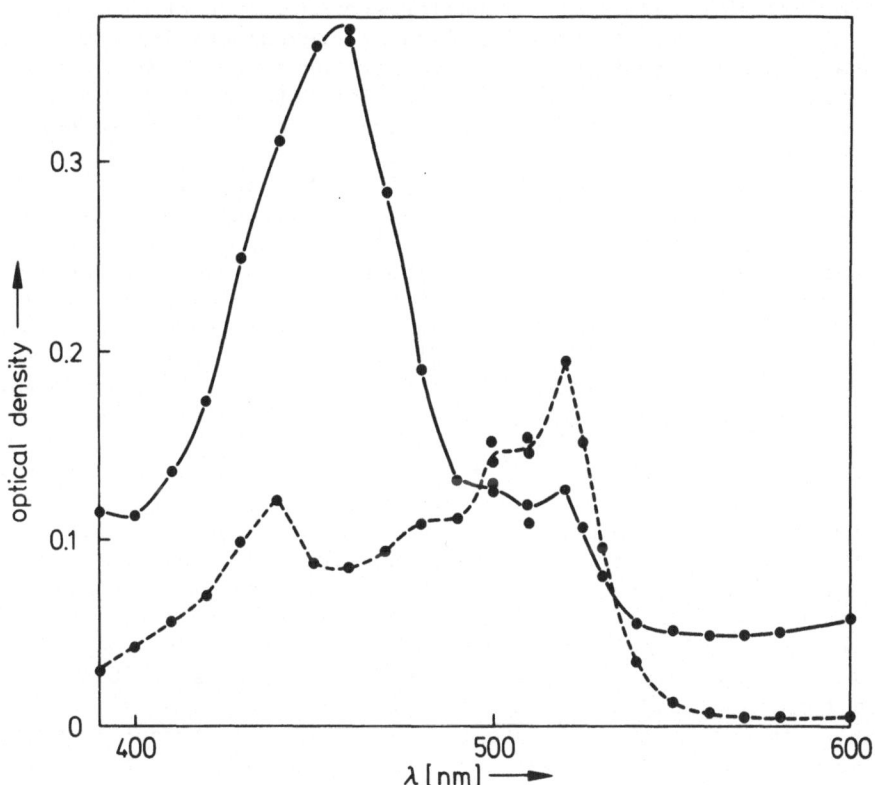

Figure 7. Transitory spectra obtained from the laser photolysis of 5×10^{-5} M PTH in deoxygenated methanol solution containing 3×10^{-3} M $Eu(NO_3)_3$. Solid line: spectra immediately after the pulse. Dashed line: spectra 300 nsec after the pulse.

that of phenothiazine cation radicals, the yield of the latter
being much larger than that observed in Eu^{3+} free solution. Oscilloscope traces such as Figure 6 can be described by first-order kinetics, plots of ln OD vs. time yielding straight lines. From the
slopes, which increase linearly with Eu^{3+} concentration, one
obtains a value for the bimolecular rate constant of the redox
reaction (8)

$$k_8(Eu^{3+}) = 4.7 \times 10^9 \ M^{-1}sec^{-1}$$

It is instructive to examine the effect of anionic micelles on
the reduction of Eu^{3+} by phenothiazine triplets. The peculiarity
of the situation encountered in anionic micellar systems lies in
the fact that multiply charged metal ions, such as Eu^{3+}, are strongly
adsorbed on the surface of the aggregates. Hence a study of this
type of reaction might enlighten us as to the nature of electron
transport processes across charged lipid water interfaces. The
electron transfer reaction in micellar systems, in contrast to the
alcoholic solutions, is not a bimolecular process obeying homogeneous second-order kinetics. It rather has to be regarded as a
summation of intramicellar events occurring between single donor
acceptor pairs. Such a reaction is expected to obey first-order
kinetics, the rate depending on the degree of covering of the
micelle by the acceptor ions and the efficiency of electron transport through the micelle-water interface. A kinetic analysis analogous to the one described for alcoholic solution yields the
interesting result that an average coverage of only one Eu^{3+} per
NaLS micelle accelerates reaction (8) to such an extent that it is
almost completed within the 15 ns duration of the laser pulse.
Hence the electron transfer from PTH^T within the micelle to adsorbed
Eu^{3+} ions must be of an extremely rapid nature. If the redox reaction
would proceed through the stage of a collision complex between
the two reactants then its rate should be relatively slow since
encounters between PTH^T and M^{n+} are improbable. In fact it has been
shown previously that diffusion-controlled intramicellar electron
transfer requires a time period of more than 10^{-6} sec.[7] A large
number of electron, charge transfer, and electrode reactions, on
the other hand, occur via tunneling of the electrons. In particular,
the work by Kuhn[8] on photoejection of electrons from chromophores
embedded in lipid monolayers has established the occurrence of
tunneling of photoexcited electrons through distances of 10 Å and
more within the lipid phase. By analogy with the latter systems
it seems reasonable to postulate that electrons can tunnel from the
PTH^T donor located within the micellar hydrocarbon phase through
the micellar Stern layer to the site of the acceptor in the aqueous
phase. The probability of these processes will depend on the height
of the energy barrier and the overlap of electronic levels of the
two reactants. These phenomena have been dealt with in a previous
paper.[4]

b)   Photo Induced Electron Transfer from a
      Donor to an Excited Acceptor Molecule

This second category of photo redox reactions to be considered
is described by Equation (2b). In the following part we will be
concerned exclusively with electron transfer reactions involving
triplet duroquinone which, in the presence of a suitable electron
donor, is reduced to durosemiquinone anion according to

$$DQ^T + D \rightarrow DQ^- + D^+ \tag{9}$$

A detailed investigation of reaction (9) in water-ethanol mixtures
has been carried out by laser photolysis technique. DQ^T was found
to abstract electrons from a variety of substrates. The rate constant for these redox processes are close to the diffusion controlled
limit (10^9-10^{10} $M^{-1}s^{-1}$) except for D = CO_3^{2-} where k equals 7 x 10^7
$M^{-1}s^{-1}$. The latter case is of particular interest and will now be
dealt with in more detail:[10]

$$DQ^T + CO_3^{2-} \rightarrow DQ^- + CO_3^- \tag{10}$$

Pronounced micellar effects on the kinetics of reaction (10) will
be demonstrated. Also, subsequent reactions of CO_3^- leading to the
formation of carbon superoxide, CO_3, and giving way to oxygen production, will be identified.
      The course of the electron transfer from triplet duroquinone
to carbonate can be readily monitored by kinetic spectroscopy.
Excitation of duroquinone by the laser pulse leads to the formation
of DQ^T with a quantum yield of practically unity. The transitory
spectrum observed immediately after the laser pulse therefore
exhibits the characteristics of the DQ^T spectrum with a broad absorption band in the wavelength region between 520 and 400 nm. During
the course of the redox reaction (10) the triplet spectrum disappears, and, concomitantly, a composite spectrum of CO_3^- and DQ^-
develops. The former species absorbs in the red wavelength region
(λ_{max} = 600 nm, ε_{max} = 1860 $M^{-1}cm^{-1}$), while the latter has an absorption maximum at 440 nm (ε_{max} = 7600 $M^{-1}cm^{-1}$).
      In the following section the kinetics of CO_3^- formation via
reaction (2) in two different media will be compared: first a water-
ethanol mixture (2/1) and second an aqueous solution of dodecyltrimethylammonium chloride (DTAC) micelles. The kinetic analysis
is performed at 600 nm, where CO_3^- contributes almost exclusively to
the transitory absorption.
      Figure 8 shows two oscilloscope traces illustrating the beha-
vior of the optical density at 600 nm after laser irradiation of
the two duroquinone carbonate solutions. In water/ethanol one
notices a gradual increase in the absorbance signal after the laser
pulse. The kinetics of this process are pseudo-first order with
respect to carbonate concentration and the kinetic analysis gives

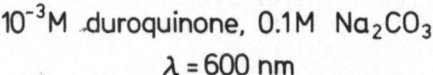

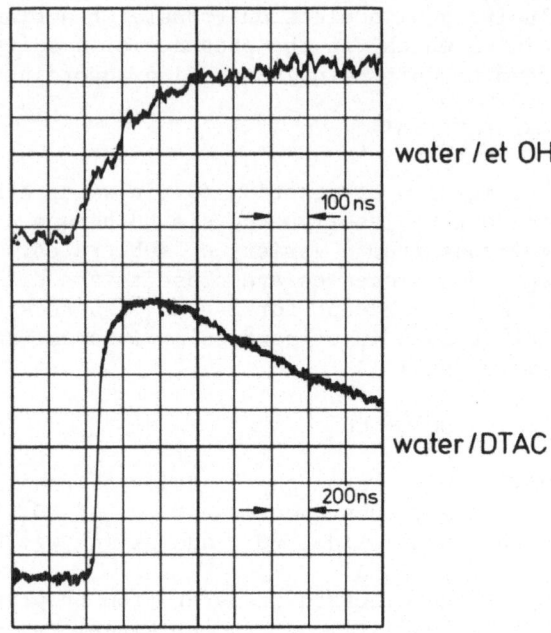

Figure 8. Oscillograms from laser photolysis investigations showing the formation of CO_3^- anion radicals in water ethanol solution and 0.03 M solution of $\bar{D}TAC$.

the rate constant $k_2 = 7 \times 10^7 \ M^{-1}s^{-1}$. The picture changes drastically when micellar solutions are considered. Here the absorbance signal at 600 nm is deflected upwards immediately during the laser pulse indicating a very rapid formation of CO_3^- anion radicals. Subsequently, a small additional growth in optical density occurs which is followed by the onset of a decay process. These effects are explained in terms of a pronounced micellar catalysis of the carbonate oxidation by duroquinone triplets.

A schematic illustration of the events occurring in the micellar system is given in Figure 9. Carbonate ions are expected to interact strongly with the positively charged amphipathic surface

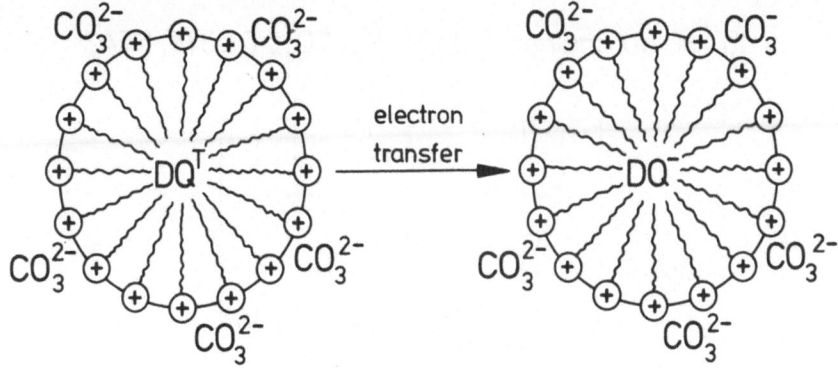

Figure 9.   Schematic illustration of an intramicellar electron transfer from CO_3^{2-} to DQ^T.

of the DTAC micelles and hence will be partially associated with the micellar Stern layer. Duroquinone, on the other hand, is solubilized in the lipoidic interior of the aggregates. Hence the redox reaction between DQ^T and CO_3^- has to be envisaged as a summation of intramicellar events. Thereby electrons will flow from the micellar surface, the site of the donor, across the lipid/water interface to the acceptor inside the aggregate. The laser photolysis results show that this process occurs with an extremely rapid rate. It should be recalled that similar high rates have been observed for the other type of intramicellar redox reaction discussed above. This effect may also be explained by assuming a tunneling mechanism for the electron transfer. It has been demonstrated recently[11] that electronic interaction between electron donor and acceptor pairs can range over distances considerably larger than collision diameters. (As a result the cross section of an electron transfer reaction can be considerably larger than the collision cross section.) These forces may still be significant at distances comparable to the micellar radius (16 Å) thus enabling rapid electron tunneling from the CO_3^{2-} donor at the surface to DQ^T in the interior of the micelle.

At the DTAC concentrations employed in our experiments a small amount of DQ is present in the aqueous phase. A minor fraction of triplet duroquinone will therefore react outside the micelles. The slower rising portion of 600 nm absorbance noted in Figure 8 is attributed to the latter bulk reaction.

10^{-3} M duroquinone, 0.1 M Na_2CO_3, 0.03 M DTAC

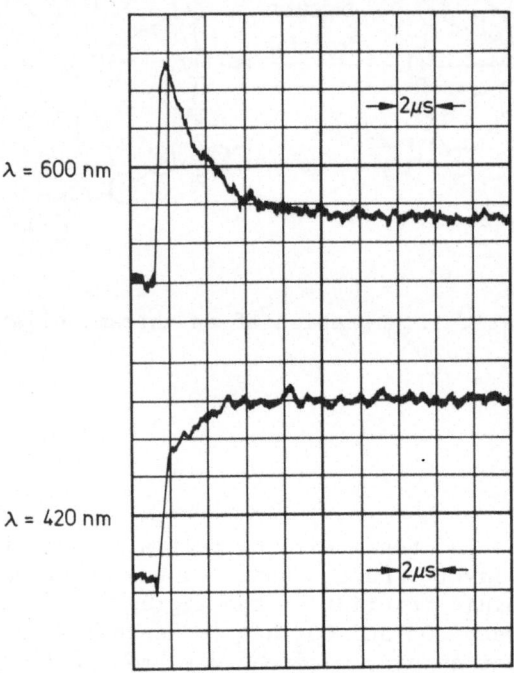

Figure 10.   Oscillograms from the laser photolysis of duroquinone in micellar DTAC solution, charge transfer from CO_3^- to DQ.

We now proceed to examine the further fate of the CO_3^--anion radical in the duroquinone/DTAC solution. The upper oscillogram displayed in Figure 10 shows the decay of the 600 nm absorption on a compressed time scale. This was found to obey first-order kinetics and was enhanced upon increasing the duroquinone concentration. This result suggested that CO_3^- is eliminated via reaction with ground state duroquinone. In order to identify the products resulting from the reaction the behavior of the transient absorption was examined in the entire wavelength region between 380 and 650 nm. The lower oscillogram in Figure 10 shows data obtained at 440 nm, the absorption maximum of DQ^-. An initial rise in absorbance is followed by a slower growth that occurs concurrently with the decay of the CO_3^- absorption at 600 nm. The spectrum in the 380 to 650 nm region present after completion of this process was found to be identical with that of durosemiquinone anion. From these results it is inferred that CO_3^- reduces duroquinone according to the equation

$$CO_3^- + DQ \rightarrow DQ^- + CO_3 \tag{11}$$

The transfer of an electron from CO_3^- to DQ must lead to the formation of carbonperoxide, CO_3, or a hydrated form of it. The question
may be raised why reaction (11) can compete so efficiently with the
back transfer of an electron from DQ^- to CO_3^- which is thermodynamically favorable. It should be noted at this point that the DQ^-/CO_3^-
radical pair on a DTAC micelle has a total spin of 1, immediately
after the electron transfer has occurred. This is a consequence of
the fact that the redox reaction (10) starts with triplet duroquinone. Therefore until spin relaxation has taken place which requires
several microseconds the reaction

$$DQ^- + CO_3^- \rightarrow DQ + CO_3^{2-} \tag{11}$$

leading to a singlet system is multiplicitly forbidden. During
this period of time CO_3^- may have escaped from the host micelle
where it was formed initially. Subsequently it is trapped in the
potential field of another micelle containing duroquinone where
reaction (11) can occur.

Carbon peroxide produced via reaction (10) decomposes according
to the overall equation

$$CO_3 \rightarrow CO_2 + \frac{1}{2} O_2 \tag{12}$$

In alkaline medium this reaction is followed by

$$CO_2 + 2 \, OH^- \rightarrow CO_3^{2-} + H_2O \tag{13}$$

Summation of the above equations yields for the overall chemical
change produced upon illumination of duroquinone/carbonate solutions
in H_2O/DTAC

$$2 \, DQ + 2 \, OH^- \xrightarrow{h\nu} 2 \, DQ^- + \frac{1}{2} O_2 + H_2O \tag{14}$$

Hence DQ^- which is stable in alkaline medium and oxygen should be
the final products of the photo reaction in this system. Results
from steady state irradiations carried out with visible light (λ
> 400 nm) support this mechanism. Spectrophotometric measurements
showed that under illumination DQ is converted into DQ^- whereas the
second product O_2 could be detected analytically.

Apparently CO_3^- exhibits electron donating as well as electron
accepting properties. While the latter have been recognized before,
this study provides the first evidence for reactions in which
CO_3^- acts as a reducing agent. This redox behavior of CO_3^- has also
recently been confirmed by short time pulse polarography experiments. It has bearings on the light induced oxygen evolution from
water _in vitro_ as well as _in vivo_.

REFERENCES

1. G. Beck, J. Kiwi, D. Lindenau, and W. Schnabel, Eur. Polym. J. 10, 1069 (1974).
2. S. A. Alkaitis and M. Grätzel, J. Amer. Chem. Soc. 98, 3549 (1976).
3. S. A. Alkaitis, G. Beck, and M. Grätzel, J. Amer. Chem. Soc. 97, 5723 (1975).
4. S. A. Alkaitis, A. Henglein and M. Grätzel, Ber. Bunsenges. Phys. Chem. 79, 541 (1975).
5. E. J. Hart and M. Anbar, "The Hydrated Electron", Wiley Interscience, New York, N.Y., 1970.
6. S. C. Wallace, M. Grätzel and J. K. Thomas, Chem. Phys. Lett. 23, 359 (1973).
7. M. Grätzel and J. K. Thomas, J. Phys. Chem. 78, 2248 (1974).
8. H. Kuhn and D. Möbius, Angew. Chem. 83, 672 (1971).
9. R. Scheerer and M. Grätzel, J. Amer. Chem. Soc., Jan. 1977.
10. R. Scheerer and M. Grätzel, Ber. Bunsenges. Phys. Chem., in print.
11. P. A. Carapellucci and D. Mauzerall, Annals of the New York Academy of Sciences, 244, 214 (1975).

RADIATION INDUCED REDOX REACTIONS IN MICELLAR SOLUTIONS

A. J. Frank[1]

Hahn-Meitner-Institut für Kernforschung Berlin GmbH

Bereich Strahlenchemie, D-1000 Berlin 39

This paper outlines some recent observations made during pulse radiolysis of fast redox reactions in aqueous solutions containing micellar aggregates. The dismutation kinetics of Br_2^- radicals in micellar cetyltrimethylammonium bromide solution is reported. Two-dimensional intramicellar surface diffusion and three-dimensional intermicellar (or solution bulk type) diffusion can be distinguished. Similarly a two-component kinetic feature is resolved in the electron-transfer reaction between triplet pyrene, located in the interior of the micelle, and Br_2^- radicals in the surrounding aqueous phase. The kinetics of the electron transfer of singlet and triplet pyrene in micelles with various radical ions (CO_2^-, CH_2O^-, CH_3CHO^-, and $CH_3COCH_3^-$) in aqueous solutions containing micelles are reported. In addition the affect of the interface potential on the reaction rates of the hydrated electron and ethanol radical with various acceptor molecules differing in their electron affinity are also considered. These results are discussed in terms of current theories on electron-transfer reactions. The movement of the electron from the aqueous phase across the electrical double layer of the micelle to an acceptor molecule solublized in the micelle is related to the favorableness of electron tunneling.

INTRODUCTION

Electron transfer across the electrical double layer of a micelle and on the micellar surface have been the subject of some recent pulse radiolysis investigations [2,3,4,5]. Employing a combination of pulse radiolysis and flash photolysis techniques, information was obtained on the mechanism for electron transfer between excited (or ground) state molecules residing in the micellar interior and radical anions formed in the aqueous phase. Rate constants for the processes exhibit a dependence on the electron affinity of the acceptor molecule and on the interface potential. These effects were interpreted as electron tunneling between the occupied electronic redox levels of the donor system D/D^- and the unoccupied levels of the acceptor system A/A^-.

In another study[5] the influence of dimensionality of reaction space on the rate constant for the dismutation reaction of the Br_2^- radical was investigated. For this system it is possible to kinetically resolve surface and bulk diffusion reactions. Randon-walk theory for two- and three-dimensional lattices with traps was invoked to explain the experimental results.

In this paper, selected examples from these investigations[2,3,4,5] will be used to illustrate general statements on electron-transfer reactions and the affect of dimensionality on reaction space in aqueous solutions containing spherical micellar aggregates of cetyltrimethylammonium bromide (CTAB) , dodecyltrimethylammonium chloride (DTAC), or sodium lauryl sulfate (SLS).

EXPERIMENTAL

The experimental arrangement has been described earlier[2,6]. Redox reactions were initiated with a 5-20 ns pulse of high energy electrons from a 12 MeV linear accelerator. Transient signals were monitored using kinetic spectroscopy with nanosecond time resolution in optical absorption.

RESULTS AND DISCUSSION

Considerations on Electron Tunneling in Heterogeneous Systems

The most important factors determining the rate of tunneling through the charged water-lipid interface of a micelle are the height and width of the potential barrier and the overlap of occupied electronic redox levels of the donor system D^+/D with unoccupied levels of the acceptor system A/A^-.

The rate constant of the reaction $D + A \longrightarrow D^+ + A^-$ is approximately given by equation $(1)^7$.

$$k = k_e \quad \text{for } \tau_e > \tau_r \tag{1a}$$

$$k = k_e \, \tau_e \, \omega \int_{-\infty}^{+\infty} K(\epsilon) \, D_{occ}^D(\epsilon) \, D_{unocc}^A(\epsilon) \, d\epsilon \quad \text{for } \tau_e < \tau_r \tag{1b}$$

where k_e is the specific rate of encounters determined from Smoluchowski's equation. The terms $D_{occ}(\epsilon)$ and $D_{unocc}(\epsilon)$ are normalized distribution functions of the occupied and unoccupied redox levels of the systems D^+/D and A/A^-. Because the donor and acceptor molecules may be located in different environments in micellar systems, a common energy is chosen in which the energy of the electron in the gas phase (or, more precisely, the vacuum level) is taken as zero. Thus ϵ is the change in free energy accompanying a Franck-Condon transition from the gas phase to an acceptor A in solution ($A + e_g^- \longrightarrow A^-$) or in the reverse direction, that is, from the donor D in solution to the gas phase ($D \longrightarrow D^+ + e_g^-$). The other terms of Equation (1) are the unreactive encounter time τ_e, the tunneling time τ_r, the tunneling attempt frequency ω (ca. 10^{15} s^{-1}), and the transmission factor $K(\epsilon)$. The transmission factor $K(\epsilon)$ is dependent on the height and width of the potential barrier. The requirement for a diffusion-controlled reaction is that the integral of Equation (1b) is larger than 10^{-4} [7].

The occupied and unoccupied distributions of redox levels of a system are equal at $\epsilon = \epsilon_F^o$ where ϵ_F^o represents the Fermi energy. The Fermi energy of the acceptor system is given by Equation (2).

$$\epsilon_F^o(A/A^-) = -EA(A) + \Delta G_s(A^-) - \Delta G_s(A) - e\Delta\psi \tag{2}$$

where EA is the electron affinity, ΔG_s is the free energy of solvation, e is the elementary charge and $\Delta\psi$ is the potential difference between the lipid phase of the micelle and the aqueous phase to which it is referenced. $\Delta G_s(A)$ is negligible compared with $\Delta G_s(A^-)$ and can be ignored. The latter term may be obtained from Born's equation[4,7]. Similarly, the Fermi energy of the donor molecule D may be determined with the aid of Equation (3).

$$\epsilon_F^o(D^+/D) = -IP(D) - \Delta G_s(D^+) + \Delta G_s(D) - e\Delta\psi \tag{3}$$

where IP is the gas phase ionization energy and the other terms have been defined. The Fermi energy is related to the standard redox potential of an aqueous solution by the relation $E_o = -(4.5\pm0.15) - \epsilon_F^o$.

According to thermodynamics, the reduction of an acceptor A by a donor D can occur if the redox potential of the system A/A^- is more positive than that of the system D^+/D. However, it is not necessarily true that the rate of the redox reaction will be enhanced as the difference between the redox levels of the system A/A^- and D^+/D are augmented. When there is weak interaction between the donor and acceptor systems as is approximately realized in the case of species separated by a water-lipid interface, the rate of reaction can be semiquantitatively correlated in terms of the overlap of the electronic redox levels of the reactants. Irrespective of thermodynamic favorability, the rate of electron transfer will be low if little overlap exists between the occupied levels of the donor system and the unoccupied levels of the acceptor system. More general theories have been developed by Marcus[8] and Levich[9].

Electron Transport Across the Water-lipid Interface[2,3]

The reactions of CTAB-solubilized singlet and triplet pyrene with CO_2^- and several α-alcohol radical anions of the general type $R_1R_2CO^-$ (CH_2O^-, CH_3CHO^- and $CH_3COCH_3^-$) were investigated. Because of their charge, the radical anions are restricted to the aqueous phase and hence electron transfer must proceed across the electrical double layer to the water-insoluble pyrene molecule located in the lipid phase of the micelle.[2]

In Figure 1 are the estimated distributions of occupied and unoccupied electronic redox levels of various redox systems of CO_2 in water and pyrene (P) in CTAB. Using diagrams such as Figure 1, conclusions can be drawn on the reactivity of species and estimates can be made of the magnitude of rate constants for electron transfers. For example, there is good overlap between the occupied levels of the redox couple CO_2/CO_2^- and the unoccupied levels of the pyrene triplet system $(^3P/P^-)_{mic}$. As a consequence, the rate for the electron transfer from CO_2^- to triplet pyrene is expected to be efficient, which it is, 5.0×10^9 $dm^3mol^{-1}s^{-1}$.

$$CO_{2aq}^- + {}^3P_{mic} \longrightarrow CO_{2aq} + P_{mic}^- \qquad (4)$$

In contrast the small overlap between the occupied levels of the CO_2/CO_2^- system and the singlet pyrene system $(^1P/P^-)_{mic}$ suggest the possibility of a very low rate constant for the reaction between CO_2^- and singlet pyrene. In fact, no reaction is observed. The relative position of the Fermi levels of $CO_2^-/CO_2^=$ and $^1P/P^-$ and the extent of overlap of the distribution functions indicate that Reaction (5) should proceed readily.

$$CO_{2aq}^- + P_{mic}^- \longrightarrow CO_{2aq}^= + P_{mic} \qquad (5)$$

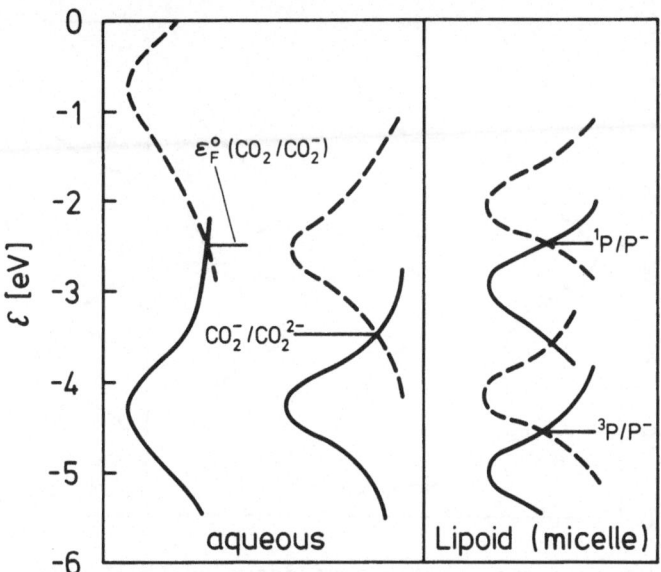

Figure 1. Estimated distributions of occupied (full lines) and unoccupied (dashed lines) electronic redox levels of various redox
systems of CO_2^- in water and of P^- in CTAB micelles ($\Delta\psi = +0.2V$)[2].

Experimentally, Reaction (5) is found to have a diffusion-controlled
rate constant, 1.8×10^{10} $dm^3mol^{-1}s^{-1}$. Reaction (6) is also expected
to occur, however, with a low rate constant because of the small
overlap of occupied levels of the $(P/P^-)_{mic}$ system with unoccupied
levels of the CO_2/CO_2^- couple.

$$P^-_{mic} + CO_{2aq} \longrightarrow P_{mic} + CO^-_{2aq} \tag{6}$$

The experimental results bear out this prediction with a rate constant of 1×10^7 $dm^3mol^{-1}s^{-1}$.

The estimated distributions of occupied and unoccupied electronic redox levels of CH_2O^-, CH_3CHO^- and $(CH_3)_2O^-$ are shown in
Figure 2 along with those of singlet and triplet pyrene. As the
Fermi energies become more positive, that is, in going from the
aldehyde radical anion to the acetone radical anion, the rate of
reaction with the singlet state is expected to increase,

$$^1P_{mic} + R_1R_2CO^- \longrightarrow R_1R_2C=O + P^-_{mic} \tag{7}$$

where $R_1R_2CO^-$ has been defined. No reaction between the singlet
state and the formaldehyde radical anion is anticipated because
the Fermi level of the radical lies below that of the singlet state.
Quite the opposite trend is followed by the triplet state.

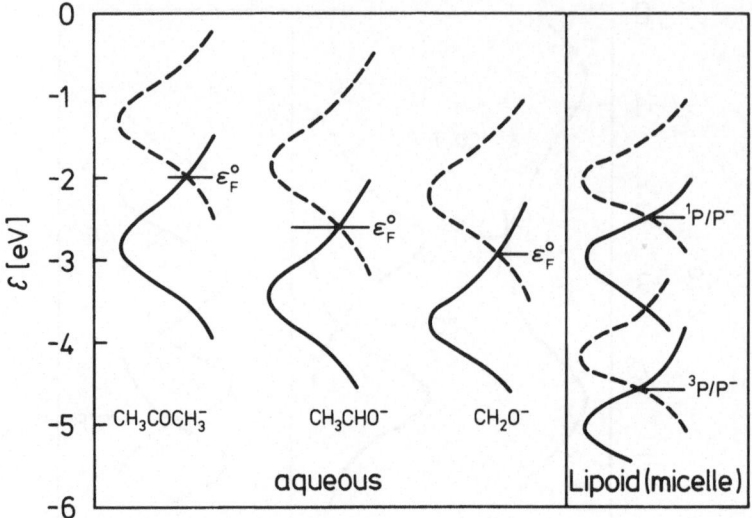

Figure 2. Estimated distributions of occupied (full lines) and un-occupied (dashed lines) electronic redox levels of various radical anions in water and of pyrene anions in CTAB micelles ($\Delta\Psi=+0.2V$)[2].

$$^3P_{mic} + R_1R_2CO^- \longrightarrow R_1R_2C{=}O + P^-_{mic} \qquad (8)$$

The rate of electron transfer from the donor radical $R_1R_2CO^-$ to the triplet is expected to <u>decrease</u> as the reduction potential of the donor species becomes more positive.

Table 1 summarizes the measured rate constants for the electron transfer reaction between the various radical anions and the singlet and triplet states of pyrene. The rate constants for reactions with the singlet state of pyrene are enhanced as the redox potentials of the radical anion become more negative in accord with thermodynamics. The opposite trend is observed for the triplet state. Both effects can be understood in terms of overlap of occupied electronic redox levels in the donor system with unoccupied levels in the acceptor system.

The redox levels of the donor and acceptor systems can be shifted relative to one another by changing the direction and magnitude of the interface potential[3]. This may be done by altering the charge of the micellar head groups or changing the concentration of the electrolyte in the aqueous phase. Predictions of the consequence of the shift are that the rate constant for electron tunneling across the interface will exhibit a dependence on the electron affinity of the acceptor system and the ionization

Table I. Rate Constants in $dm^3mol^{-1}s^{-1}$ of Reactions of Various Radical Anions with Singlet and Triplet Pyrene in CTAB Micelles at a pH of 13^2

Radical Anion	CH_2O^-	CH_3CHO^-	CO_2^-	$CH_3COCH_3^-$
Rate Constant for Reaction with:				
Singlet Pyrene	$-$	1.7×10^8	$-$	2.3×10^9
Triplet Pyrene	1.8×10^{10}	8.0×10^9	5.0×10^9	2.3×10^9
$E_o(D/D^-)$	-1.6	-1.8	-2.0	-2.5
$\varepsilon_F^0(D/D^-)$	-2.9	-2.7	-2.5	-2.0

potential of the donor system as may be inferred from Equations (2) and (3). The former possibility was examined for the reaction of the hydrated electron and ethanol radical with a variety of acceptors whose gas phase electron affinities ranged from 0.6 to 6.5 eV.

$$e_{aq}^- + A_{mic} \longrightarrow A_{mic}^- \qquad (9)$$

$$CH_3CHOH + A_{mic} \longrightarrow A_{mic}^- + H_{aq}^+ + CH_3CHO \qquad (10)$$

Representative examples of the oscillograms for the system studied are those of 1,2,4,5-tetracyanobenzene (TCNB) in aqueous solutions of CTAB micelles shown in Figure 3. The signal of the TCNB anion (λ_{max} = 460 nm) is composed of a fast and a slow component. The growth kinetics of the fast component matches the first-order decay of the electron absorption at 600 nm. The slower component at 460 nm corresponds to the reaction of the ethanol radical with TCNB.

In Figure 4 is shown a plot of the rate constants for the reactions of the hydrated electron with various acceptors (TCNB, tetranitromethane, p-chloranil, nitroanthracene and pyrene) against the interface potential. For cationic micelles of high surface potential ($\Delta\Psi$ = 0.5 V), the rate constants show little dependence on the electron affinities of the acceptors and lie near the diffusion-controlled limit, about 3×10^{10} to 10^{11} $dm^3mol^{-1}s^{-1}$. As the interface potential becomes more negative, the rate constants become more dependent on the nature of the acceptor. At a surface potential of -0.5 v, the values of the rate constants vary over three orders of magnitude. Acceptors like 1,2,4,5-tetracyanobenzene and tetranitromethane which have similar electron affinities (1.8 ± 0.2 eV) display the least variation. Presumably,

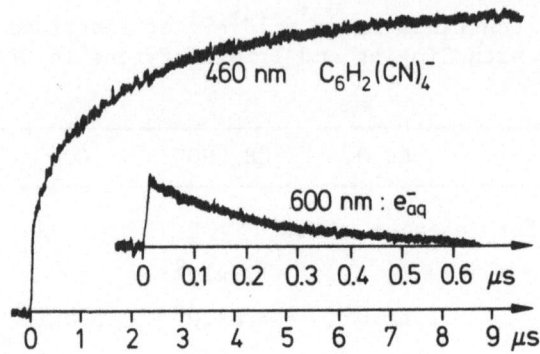

Figure 3. Absorption versus time curves for the decay of the hydrated electron and the build-up of the anion of tetracyanobenzene for CTAB micelles, 10^{-1} mol dm^{-3} Na_2SO_4 and 10^{-4} mol dm^{-3} $C_6H_2(CN)_4$[3].

at this electron affinity, the relative position of the Fermi levels of the redox couples A/A^- and aq/e_{aq}^- corresponds to large overlap of the unoccupied and occupied redox levels of the respective systems (cf. Figure 5).

Table II lists the rate constants for the reaction of the hydrated electron and ethanol radical with various acceptors at

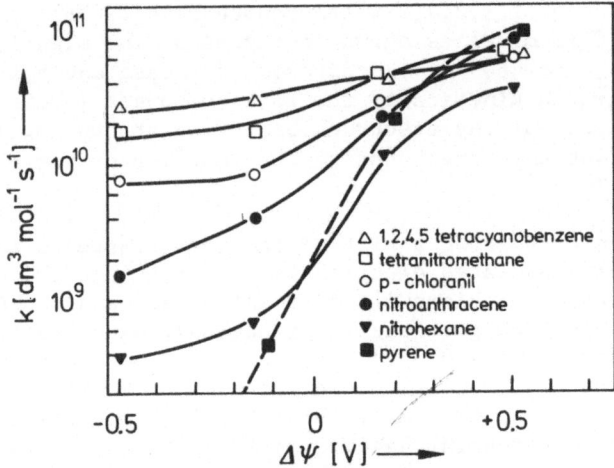

Figure 4. Rate constants of the hydrated electron with various acceptors as a function of the potential difference between the lipid part of the micelle and the aqueous phase[3].

Table II. Rate Constants k in $dm^3mol^{-1}s^{-1}$ of the Reactions of the Hydrated Electron and the Ethanol Radical with Acceptors in Various Micelles[3]

Acceptor	CTAB or DTAC Micelles				SLS Micelles			
	+ 0.5 V		+ 0.15 V		- 0.15 V		- 0.5 V	
	e^-_{aq} $k \times 10^{-10}$	CH_3CHOH $k \times 10^{-9}$	e^-_{aq} $k \times 10^{-10}$	CH_3CHOH $k \times 10^{-9}$	e^-_{aq} $k \times 10^{-10}$	CH_3CHOH $k \times 10^{-9}$	e^-_{aq} $k \times 10^{-10}$	CH_3CHOH $k \times 10^{-9}$
TNM*	7.7	3.5	5.0	3.3	1.8	3.8	1.8	3.5
TCNB*	7.8	4.1	4.2	4.0	3.0	3.8	2.6	3.3
TCBQ*	6.3	3.3	3.0	3.1	8.7	3.4	7.7	2.9

*The symbols are defined as follows: TNM = Tetranitromethane, TCNB = 1,2,4,5-Tetracyanobenzene, TCBQ = 1,2,4,5-Tetrachlorobenzoquinone

different interface potentials. In the case of the ethanol radical,
the rate constants of the charge transfer were independent of the
surface charge of the micelles and the electron affinity of the
acceptors. The values of the rate constants (ca. 3.5×10^9 dm^3
$mol^{-1}s^{-1}$) are typical of homogeneous diffusion-controlled reactions
of neutral free radicals with small molecules. Apparently the activation energy is small for the movement of a neutral radical
from the aqueous phase to the lipid interior of the micelle.

The rate constants of the hydrated electron as a function of
the free energy ΔG of Reaction (9) are plotted in Figure 5. The
free energy ΔG was determined from the difference in Fermi levels
of the acceptor and donor systems: $\varepsilon_F^0(A/A^-) - \varepsilon_F^0(aq/e_{aq}^-)$. Congru-
ous with considerations of the redox level distributions[7] and the
Marcus theory[8], the rate constants become larger as ΔG becomes
more negative, reach the diffusion-controlled limit at about -1 eV
and then decrease. The diffusion-controlled limit occurs at electron affinities between 1.6 eV and 2.0 eV; at higher or lower
electron affinities, the rate constant is less. As the acceptors
used belong to different classes of compounds, however, more detailed conclusions should not be made. Differences in the hydrophobicity of the acceptors will also determine their relative
position in the micellar interior, an effect that will influence
the rate of reaction.

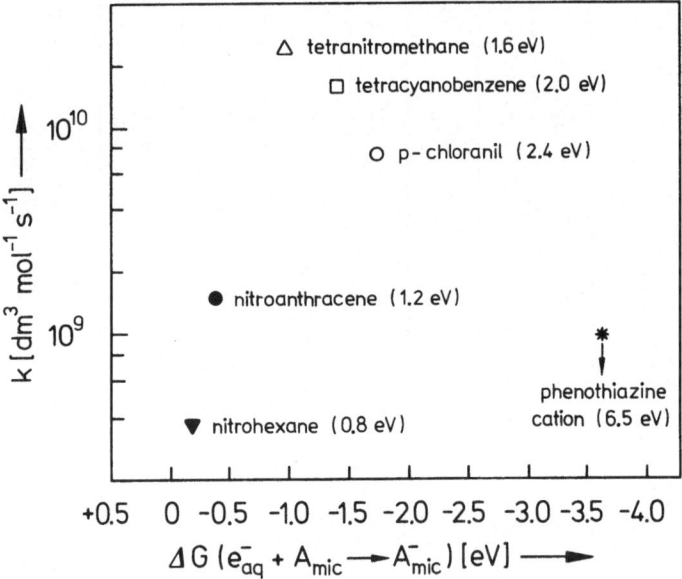

Figure 5. Rate constants as a function of ΔG for the reaction
$e_{aq}^- + A_{mic} \longrightarrow A_{mic}^-$. The values in parantheses are the electron
affinities of the species[3].

The invariance of the rate constants for the reactions of the hydrated electron with acceptors in cationic micelles (Figure 4) must reflect the long residence time of the electron in the positive potential field of the micelle ($\Delta\Psi$ = 0.5 V). Consequently, the time required for electron tunneling may vary over several orders of magnitude and still not influence significantly the measured rate constant. When the tunneling time is longer than the duration of an encounter, the rate constant will be lower than the diffusion-controlled limit. In this context as the surface potential becomes more negative, the rate of reaction will depend more on the tunneling time.

Electron Transport Across the Lipid-water Interface[4]

Up until now the discussion has centered on electron transfer across the water-lipid interface. In this section we investigate the transport of charge in the reverse direction, that is, electron movement from a stable donor D in the lipid phase of the micelle to a transitory acceptor ion A in the aqueous phase. In particular, the oxidation of triplet pyrene in cationic CTAB micelles by Br_2^- radicals in the aqueous phase is examined[4].

$$^3P_{mic} + Br_{2aq}^- \longrightarrow P_{mic}^+ + 2Br_{aq}^- \tag{11}$$

The effect of the triplet energy (2.1 eV) is to make the triplet state of pyrene a better reductant than the ground state which is unable to reduce Br_2^-. Earlier in this paper it was pointed out that the triplet state of pyrene can, unlike the ground state, reduce CO_2^- and H_2CO^-. Quite generally, depending upon the nature of the other reactant, a molecule in the triplet state can be both a better reductant and oxidant than one in the ground state. The triplet state of a molecule will have both a lower ionization potential and a higher electron affinity than that of the ground state by the amount of the triplet state energy.

The reactions of Br_2^- in aqueous solutions containing CTAB micelles in the absence of pyrene will be discussed in the subsequent section. Here we examine theoretical predictions for the redox reactions involving the pyrene and Br_2^- systems in CTAB micellar solution and consider the unusual kinetic behavior of the reactants.

The experiment was undertaken to examine the possibility that triplet pyrene could reduce Br_2^- as was predicted from the calculations of redox levels shown in Figure 6. It can be seen in the figure that only triplet pyrene and not the ground state is

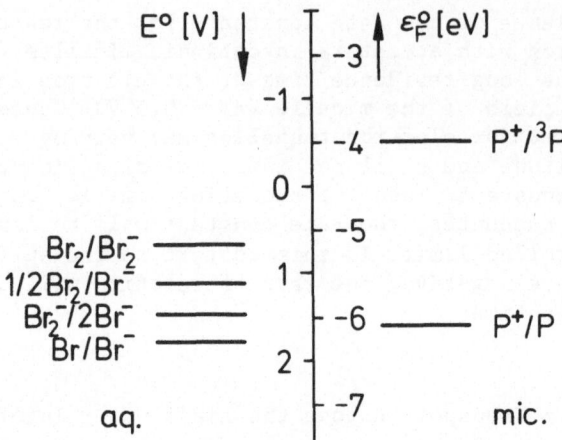

Figure 6. Standard redox potentials, E_o, and Fermi energies, ε_F^o, of various systems in aqueous solution and in the lipid part of a CTAB micelle. P:pyrene[4].

expected to reduce Br_2^-. Also the cation of pyrene is not anticipated to back react with the bromide anion to form neutral pyrene and Br atoms. However, it is feasible, based on the redox levels in the diagram, that P^+ can react with Br_2^- via Reaction (12).

$$P^+_{mic} + Br^-_{2aq} \longrightarrow P_{mic} + Br_{2aq} \tag{12}$$

These predictions were confirmed by experiments.

In Figure 7 are displayed the experimental results. The decay of the pyrene triplet in Figures 7a and 7b shows a fast and a slow component which we attribute to Reaction (11). The fast component follows first-order kinetics with a rate constant of $2.3 \times 10^6 s^{-1}$ and the slow one obeys second-order kinetics with a rate constant of $1 \times 10^9 dm^3 mol^{-1} s^{-1}$. Kinetically matching the fast component decay is the growth of the pyrene cation ($\lambda_{max} = 448$ nm). About 13% of the initially formed Br_2^- radicals are consumed during the decay of the fast component while at the same time about 16% of the micelles contain triplet molecules. The near agreement of these numbers together with the two-component kinetic features discussed above support the following explanation: Initially Br_2^- radicals are formed homogeneously in the aqueous phase but are rapidly trapped in the positive potential field of the micelles. In the event that a Br_2^- radical meets a triplet containing micelle on the first encounter, the reaction will take place. A much longer time would be required for a reaction to occur if the first encounter involved a triplet-free micelle because of the long residence time of Br_2^- in the vicinity of the positively charged

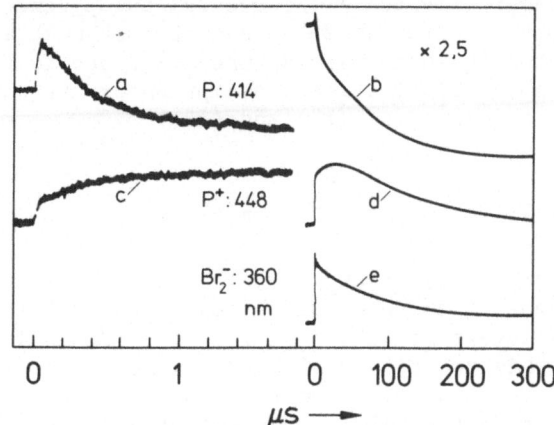

Figure 7. Absorption versus time curves for the pyrene anion (P^-), pyrene cation (P^+) and Br_2^- in solution containing 10^{-2} mol dm^{-3} CTAB, 2×10^{-2} mol dm^{-3} NaBr and 10^{-4} mol dm^{-3} pyrene.

micelle, at least 2 μs[5].

    Several additional arguments support this mechanism. For example, when the ratio of micelle to triplet pyrene is increased, the fraction of the Br_2^- decay due to the fast component decreases indicating a smaller probability that the first encounter will involve a triplet containing micelle. The time constant for the fast process is shown to be determined by the time required for Br_2^- to react with triplet pyrene and not the time for it to diffuse to the micelle. For instance, the time constant is invariant when the micelle concentration is changed while the ratio of pyrene to micelle is kept constant. Furthermore the conclusion is corroborated by an estimation of the time required for diffusion of Br_2^- radicals to CTAB micelles. A diffusion-controlled rate constant k_e is calculated to be 2.4×10^{11} dm^{3}mol^{-1}s^{-1} using Smolochowski's equation corrected for charged encounters. When k_e is substituted into the expression $\tau_e = \ln 2/k_e$ [micelle], a half-life is obtained which is about 5 times shorter than the experimental value supporting the premise that the diffusion time to the micelle is not the rate determining step. Since tunneling occurs more rapidly than the average diffusion time of triplet pyrene to the surface, the latter time determines the life-time of the fast component. As the residence time of Br_2^- on the micelle is at least 2 μs[5], substantial time is available for the reaction to take place. This is quite a remarkable difference between chemistry of ionic micelles and ordinary reaction kinetics in solution where encounter times are generally much shorter than the time between subsequent encounters.

On a longer time scale (Figure 7d) the pyrene cation signal passes through a maxium at 35 μs and then at much longer times decays with about the same rate as the Br_2^- trace in Figure 7e. The processes responsible for the consumption of the cation radical are ascribed to Reactions (12) and (13).

$$P_{mic}^+ \longrightarrow P_{aq}^+ \xrightarrow{\text{H}_2\text{O}} product \tag{13}$$

The reaction $P_{mic}^+ + Br_{aq}^- \longrightarrow P_{mic} + Br_{aq}$ is ruled out for several reasons other than that based on the relative positions of the redox levels for the $(P^+/P)_{mic}$ and (Br/Br^-) couples. To begin with, this reaction will lead to the chain oxidation of triplet pyrene which will mean that the amount of triplet consumed will be less than the initial concentration of Br_2^-, whereas the reverse is found. Also the possibility of the reaction is dismissed on the basis of a laser photolysis study of pyrene in micelles of DTAC which indicates a half-life for the cation of 19 μs. In this system the reaction can only be due to the hydrolysis of the cation (Reaction 13), ejected from the micelle interior, since no Br^- species is present and the cation cannot oxidize the Cl^- counterions. Reaction (12) may also contribute to the disappearance of the cation. Thermodynamically the reaction is expected to occur because the redox potential of the couple Br_2/Br_2^- ($E^o = 0.68$ V) is more negative than that of the system P^+/P ($E^o = 1.6$ V). Statistically it can be shown that the probability of two Br_2^- radicals and one triplet pyrene molecule being associated with the same micelle is about 7%. Also the time required for a Br_2^- radical to diffuse from one micelle to a second may be as short as 2 μs[5]. Thus Reaction (12) contributes to the decay of the cation signal.

Effects of a Reduction of Dimensionality on the Reaction Rates: Intramicellar Surface Reaction Versus Solution Bulk Type Reactions[5]

Investigations that have been discussed up to this point have involved electron movement across the electrical double layer of the micelle in the absence of bond breakage. In this section electron transfer on the micellar surface is described where bond rupture does take place. In particular, the dismutation reaction of Br_2^- radicals in aqueous solution of CTAB micelles is examined[5]. The CTAB/Br_2^- system represents the first example in which two components of a diffusion-controlled reaction differing in dimensionality can be distinguished experimentally. The components were attributed to two-dimensional intramicellar surface diffusion and three-dimensional intermicellar or bulk-type diffusion. A schematic diagram of the three processes is outlined in Figure 8.

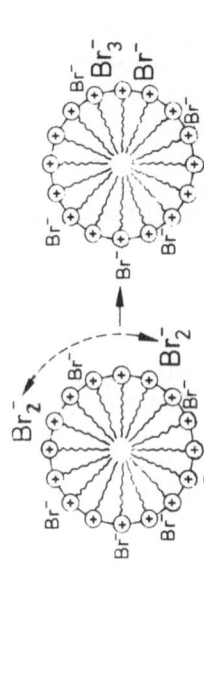

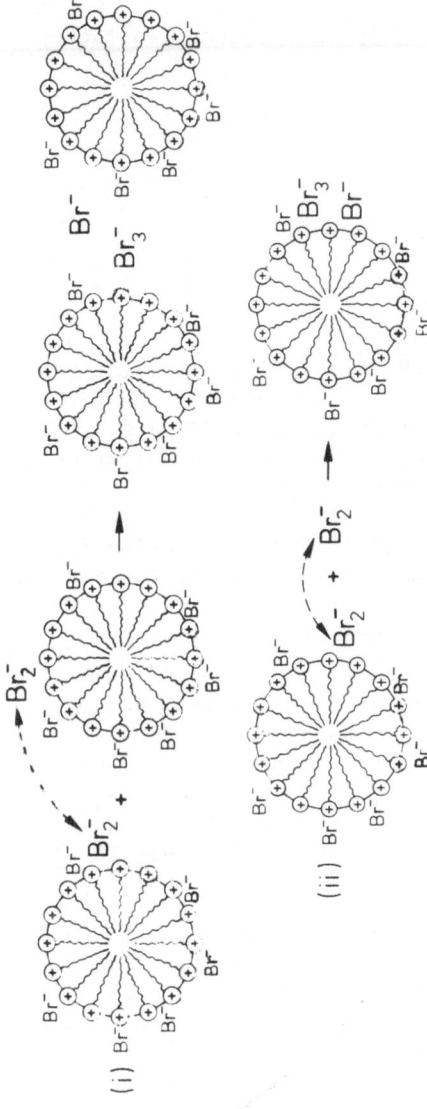

Figure 8. Schematic illustration of possible pathways for the dismutation reaction of Br_2^- ion radicals in micellar CTAB solution[5].

Irradiation of N_2O saturated solutions containing NaBr either with or without CTAB micelles initiates a series of reactions leading to the formation of 5.5 Br atoms for each 100 eV of energy absorbed [10,11]. Subsequent reactions of Br atoms produce Br_2^- radicals.

$$Br + Br^- \rightleftharpoons Br_2^- \qquad (14)$$

followed by the dismutation reaction and the resulting production of tribromide anions.

$$Br_2^- + Br_2^- \longrightarrow Br_3^- + Br^- \qquad (15)$$

In micelle-free solutions these reactions have been well-characterized [10,11,12]. At the bromide ion concentration (0.02 mol dm^{-3}) employed in our experiment, the forward reaction of Step (14) is completed during the electron pulse (5 - 20 ns pulse width) from the accelerator and the contribution of the back reaction is practically negligible.

The equilibrium of Reaction (16) is established within about 100 ns under our experimental conditions.

$$Br_3^- \rightleftharpoons Br_2 + Br^- \qquad (16)$$

In Figure 9 the kinetic behavior of Br_2^- radicals (λ_{max} = 360 nm) in the presence and absence of CTAB is shown. The oscillograms in the left column of Figure 9 show the smooth decay of Br_2^- in micelle-free solution. The kinetics corresponds to second order and has a rate constant of 1.8×10^9 dm^{3}mol^{-1}s^{-1} (after correcting for the salt effect) which agrees with earlier determinations [10,11]. In the presence of CTAB micelles, the decay of Br_2^- occurs in two steps. The first step is completed in about 2 µs and follows first-order kinetics with a rate constant of 2.1×10^6s^{-1}. The slower decay occurs over a period of 100 µs with a second-order rate constant of 1.6×10^9dm^{3}mol^{-1}s^{-1} which is similar to that for the micelle-free solution. Concurrent with and matching the decay kinetics is the growth of the tribromide anion signal at 270 nm (Figure 9, right column).

In addition to the kinetic features which distinguish micelle from micelle-free solutions, a conspicuous difference exists in the transitory spectrum of the Br_2^- radical and the tribromide anion. On the basis of Reactions (15) and (16), one-eighth of the initially produced Br_2^- radicals should be converted into Br_3^- anions in aqueous solution. Although this is indeed observed in aqueous NaBr solutions in the absence of micelles, the final yield of Br_3^- in the presence of micelles corresponds to almost one half of the

0.02 M NaBr, N_2O saturated

H_2O 0.005 M CTAB, H_2O

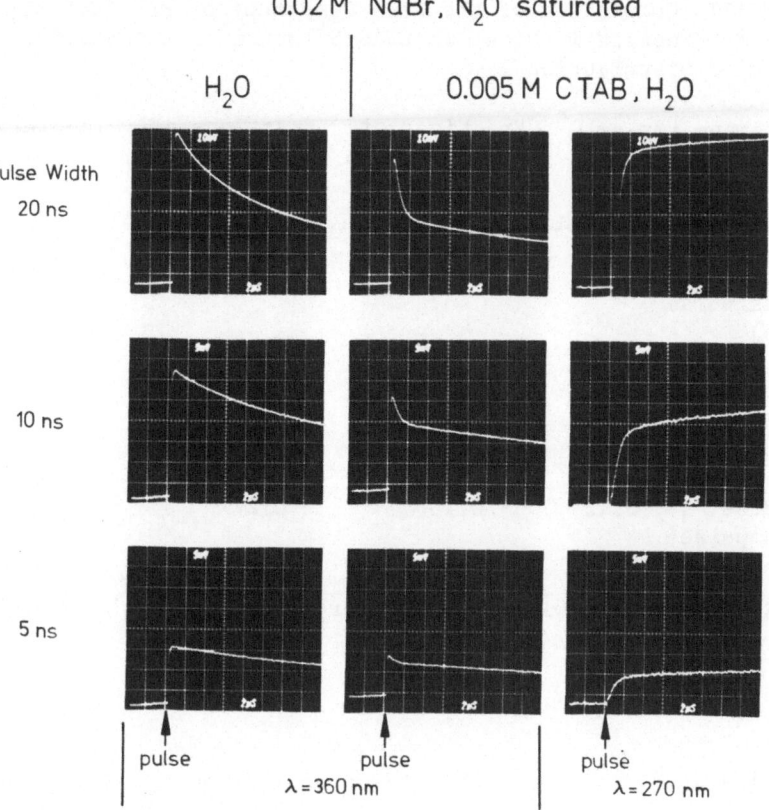

Figure 9. Oscilloscope traces of the kinetics of Br_2^- radicals (left and middle columns) and Br_3^- ions (right column). Left column: Aqueous 0.02 dm^3mol^{-1} NaBr saturated with N_2O, wavelength - 360 nm; Middle column: Aqueous 0.02 dm^3mol^{-1} NaBr and 5 x $10^{-3}dm^3mol^{-1}$ CTAB saturated with N_2O, wavelength - 360 nm; Right column: Aqueous 0.02dm^3mol^{-1} NaBr and 5 x $10^{-3}dm^3mol^{-1}$ CTAB saturated with N_2O, wavelength - 270 nm. Upper row: Abscissa - 2 μs per large division, ordinate - 5% absorption change per large division, pulse width - 20 ns; Middle row: Abscissa - 2 μs per large division, ordinate - 3% absorption change per large division, pulse width - 10 ns; Lower row: Abscissa - 2 μs per large division, ordinate - 3% absorption change per large division, pulse width - 5 ns^5.

initially produced Br_2^- concentration indicating that the equilibrium situation in Reaction (16) lies far to the left. Evidently Br_3^- is formed at a site where there is a high local Br^- concentration which we suggest is at the surface of the micelle, that is, the Stern layer.

Table III. Comparison of the Fast Component of Br_2^- Radicals and Br_3^- Ions Kinetics in Micellar CTAB Solution with a Statistically Predicted Distribution[5].

Experimental runs	$[Br_2^-] \times 10^6$ (mol dm^{-3})	Fractions of Br_2^- decay via fast component	Fractions of Br_3^- growth via fast component	Theoretical fraction*
High dose (20 ns pulse width)	26.9	0.61	0.47	0.55
Medium dose (10 ns pulse width)	7.32	0.30	0.35	0.32
Low dose (5 ns pulse width)	3.12	0.16	0.27	0.30

*Ratio of micelles with 2 or more Br_2^- radicals to those with 1 or more Br_2^- radicals.

Table III indicates the affect of dose on the decay kinetics of Br_2^-. It can be seen that as the pulse width is decreased, the initial concentration of Br_2^- is diminished as is the fraction of Br_2^- decay (or Br_3^- growth) via the fast process. A comparison of the last three columns in Table III will show that the fraction of Br_2^- consumed (or Br_3^- formed) correlates reasonably well with the statistical prediction of the ratio of micelles with 2 or more Br_2^- radicals to those with 1 or more Br_2^- radicals.

An explanation of the data is pictorially presented in Figure 8. The initially fast Br_2^- decay is presumed to occur from surface reactions of Br_2^- radicals on a single host micelle (process a) while the slow component (processes a and b) represents inter-micellar, micellar-bulk, and bulk reactions. The validity of the interpretation is further examined by application of Montroll's theory of random walk on two- and three- dimensional lattices with traps[13,14].

The problem as formulated considers one Br_2^- radical, which may be bound to a micelle, fixed at the center of a coordinate system. The calculation involves finding the average number of steps $\langle n \rangle_3$ required for a second radical present in bulk solution or on another micelle to meet the fixed one and be "trapped" with subsequent formation of Br_3^- and Br^-. Similarly in the two-dimensional situation where both radicals are on the same micellar surface, one radical is again assumed fixed at some point and the average number of steps $<n>_2$ is determined for the second Br_2^- radical to be "trapped" by the first one. Both problems have been solved by Montroll for two- and three-dimensional lattices containing traps[13,14]. Under our experimental conditions, we determine that $<n>_3 \approx 3.4 \times 10^4$ and $<n>_2 \approx 82$ for the case of a 20 ns pulse (Table III). Since the time for reaction, $<\tau>$, of two Br_2^- radicals in two- and three-dimensional space is equal to the average number of diffusion jumps required for trapping, $<n>$, multiplied by the time needed for a diffusion jump, t, the half lifetimes of the dismutation reactions can be semi-empirically assessed. In terms of the ratio of the rate constants of the processes, the derived relation is

$$\frac{k(a)}{2k(b + c)} = (ln2) \, [Br_2^-]_0 \, \frac{<n>_3}{<n>_2} \qquad (17)$$

where the labels a, b, and c refer to the pathways indicated in Figure 8. Substitution of the values for $<n>_3$ and $<n>_2$ and $[Br_2^-]_0 =$ 2.69 × 10⁻⁵ mol dm⁻³ (Table III) yields a ratio of 7.7 × 10⁻³. In view of the approximations used in the random-walk model, this is in reasonably good agreement with the experimental ratio of the rate constants for the fast and slow components of the dismutation reaction which is determined to be 1.5 × 10⁻³.

Clearly the reduction of dimensionality in which diffusion takes place from three-dimensional space to two-dimensional surface diffusion has a dramatic effect on the probability of reaction. This may be the way, as Adam and Delbrück[15] observed that organisms are able to cope with some of the problems of timing and efficiency, in which small numbers of molecules and their diffusion are involved.

REFERENCES

1. Present address: Lawrence Berkeley Laboratory, University of California, Berkeley, California, U.S.A.
2. A.J. Frank, M. Grätzel, A. Henglein, and E. Janata, Ber. Bunsenges physik. Chem, 80, 294(1976).
3. A.J. Frank, M. Grätzel, A. Henglein, and E. Janata, Ber. Bunsenges physik. Chem, 80, 547(1976).
4. A.J. Frank, M. Grätzel, A. Henglein, and E. Janata, (1976), Intern. J. Chem. Kinetics, in press.

5. A. J. Frank, M. Grätzel, and J. J. Kozak, J. Amer. Chem. Soc. 98, 3317 (1976).

6. M. Grätzel, A. Henglein, and E. Janata, Ber. Bunsenges. physik. Chem. 79, 475 (1975).

7. A. Henglein, Ber. Bunsenges. physik. Chem. 79, 129 (1975).

8. R. A. Marcus, Ann. Rev. Phys. Chem. 15, 155 (1964).

9. V. G. Levich in "Physical Chemistry - An Advanced Treatise", H. Eyring, D. Henderson and W. Jost, Editors, Vol. IXB, p. 985, Academic Press, New York, 1970.

10. B. Cercek, M. Ebert, C. W. Gilbert, and A. J. Swallow, in "Pulse Radiolysis", M. Ebert, J. P. Keene, A. J. Swallow and J. H. Baxendale, Editors, p. 83, Academic Press, London, 1965.

11. H. C. Sutton, G. E. Adams, J. W. Boag, and B. D. Michael, ibid., p. 61.

12. D. Wong and B. DiBartolo, J. Photochem. 4, 249 (1975).

13. E. W. Montroll, J. Math. Phys. 10, 753 (1969).

14. E. W. Montroll, in "Proceedings of the International Conference on Statistical Mechanics", Suppl. to J. Phys. Soc. Japan, 26 (1969).

15. B. Adam and M. Delbrück in "Structural Chemistry and Molecular Biology", A. Rich and W. Davidson, Editors, W. H. Freeman and Co., San Francisco (1968).

RADIATION-INDUCED PROCESSES IN NONIONIC MICELLES

K. Kalyanasundaram and J. K. Thomas

Department of Chemistry and Radiation Laboratory

University of Notre Dame, Notre Dame, Indiana 46556

The static and dynamical properties of nonionic micelles (Triton X-100, Igepal CO-630 and Brij-35) in aqueous solution have been investigated by pulsed ^{1}H, ^{13}C NMR relaxation, fluorescence probing and pulse radiolysis techniques. Chemical shifts and spin lattice relaxation times presented for the various resolved resonances in the proton and proton-decoupled ^{13}C NMR spectra provide detailed information on the nature and segmental mobility of hydrocarbon chains in micellar core and that of ethylene oxide units in the palisade layer. The permeability of these nonionic micelles with respect to various species (ionic and nonionic) has been investigated by examining the dynamics of quenching of fluorescence emitted by "external" probe such as pyrene and "built in" phenoxyl unit. The basic photophysical features such as UV absorption, fluorescence lifetime and quantum yields for phenoxyl chromophore are also reported and these are used to gain information on the environment around these probes. Efficient excitation singlet energy transfer between phenoxyl unit and pyrene (solubilized in micellar core) has been observed. The 347.1 nm ruby laser photolysis of pyrene solubilized in these nonionics indicates efficient biphotonic photoionization of pyrene in these systems. Pulse radiolysis studies with solubilized pyrene and biphenyl indicate that hydrated electrons (e^-_{aq}) enter nonionic micelles fairly efficiently. Finally microviscosity data, derived from fluorescence depolarization studies with 2-methylanthracene are also presented.

INTRODUCTION

Molecular studies of self-association of amphiphilic molecules in aqueous and non-aqueous media to micellar aggregates is a topic of increasing interest. These systems are similar to other multimolecular aggregates such as biomembranes,[1] while the micellar catalysis is in many ways analogous to enzymatic catalysis.[2] Studies on the micelles formed by ionic surfactants have been quite extensive and their structures well characterized. However, there have been relatively few studies of their nonionic counterparts.[3] In our earlier studies, the static and dynamical properties of ionic micelles were investigated by pulsed NMR, laser photolysis and pulse radiolysis techniques. In this study these techniques have been extended to the case of nonionic micelles formed from polyethyleneoxyalkylphenols and alkyl ethers.

In this study, in particular, the static and dynamical properties of micelles, formed by nonionic surfactants p-di-t-butyl polyoxyethylene (9.5) ether (Triton X-100), p-nonyl phenoxy-polyoxyethylene (9) ether (Igepal CO-630) and polyoxyethylene (23) dodecanol (Brij 35) have been investigated by means of ^{1}H, ^{13}C NMR, time-resolved fluorescence and pulse radiolysis techniques.

To facilitate discussion, Figure (1) represents the currently accepted model of a nonionic micelle. Micelles formed by Igepal CO-630 and Triton X-100 are comparatively large and approximately spherical (r = 5nm) with aggregation numbers around 150. On the other hand, Brij 35 forms somewhat compact rod shaped micelles (r = 2nm, ℓ = 18 nm) made up of about 40 surfactant molecules. The surfactant Triton X-100 has a highly branched di-t-butyl short hydrophobic core, while to ethylene oxide units of the head group have varying lengths, the average chain length being about 9.5 ethylene oxide (EO) units. An interesting feature of the micelles formed by Igepals and Tritons is that these surfactants have a "built-in" chromophore phenoxyl unit at approximately the midpoint of the molecule. Thus the intriguing possibility exists of using this built-in probe to study the dynamic features of these micelles.

In the first part chemical shifts and spin-lattice relaxation times (T_1 values) are reported for various resolved resonances in the proton and proton decoupled carbon-13 nmr spectra of aqueous solutions of the above nonionic detergents. These data indicate a gradient in the segmental mobility of the hydrocarbon chains as well as the ethylene oxide units. The permeability of the non-ionics with respect to various ionic and neutral species has been investigated by kinetic analysis of fluorescence quenching studies. Fluorescence emitted by probes present either as a built-in (phenoxyl) probes or introduced externally (pyrene) is quenched by molecules of a nonionic micelle. Studies of the basic photophysical

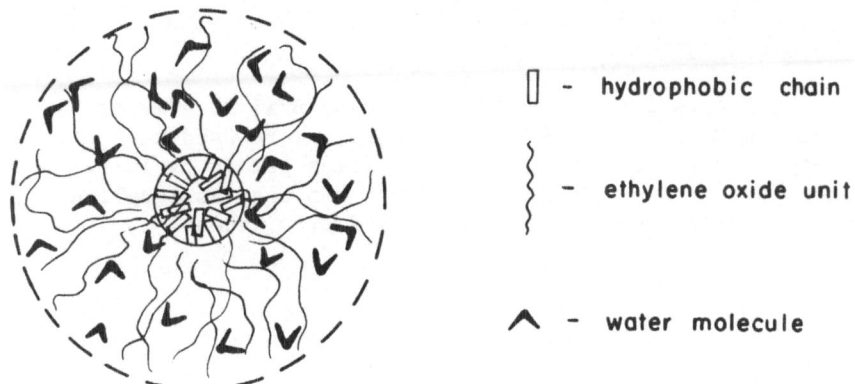

hypthobic chain — *(see figure labels)*

Figure 1. Model of a nonionic micelle.

features of the built-in phenoxyl probe provide information on the
environment around the chromophore in these nonionic micelles.
Solubilization of pyrene adjacent to the phenoxyl unit is established
by the efficient singlet excitation energy transfer studies. The
unique solubilization site for pyrene in the hydrophobic core of
the micelle leads to interesting photochemical consequences. For
example, the 347.1 nm ruby laser photolysis of pyrene solubilized
in the nonionics gives rise to efficient photoionization by a
biphotonic mechanism. Similarly, hydrated electrons produced in
the aqueous phase by pulse radiolysis enter nonionic micelles
relatively efficiently. These studies are supplemented by microviscosity data obtained from fluorescence depolarization studies
with 2-methylanthracene.

EXPERIMENTAL SECTION

   High resolution ^{1}H, ^{13}C NMR spectra and T_1 measurements were
carried out on a Varian XL-100 NMR spectrometer equipped with
Nicolet TT-100 Fourier transform accessories. The T_1 measurements
employed the inversion-recovery method of Vold et. al.[4] Laser
photolysis studies were carried out with a Korad frequency doubled
Q-switched ruby laser. The 347.1 nm UV pulse had a half width of
15 ns with an energy output of ≃200 mJ. Transient absorption and
fluorescence were detected by fast kinetic spectroscopy, a detailed
description of which has been given elsewhere.[5] Steady-state
fluorescence and fluorescence depolarization studies were carried
out on an Aminco-Bowman spectrophotofluorimeter. Pulse radiolysis
studies were carried out with 10 nsec electron pulses from a 10
MeV Linac. Measurements of very short fluorescence lifetimes

as observed with the phenoxy groups of Igepals and Tritons were
carried out with 1.5 ns Cerenkov light pulses. These pulses were
produced by pulse electron irradiation of a quartz bead with a
2.5 MeV Van de Graaff generator.

Nonionic detergents Triton X-100 (Rohm-Haas) Brij 35 (Pierce
Chemicals), Igepals CO-630 and CO-880 (GAF Corp.) were used as
supplied. Pyrene (Kodak) was passed through silica gel in cyclohexane and then recovered. D_2O (Merck, 99.9%) was used as supplied.
2-methylanthracene (Aldrich) was recrystallized from ethanol
before use.

RESULTS AND DISCUSSION

Proton NMR Relaxation Studies

Nuclear Magnetic Resonance Studies have proved to be an
efficient method[10-13] for investigating structural dynamics in ionic[6-9]
and nonionic micellar solutions. Changes in chemical shifts,
line widths and spin-lattice relaxation times give detailed information concerning the segmental mobility of various parts of the
hydrocarbon chains of the micelles. Figure (2) presents typical
proton NMR spectra of micellar solutions of Igepal CO-630 in D_2O.
The assignments for the various proton signals are also indicated
in this figure. The spectrum shows partially resolved resonances
for the alkyl and aryl protons, with the ethylene oxide protons
appearing as a broad band, consisting of several overlapping
resonances. The spin-lattice relaxation times (T_1 values) for
various resolved resonances have been measured and these results
are summarized in Table I. The observation that the ethylene oxide
band consists of several overlapping resonances is clearly demonstrated in Figure (3). This figure presents a series of partially
relaxed fourier transform (PRFT) spectra[4] for this band. It is
noted that while the proton on the high field side relax quite
rapidly, the proteon on the low field side relax much more slowly.
(cf. for example, the spectrum corresponding to a delay of 250
nsec between the 180° and 90° pulses). In fact, under high
resolution (100 MHz) one can identify as many as six resonance
lines and the T_1 values for these lines can also be measured. This
type of behavior in Triton X-100 micelles has, in fact, been noted
earlier by Podo et. al.[12] and quite recently by Riberio and
Dennis.[13] Chemical shift assignments indicate that the protons on
the low field side arise from EO units at the end of the chain
while those of high field side arise from EO_3 units adjacent to the
phenoxyl unit. Earlier detailed studies[12,13] with Triton X-100
have clearly established that the micellar core is free of water.
Our own NMR and UV spectral studies (see later) also indicate this
type of behavior for the Igepals (nonyl phenoxy polyethylene

oxyethanols). Apparently this type of behavior, viz. a micellar core free of water, seems to be a general feature of the nonionic micelles.

Table I. Proton Spin-Lattice Relaxation Times (T_1 Values) for Nonionic Micelle Igepal CO-630 (4 mM) in D_2O

Peak	δ(ppm)	T_1 (ms)	Peak	δ(ppm)	T_1 (ms)
a	0.58	230	e	3.65	240
b	0.64	200	f	3.80	130
c	1.00	95	g	6.64	200
d	(3.40–3.60)	240,280,350 505,515,545	h	6.92	165

The spin-lattice relaxation data, presented in Table I, also provide interesting data on the segmental mobility of the alkyl chains as well as those of ethylene oxide units. In multimolecular assemblies such as micelles or membranes, the major mechanism for spin-lattice relaxation relaxation of proton spins is one of scalar dipole-dipole coupling between adjacent spins. In this case, T_1, the relaxation time is given by[14,15]

$$\left(\frac{1}{T_1}\right) = N \cdot h^2 \gamma_1^2 \gamma_2^2 \frac{1}{\gamma_{12}^6} \tau_{eff} \qquad (1)$$

where h = reduced Planck's constant, N = number of directly bonded protons, γ_1 and γ_2 the gyromagnetic ratio for the two spins involved, r_{12} internuclear distance and τ_{eff} the effective correlation time. If one treats the tumbling spheres hydrodynamically, τ can be related by Stokes law, to the viscosity of the medium in which the spins tumble:

$$\tau = \frac{4\pi \eta a^3}{3} \qquad (2)$$

Here a refers to the radius of the tumbling sphere and all other symbols designate familiar parameters and constants. To a first approximation, the overall reorientation of micelles are slow compared to the internal modes, it is then reasonable to assume that τ_{eff} is determined predominantly by the rotational mobility of the various segments of the chain. Thus, measurements of T_1

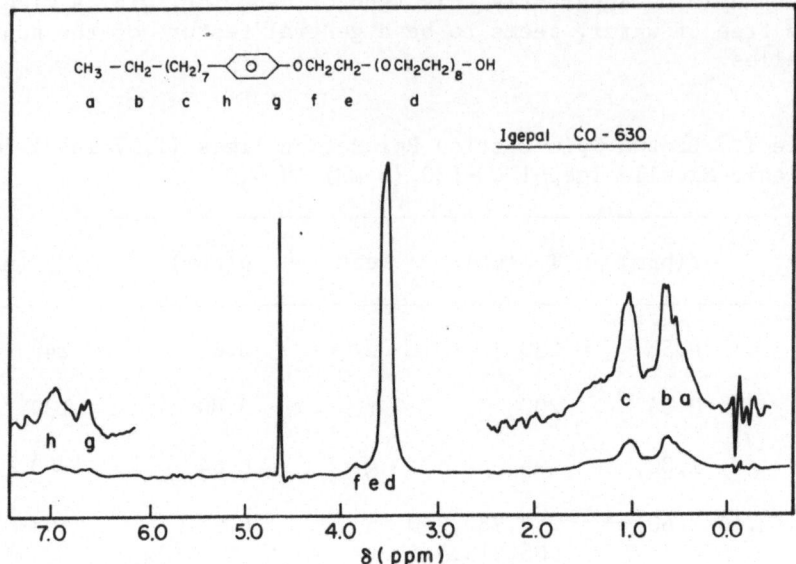

Figure 2. ^{1}H NMR spectrum of 5 mM Igepal CO-630 in D_2O.

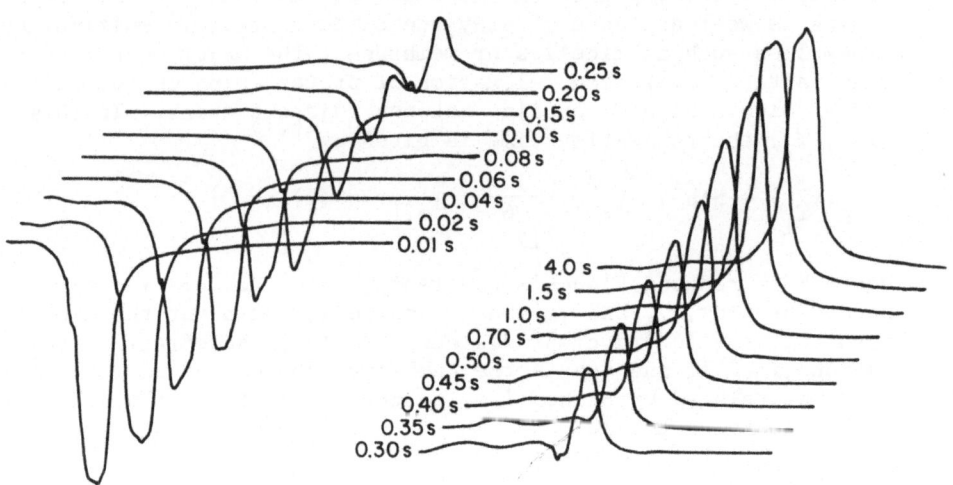

Figure 3. Partially relaxed fourier-transform spectra for the ethylene oxide band in Igepal CO-630 pmr spectrum.

values for various protons provide a direct method of determining the viscosity surrounding a given spin along the chain and thus the segmental mobility of the hydrocarbon chain inside the micelle.

The relative magnitude of the T_1 values for various protons (alkyl, aryl and methylenes of the EO units) are considerably less than that observed in pure hydrocarbon liquids. This indicates that the micellar interiors in these nonionic micelles are more rigid than in hydrocarbon liquids. Furthermore, the gradation in T_1 values observed indicates a gradient in the segmental mobility of the various units. As one moves farther down towards the inner core from the phenoxyl unit, the freedom of motion for the alkyl chain increases. This is also the case for the ethylene oxide units, where the EO units towards the end of the hydrophilic chains possess maximum freedom of motion. The above picture of segmental mobility for the Igepal CO-630 micelle is consistent with similar results reported recently for the Triton X-100 micelles.[12]

Table II. Carbon-13 Spin-lattice Relaxation Times (T_1 Values) for Nonionic Micellar System: Triton X-100 (4 mM) in D_2O.

Peak	δ (ppm)	T_1 (ms)	Peak	δ (ppm)	T_1 (ms)
a	32.4	380	g	128.0	480
b	57.6	320	h	158.0	2000
c	32.8	1800	i	68.0	350
d	38.0	1850	j	70.8	380
e	143.0	1900	k	61.2	1100
f	115.0	460	l	72.8	1000

Carbon-13 NMR Relaxation Studies

As was shown earlier with ionic micelles,[6,16] a major shortcoming of the proton NMR studies is the fact that the methylene signals are an average comprised of contributions from different methylene groups of the alkyl chain. With Carbon-13 NMR it is possible to detect the motions of the individual carbon atoms, forming the backbone of the surfactant molecule. The proton-

decoupled ^{13}C NMR spectra of micellar solutions of a typical nonionic surfactant Triton X-100 is presented in Figure (4). The C-
13 spectrum shows more resonances than is observed with the proton
spectrum. The chemical shift assignments, relative to TMS, for
the various lines are also indicated in the figure. The assignments are based on earlier assignments for pure hydrocarbon
liquids[17,18] ionic micelles[8,16] and on model compounds. The chemical
shifts and spin-lattice relaxation times for various carbon resonances in micellar solutions of the nonionic micelle Triton X-100
are summarized in Table II.

The carbon-13 spin-lattice relaxation data, presented in the
table, provide more detailed information on the segmental mobility
of the fatty acid chains as well as that of ethylene oxide units
in the nonionic micelles. As with the case of proton relaxation
times, the C-13 T_1 values for alkyl chain carbon atoms are considerably less than those observed in pure liquids and ionic
micelles. This indicates that the micellar core in a nonionic
micelle is considerably more rigid in the other systems. A
gradient in the segmental mobility of the chain (both the alkyl as

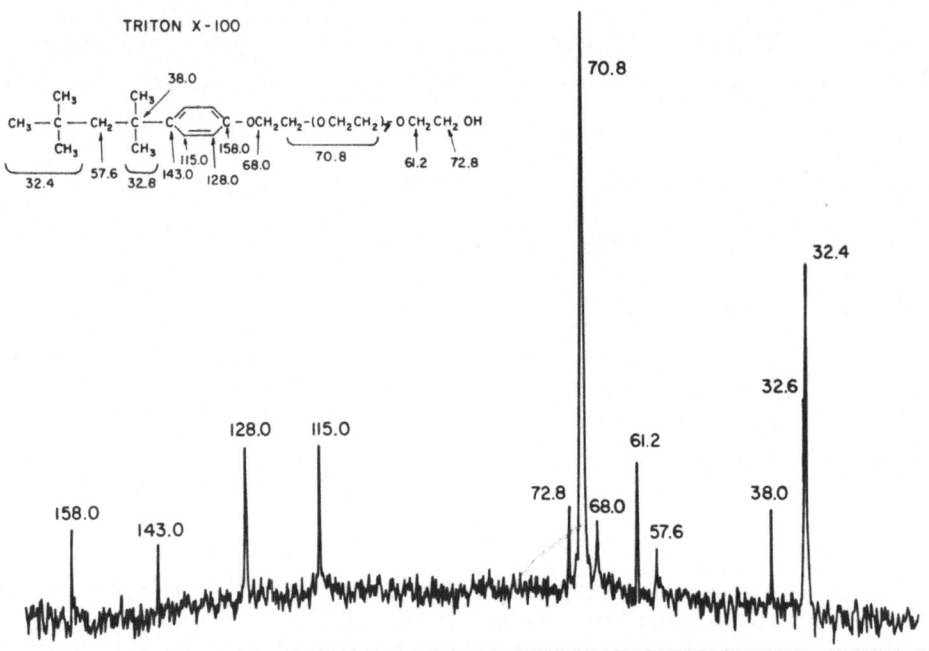

Figure 4.   Carbon-13 NMR spectrum of micellar Triton X-100 solutions
in D_2O.

well as the ethylene oxide units) is indicated by the increase in
the T_1 values. This is noted as one moves either towards the
inner core or towards the surface, with respect to the phenoxyl
unit. UV absorption spectral studies reported in the latter
sections are also in accord with such a picture.

Fluorescence Probe Studies with Pyrene

Several dynamical features of ionic micelles, viz. permeability
of various molecules through the micelle and counter ions binding
to the micellar surface, have been studied recently by fluorescence
probe analysis utilizing pyrene.[3] Fluorescence probes such as
pyrene are strongly hydrophobic in nature and dissolve in the
interior hydrophobic core of the micelle. In the absence of
quenching molecules, the pyrene monomer fluorescence is long lived
(τ_{fl} >300 ns). However, in the presence of certain molecules
called quenchers, the fluorescence decay may be greatly enhanced.
The rate at which the quencher enters the micelle and/or the
fluorescing probe diffuses in the micellar core determines the
kinetics of the quenching process. Hence, a kinetic analysis of
the fluorescence decay in the absence and in the presence of
various quenchers yields information about the permeability of the
micelle with respect to the quencher. It is also possible to
monitor the movement of the fluorescent probe in the hydrocarbon
interior of the micelle. NMR and UV spectral data suggest that
the hydrophobic core of the nonionic micelle is free of water.
Thus it is reasonable to assume that pyrene is solubilized in the
inner core of the micelle. It will be shown later than such a
hypothesis is supported by (i) efficient formation of pyrene
excimer; (ii) excitation singlet energy transfer studies between
pyrene and the built-in probe phenoxyl indicate an upper limit for
the distance between pyrene and phenoxyl to be 1.5 nm (15Å) and
(iii) and quenching rate constants for pyrene and phenoxyl are
very similar. Hence, a reasonable model for the solubilization
site of pyrene in nonionic micelles locates the probe in the
hydrophobic core in the vicinity of the phenoxyl unit.

Figure (5) presents typical fluorescence traces obtained in
the laser photolysis of pyrene in Triton X-100 micelles in the
absence and in the presence of quenchers oxygen, I^- and Tl^+. In
the absence of quenchers, the pyrene monomer fluorescence is long
lived with $\tau_f \simeq 320$ ns in nonionic micelles Triton X-100 and
Igepal Co-630. The fluorescence decay is enhanced in the presence
of quenchers and the quenching efficiency is described by

$$k = k_o + k_q [Q] \qquad\qquad (3)$$

where k = observed rate constant for monomer fluorescence decay,

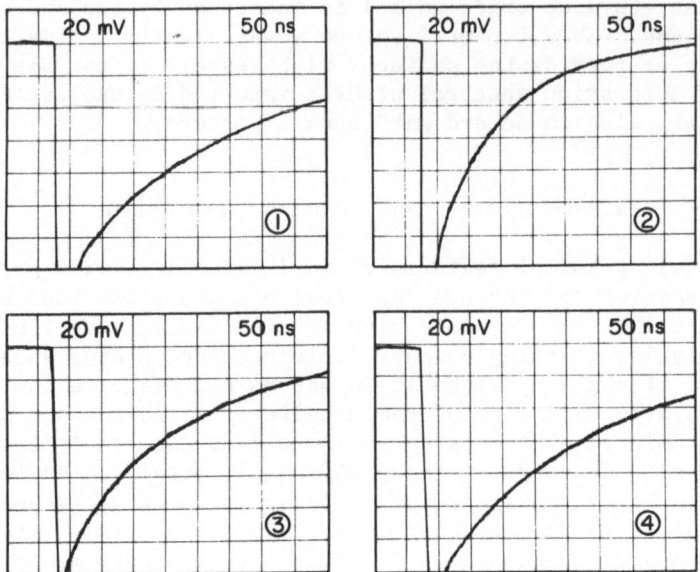

Figure 5. Decay of pyrene (10 μM). Monomer fluorescence at λ =
400 nm in 4 mm Igepal CO-630 in the absence and presence of various
quenchers (i) deaerated solution; (ii) oxygenated solution; (iii)
$[I^-] = 4$ mm; (iv) $[Tl^+] = 4$ mM.

k_o and k_q are rate constants for the decay in the absence and in
the presence of quencher Q respectively and [Q] is the concentration of the quencher. The fluorescence decay of pyrene is
exponential in the presence and absence of quenchers. Table III
summarizes the data obtained from the laser photolysis studies of
pyrene monomer fluorescence in the presence of various ionic and
neutral quencher molecules. In general, the nonionic micelles
provide a mild inhibition for the movement of quenching species to
the probe. For a given quencher, the quenching rates in the
spherical micelles are higher than those in the larger rod-shaped
aggregates. Oxygen diffusion, however, is more efficient in the
spherical micelles as is indicated by the lower quenching rates
observed in the presence of nonionic micelles. The micelles
inhibit the quenching rates of pyrene fluorescence by I^- and
nitromethane to approximately the same extent. A distinctive
feature of the data presented in Table III is the abnormally low
values obtained for the quenching rate constants for positively
charged quenchers like Tl^+ and Cu^{2+}. The very low quenching rates
obtained with the positively charged quenchers can be rationalized
in terms of adsorption of these ions in the ether units of the
ethylene oxide coils. It is well known[19,20] that cations such as
Tl^+ are strongly adsorbed by the ether linkages in polyoxygenated

Table III. Summary of Fluorescence Quenching Efficiency in Various Nonionic Micellar Systems.

Quenchers	k_q (Pyrene) $(\times 10^{-8} M^{-1} s^{-1})$				k_q (Phenoxyl) $(\times 10^{-8})$
	CH_3OH	Igepal CO-630	Triton X-100	Brij-35	Triton X-100
Oxygen	200	55 (3.6)	57 (3.6)	24 (8.3)	–
CH_3NO_2	36	5.3 (6.8)	5.1 (6.8)	17 (2.1)	9.0 (4.0)
I^-	30	4.0 (7.5)	6.9 (4.5)	5.4 (5.7)	
Tl^+	50	1.6 (31)	2.2 (24)	2.7 (19)	1.6 (31)

Note: The numbers in parenthesis after the quenching rate constants refer to the ratio of the rate constants with reference to those in homogeneous sytems, methanol.

compounds such as crown ethers. Uncomplexed Tl^+ ions in aqueous solutions fluoresce strongly with a fluorescence maximum around 358 nm and quantum yield 0.17. Complexing of Tl^+ by ether quenches the fluorescence. The fluorescence spectra of Tl^+ ions in water and in the presence of varying concentrations of Brij 35 are presented in Figure (6). Addition of increasing amounts of Brij 35 decreases the thallous ion fluorescence intensity, indicating the complexing of Tl^+ ions in the ethylene oxide coils of Brij 35.

UV Absorption and Fluorescence Studies with Phenoxyl Chromophore

As indicated earlier, a distinctive feature of nonionic surfactants such as Igepal CO-630 or Triton X-100 is that these detergent molecules possess a phenoxyl group with a distinct UV absorption and fluorescence spectra. A few studies have already[21-23] been made with a phenoxyl group of this type of amphiphiles. Both the UV absorption and fluorescence spectra show distinct concentration dependence.

Aqueous solutions of the nonionic surfactant Igepal CO-630 below the critical micelle concentration have an UV absorption spectrum with maximum around 275 nm ($\varepsilon = 1100$) and a shoulder around 281 nm ($\varepsilon = 910$). Increasing the concentration above CMC leads to a strong perturbation in the absorption spectra. The intensity of the spectrum increases, a red shift is also noted and some fine structure also appears. The variation of the molar

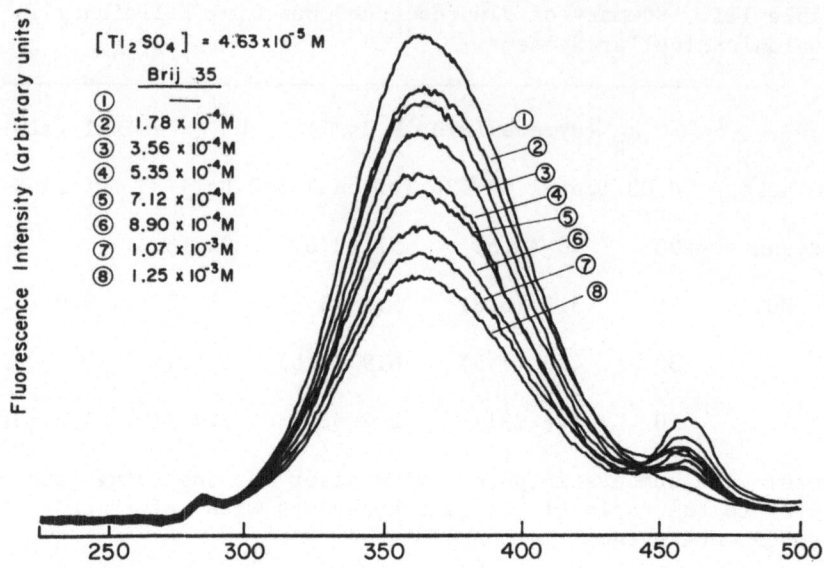

Figure 6.   Quenching of Tl$^+$ fluorescence of Brij 35 (λ_{excit} 220 nm).

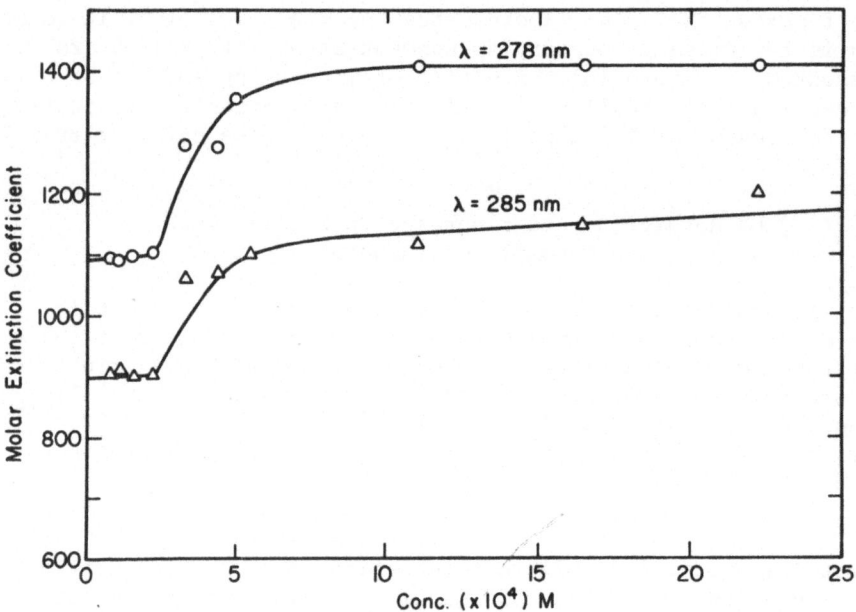

Figure 7.   Concentration dependence of UV absorption maxima and molar extinction coefficient for Triton X-100.

extinction coefficient with Triton X-100 concentration is presented
in Figure (7). The break in the shape of the curves occurs at ca.
2×10^{-4} M. This value is very close to that determined for the
CMC of Triton X-100.

Further insight into the environment of the phenoxyl chromophore inside the nonionic micelles can be obtained by comparison
of the absorption maxima and extinction coefficient in micellar
solutions to those observed in pure solvents. In solvents such as
methanol or dioxane, the phenoxyl group absorption spectrum is
independent of the surfactant concentration. In mildly polar
solvents such as methanol (D = 33), the UV absorption spectrum has
maxima at 278 nm (ε = 1495) and at 284 nm (ε = 1200). In dioxane
the corresponding values are 278 nm (ε = 1730) and 284 nm (ε =
1390). The molar extinction coefficients and absorption maxima in
micellar solutions are comparable to those observed in methanol
suggesting that the phenoxyl group in these nonionic micelles, on
the average, experiences environments similar to that in methanol.
Presumably the phenoxyl group serves as a barrier between the
hydrophobic inner core and the hydrophilic ethylene oxide palisade
layer. The similarity of the UV spectra in micellar and methanol
solutions strongly suggests water penetration of the micelle up to
the phenoxyl unit. It is pertinent to mention here that this
"distinctive" perturbation of the UV absorption of phenoxyl chromophore with the onset of micellization has been successfully developed
into a differential spectrophotometric method for determination of
CMC[23] and has also been used to study the kinetics of micelle
dissociation equilibria.[24]

The concentration dependence of fluorescence spectra for
aqueous solutions of Triton X-100 is presented in Figure (8).
At concentrations well below CMC and up to CMC, the fluorescence
spectrum consists of a broad band with a maximum at 302 nm. At
concentrations well above CMC, an additional band appears with a
maximum around 340 nm. The intensity of this band increases with
increasing concentration of the surfactant. This type of behavior,
i.e. development of a new emission band around 340 nm, has been
reported earlier[22] for a similar nonionic detergent and has been
attributed to excimer emission. As the absorption spectrum is <u>not</u>
independent of the surfactant concentration, it is difficult to
unambiguously attribute this new spectral band to an excimer.[25] An
alternative explanation is to cite absorption and fluorescence
from dimeric species, at high concentration, as distinct from
monomeric species at low concentrations.

The fluorescence lifetimes for the phenoxyl group emission in
nonionic micellar systems were determined using the 1.5 ns Cerenkov
pulses from a 2.3 MeV Van de Graaff. The measured fluorescence
lifetimes for phenoxyl emission in Igepal CO-630 ndTriton X-100

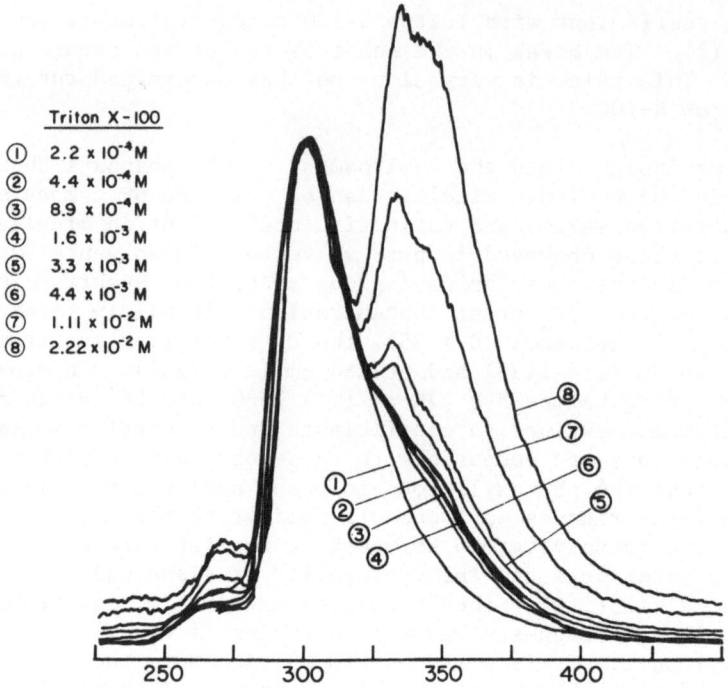

Figure 8. Concentration dependence of phenoxyl group fluorescence
in aqueous solution of Triton X-100.

(4 mM aqueous solution) are 3.6 and 4.0 ns, respectively. The
quantum yields for this fluorescence (relative to anisole quantum
yield in cyclohexane of $\Phi = 0.29$) are 0.34 and 0.32 for Igepal CO-
630 and Triton X-100, respectively. Above CMC, the fluorescence
lifetimes were found to be independent of concentration. Fluorescence depolarization with 2-methyl-anthracene indicates the
microviscosity for the inner hydrophobic core of Triton X-100
micelles to be of the order of 35 cP.

    The quenching of phenoxyl fluorescence by external additives
was also studied by monitoring the fluorescence intensity in a
spectrofluorimeter. The quenching rate constants for various
quenchers, derived from Stern-Volmer slope for phenoxyl fluorescence
quenching in Triton X-100 are included in Table III. In general,
the quenching rates are smaller for pyrene as compared to phenoxyl
group indicating a deeper location of pyrene in the inner core.

Energy Transfer Within Micelles

The unique solubilization of pyrene in the nonpolar inner core, adjacent to the phenoxyl chromophore, can be used to study energy transfer inside the micelle. A strong overlap of the phenoxyl fluorescence with the pyrene absorption spectrum occurs and singlet-singlet excitation energy transfer readily takes place from phenoxyl to pyrene. Figure (9) presents data for this type of singlet energy transfer from phenoxyl chromophore to pyrene solubilized in the micellar core. In the absence of "acceptor" pyrene molecules, the excited phenoxyl group fluoresces with a maximum around 300 nm. In the presence of increasing amounts of pyrene, the donor fluorescence decreases with a concomittant increase in the (acceptor) pyrene fluorescence. (Based upon the optical densities at the wavelength of excitation for the highest concentration of pyrene and phenoxyl, it is extimated that direct excitation of pyrene is less than 5%.) According to Forster's theory,[26] the interchromophore distance R_o (distance at which the rate of transfer is equal to the spontaneous rate of fluorescence decay) is related to the overlap integral of the donor fluorescence and acceptor absorbance by:

$$R_o^6 = \frac{9000k^2 Q_D \cdot \ln 10}{128\pi^5 n^4 N} \int \frac{f_D(\nu)\varepsilon_A(\nu)d\nu}{\nu^4} \qquad (4)$$

where k^2 = orientation factor; n = refractive index of the medium at λ_{excit}; N = Avogadro's number; Q_D = quantum yield of donor. If one defines the energy transfer efficiency T by

$$T = (1 - \frac{Q_T}{Q_o}) \qquad (5)$$

where Q_o and Q_T are quantum yields of donor in the absence and in the presence of transfer, respectively, then R, the distance between the two fluorescent chromophores, is given by

$$R = R_o(\frac{1}{T} - 1)^{1/6} \qquad (6)$$

Thus, by measuring the energy transfer efficiency T for the maximal energy transfer, it is possible to get an estimate for R. In the present case of energy transfer between the phenoxyl group and pyrene with an estimated R_o = 24.67Å and Q_D = 0.34, an upper limit for the distance between the two chromophores is estimated to be about 15Å. This distance is essentially the length of the hydrocarbon chain in Igepal CO-630, suggesting that pyrene is distributed uniformly in the micellar core.

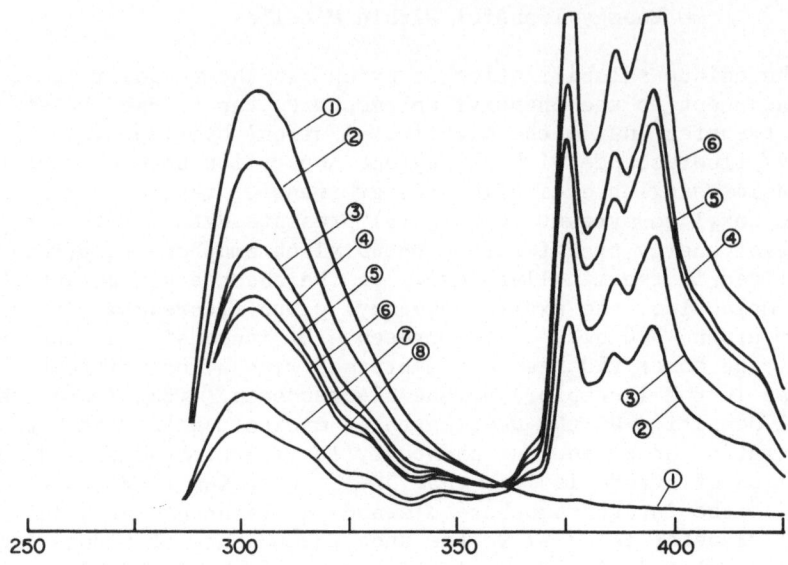

Figure 9.   Excitation singlet energy transfer from phenoxyl group to pyrene in nonionic micelle Igepal CO-630.

Laser Photoionization and Pulse Radiolysis Studies
in Nonionic Micelles

Photoionization of aromatic hydrocarbons and subsequent electron transfer reactions in bioaggregates simulates events that may play a major role in processes such a photosynthesis and electron transport in membranes. Earlier studies[27,28] have shown that several polycyclic hydrocarbons such as pyrene are photoionized efficiently in aqueous ionic micellar solutions by one and two photon processes. Pulse radiolysis studies[29,30] have shown that micellar surface charge has a pronounced catalytic (or inhibitive) effect of the reactions of hydrated electrons (e_{aq}^-). It is pertinent to study similar photoionization and electron-transfer reactions in nonionic micelles, as these systems are of intermediate complexity between ionic micelles and phospholipid dispersions.

Pyrene, in aqueous micellar solutions of nonionic surfactants such as Igepal CO-630 or Triton X-100, is photoionized by intense 10 nsec pulses of 347.1 nm light from a Q-switched ruby laser.

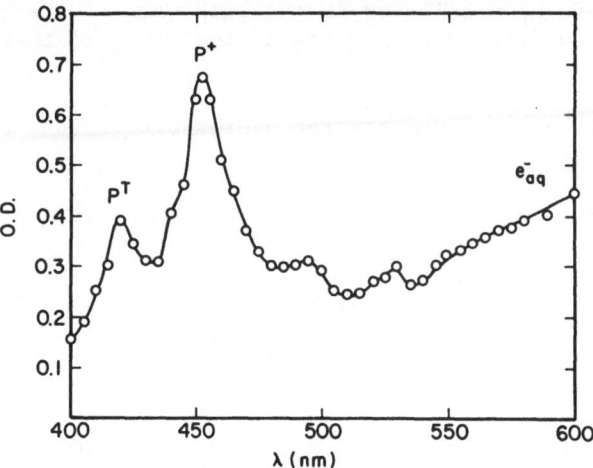

Figure 10. Transient absorption spectrum in the laser photolysis of pyrene (10^{-4} M) in Igepal CO-630 (4 mm) micelles, spectrum immediately after the pulse.

The pyrene cation is observed in the micellar phase and the electron as the hydrated electron in the aqueous phase. Figure (10) presents the transient absorption spectrum obtained in the laser photolysis of pyrene in nonionic micelles such as Igepal CO-630. The spectrum shows two peaks at 420 and 455 nm and a broad shoulder rising towards the red. The latter absorption is removed by typical electron scavengers such as N_2O and is thus attributed to the hydrated electron. The peaks at 420 and 455 nm are assigned to the pyrene triplet and pyrene cation respectively. The small hump around 490 nm arises due to pyrene singlet and triplet absorptions. These assignments are based on the earlier assign-ments for these peaks in the laser photolysis of pyrene in homo-geneous[31] and in ionic micellar systems.[32] By comparison with photoionization in pure liquids or ionic micelles, the formation of hydrated electrons in the present system occurs by a biphotonic photoionization.

$$Py \xrightarrow{2h\nu} Py^+ + e^-_{aq} \qquad (7)$$

Biphotonic photoionization is supported by the observation that the yield of hydrated electrons is proportional to the square of the laser light intensity. The solvation of the photoejected electron can occur either in the water present in the ethylene oxide core of the micelle or in the bulk water. The present study does not allow distinguishing one or the other of these two pathways.

Table IV. Rate Constants for the Hydrated Electron Reactions.

Reaction	Igepal CO-630	Triton X-100
e^-_{aq} + Surfactant (> CMC)	1.7×10^7	3.8×10^7
e^-_{aq} + Pyrene	1.7×10^9	2.0×10^9
e^-_{aq} + Biphenyl	3.8×10^9	4.0×10^9

A complementary situation to the above mentioned process, viz. the ease of entry of a hydrated electron into the micelle, can be followed by pulse radiolysis techniques. Earlier studies with ionic micelles[29] have shown that while the positive charge on the cationic micelles enhanced the rate of entry of e^-_{aq} into the micelle, the large negative charge on anionic micelles reduced the ease of entry of e^-_{aq} into the micelles. Table IV summarizes the rate constants for the reaction of e^-_{aq} with surfactant as well as with pyrene and biphenyl solubilized in the interior of the nonionic micelles. While the reactivity of e^-_{aq} with micellized nonionic surfactants is quite low, the e^-_{aq} enters the nonionic micelles fairly efficiently as is indicated by the high rate constants observed for the reactions of e^-_{aq} with solutes pyrene and biphenyl. The aqueous rate constants for e^-_{aq} with pyrene and biphenyl are 1.1×10^{10} and 5.0×10^9, respectively.[33] The presence of nonionic micellar interface thus appears to provide a mild inhibition for the reaction (e^-_{aq} + pyrene) while the reaction (e^-_{aq} + Φ_2) is not significantly altered. The higher reactivity of e^-_{aq} with biphenyl could be due to a more polar solubilization site for biphenyl compared to pyrene. There are no other data to support this suggestion. Nevertheless, the data suggest absence of large electrical charge barriers as observed with ionic micelles.

ACKNOWLEDGEMENT

The Radiation Laboratory of the University of Notre Dame is operated under contract with the U.S. Energy Research and Development Administration. This is ERDA Document No. COO-38-1048.

REFERENCES

1. C. Tanford, "The Hydrophobic Effect: Formation of Micelles and Biological Membranes," Wiley-Interscience, New York, 1973.
2. J. H. Fendler and E. J. Fendler, "Catalysis in Micellar and Macromolecular Systems," Academic Press, New York, 1975.
3. M. Gratzel and J. K. Thomas, in "Modern Fluorescence Spectroscopy," E. L. Wehry, Editor, Plenum Press, New York, to appear.
4. R. L. Vold, J. S. Waugh, M. P. Klein and D. E. Phelps, J. Chem. Phys. 48, 3831 (1968).
5. R. McNeil, J. T. Richards and J. K. Thomas, J. Phys. Chem. 74, 2290 (1970).
6. J. Clifford and B. A. Pethica, Trans. Faraday Soc. 60, 1453 (1964); 61, 182 (1965).
7. J. Clifford, ibid. 61, 1276 (1965).
8. E. Williams, B. Sears, A. Allerhand and E. H. Cordes, J. Amer. Chem. Soc. 95, 4871 (1973).
9. R. T. Roberts and C. Chachaty, Chem. Phys. Lett. 22, 348
10. L. M. Corkill, J.F. Goodman and J. Wyer, Trans. Faraday Soc. 65, 9 (1969).
11. C. J. Clemett, J. Chem. Soc, A, 2251 (1970).
12. F. Podo, A. Roy and G. Nemethy, J. Amer. Chem. Soc. 95, 6164 (1973).
13. A. A. Ribeiro and E. A. Dennis, Chem. Phys. Lipids 14, 193 (1975); Biochem. 14, 3746 (1975).
14. A. Abragam, "Principles of Nuclear Magnetism," Oxford University Press, London, 1961.
15. A. Carrington and A. D. McLachlan, "Introduction to Magnetic Resonance," Harper and Row, New York, 1967.
16. K. Kalyanasundaram, M. Gratzel and J. K. Thomas, J. Amer. Chem. Soc. 97, 3915 (1975).
17. J. R. Lyerla, H. M. McIntyre and D. M. Torchia, Macromolecules 7, 11 (1974).
18. G. C. Levy, R. A. Komoroski and J. A. Halstead, J. Amer. Chem. Soc. 96, 3456 (1974).
19. G. Cornelius, W. Gartner and D. H. Haynes, Biochem. 13, 2052 (1974).
20. T. Platzner, M. Gratzel and J. K. Thomas, (1974) unpublished results.
21. W. B. Gratzer and G. H. Beaven, J. Phys. Chem. 73, 2270 (1969).
22. S. Ikeda and G. D. Fasman, J. Polym. Sci. 8, 991 (1970).
23. A. Ray and G. Némethy, J. Phys. Chem. 74, 809 (1971).
24. J. Lang and E. M. Eyring, J. Polym. Sci. 42(10), 89 (1972).
25. Th. Forster, Angew. Chem. Intl Edit. 8, 333 (1969).
26. Th. Forster, Disc. Faraday Soc. 27, 7 (1959).
27. S.C. Wallace, M. Gratzel and J. K. Thomas, Chem. Phys. Lett. 23 359 (1973).

28. M. Gratzel and J. K. Thomas, J. Phys. Chem. 78, 2248 (1974).
29. M. Gratzel, J. J. Kozak and J. K. Thomas, J. Chem. Phys. 62, 1632 (1975).
30. M. Gratzel, J. K. Thomas and L. K. Patterson, Chem. Phys. Lett. 29, 393 (1974).
31. J. T. Richards, G. West and J. K. Thomas, J. Phys. Chem. 74, 4137 (1970).
32. M. Gratzel, K. Kalyanasundaram and J. K. Thomas, J. Amer. Chem. Soc. 97, 7869 (1974).
33. S. C. Wallace and J. K. Thomas, Radiat. Res. 54, 49 (1973).

RADIATION-INDUCED PEROXIDATION IN FATTY ACID SOAP MICELLES

L. K. Patterson and
J. L. Redpath

Department of Medical Physics
Michael Reese Hospital
Chicago, Illinois

The effects of aggregation and of added radical scavengers on radiation-induced peroxidation in fatty acid soap solutions have been examined using linoleic, linolenic and arachidonic acids. Conjugate formation monitored by UV absorption at 232 nm was used as a measure of initial hydroperoxide formation. The dependence of hydroperoxide yield on soap concentration in these systems and yields in mixed micelles suggest that configuration of the unsaturate moieties may play a significant role in determining peroxidative reaction chain length. Evidence is given from several competition measurements and from effects of added $S_2O_8^{2-}$ that the predominant chain initiating radical in these systems is OH. Solubilization of alcohols, especially tert-butanol, by soap aggregates results in little reduction of conjugate diene formation indicating that alcohol radicals within the micellar pseudo-phase may initiate peroxidative processes. Introduction of α-tocopherol into linoleate aggregates sharply inhibits hydroperoxide yield even at molecular concentrations relative to soap of $1:10^4$. Irradiation of soap solutions (1×10^{-2} M) containing low concentrations of α-tocopherol (2.5×10^{-6} M) eventually results in loss of antioxidant behavior and the onset of prooxidant activity.

INTRODUCTION

A broad variety of unsaturate moieties may be found in the different members of the surfactant class known as fatty acids. Several of these moieties are present in biological lipid systems, serving - at least in part - to determine physical characteristics in the lipid regions of membranes. In such structures, unsaturate groups exist predominantly in nonconjugated cis-configuration; they introduce "kinks" into the geometry of the hydrocarbon chain and perturb the paracrystalline close packing of molecules which occurs with saturated hydrocarbon chains.[1] A consideration of the melting points of pure C_{18} fatty acids in relation to the nature of their unsaturate groups illustrates this point: stearic, $70^\circ C$; oleic (cis- Δ^9), $16.3^\circ C$; elaidic (trans- Δ^9), $45^\circ C$; linoleic (cis, cis- $\Delta^{9,12}$), $-5^\circ C$; linolenic (cis, cis, cis- $\Delta^{9,12,15}$), $-11^\circ C$.[2] Introduction of cis-unsaturated fatty acids into a micellar core region or their phospholipid derivatives into a membrane bilayer will then increase fluidity of the associated microenvironment.[3] However, the possibility should not be overlooked that such packing characteristics of fatty acid groups may also influence important reaction processes such as peroxidative degradation of lipids.

While processes by which fatty acids may be degraded in the presence of oxygen have been subject to investigation for some time,[4,5] the relatively recent implication of lipid peroxidation in various medical disorders[6,7,8] and in radiation induced biomole- cular damage[9,10] suggest that detailed understanding of molecular parameters which determine the nature and extent of such peroxida- tion could be of marked interest. Initiation of oxygen related degradation in fatty acids by ionizing radiation has shown that, under certain conditions, the mole quantities of product far exceed those of initial radical formed.[11] Such findings are indicative of chain reaction processes, and a general mechanism has been suggested by Bateman for olefin peroxidation from which analogy may be drawn for unsaturated fatty acid groups.[12] Here H is an α-methylenic hydrogen atom associated with an unsaturate moiety. In systems with nonconjugated bonds it is presumably an allylic hydrogen on the carbon between bonds.

Initiation: Production of R$\cdot$ from RH by hydrogen abstraction r_1
Propagation: R$\cdot$ + O_2 $RO_2\cdot$ k_2
                        $RO_2\cdot$ + RH RO_2H + R$\cdot$ k_3
                            2R$\cdot$ $\Big\{$ Non-initiating or k_4
                        R$\cdot$ + $RO_2\cdot$ propagating k_5
                          $2RO_2\cdot$ products k_6

From photochemical data with ethyl linoleate, values for k_2, k_4, k_5, and k_6 were determined to be in the range of $1-5 \times 10^7$ M^{-1} sec^{-1} while k_3 was determined to be 50 M^{-1} sec^{-1}. With k_3 being the apparent rate determining step in propagation, extent of peroxidation in a given system will be a function of several parameters: lability of the methylenic hydrogens, local concentration of RH in a membrane, micelle or emulsion, and probability for favorable alignment of $RO_2 \cdot$ with an adjacent RH.

Several indices of peroxidation yield in fatty acid and lipids have been used by various workers.[13] Among the most accessible of these is the production of conjugated dienes from nonconjugated moieties accompanying processes involving the allylic hydrogens. Such dienes may be detected by UV absorption at 232 nm. An extinction coefficient of 3×10^4 $M^{-1}cm^{-1}$, based on ε for 9:11-octadecadienoic acid, is generally used for the associated hydroperoxide.[14] In linoleate ester systems a 1:1:1 correspondence has been noted for consumption of oxygen, hydroperoxide yield and conjugated diene formation.[15] Wills and Rotblat have extended UV measurements to irradiated linolenic acid solutions and found a linear relationship between radiation-induced diene and thiobarbituric acid peroxide values.[16] Brice has used conjugated diene absorption in analysis for arachidonic acid following isomerization.[17] In using dienes as a measure of peroxide, it should be noted that this parameter may fail to accurately reflect peroxide content as secondary reactions take place.

There are marked differences between the environments of the fatty acid grouping in a biological membrane and in a simple micelle which may reasonably be expected to influence the course of reaction. The radii of curvature which may have a large influence on interactions of adjacent molecules are significantly different; lipids are but one component of the membrane system which contains other species such as lipophylic proteins and steroids. However, micelles do provide a hydrocarbon microenvironment which exhibits a degree of molecular ordering yet provides large surface-to-volume ratios for contact with the aqueous phase and radicals generated in it. While radiation-induced peroxidation has also previously been examined in pure fatty acids, the uniformity of oxygenation may be questioned. In micellar solution, however, equilibrium with O_2 is rapid and complete.[11] In the present study, micelles of linoleate, linolenate, and arachidonate soaps have been employed as model systems for investigation of the roles played by molecular ordering and geometry in peroxidation. The concept has been employed here that a relationship between production of conjugated groups and hydroperoxide formation exists; however, conclusions drawn are based largely on _relative_ yields of conjugated dienes for each soap under varying solution conditions.

EXPERIMENTAL

Fatty acids used in this study were obtained from Nu-Chek
Prep and all exceed 99% in purity. One sample of linoleic acid
from Calibiochem (A grade) was used for comparative purposes and
similar interdependences of dose, concentration and conjugate
yield discussed below were found for samples from the two sources.
The α-tocopherol was also furnished by Calbiochem. Unless otherwise specified, solutions were prepared in 0.02M PO_4 buffer and
adjusted to a pH of 10.4 - 10.7. Triple distilled water was used
in preparation of all solutions. Steady state irradiations were
carried out with a Co^{60} source delivering 55 rads/minute to samples
under study. Where G-values for conjugate at 232 nm are reported,
an extinction coefficient of 3×10^4 $M^{-1}cm^{-1}$ was assumed based on
values from the literature quoted above. Critical micelle concentrations were determined by the capillary rise and hanging drop
methods.

RESULTS AND DISCUSSION

Effect of Soap Concentration on G-Values
of Conjugated Diene Production

A dependence of conjugated diene yield on solute concentration
in irradiated linoleic acid solutions has been reported by Mead[14]
and by Gebicki and Allen;[15] the latter authors identified micelle
formation as responsible for structures favorable to enhanced yield.
Radiation-induced conjugated diene formation has been measured
here in linoleate, linolenate and arachidonate solutions as a function of concentration and dose. Figure 1 gives the dependence of
UV absorption at 232 nm on dose for several linolenate concentrations; a linear relationship may be seen in each case. Similar
findings were obtained for arachidonate. Earlier results, reported
for linoleate by Gebicki and Allen[15] were confirmed though yields
in the present study were somewhat higher. From the slopes of such
curves, G-values for diene formation and, hence, hydroperoxide are
obtained. These are plotted as functions of soap concentration in
Figure 2. It may be seen that the onset of hydroperoxide enhancement with increasing soap concentration becomes less sharp as the
number of cis double bonds in the chain increases. Further, though
the yields are all comparable at the lower concentrations, where the
soap is in monomeric form, yields at high concentration appear to
be less with increasing number of cis groups. Both of these effects
suggest a potential relationship between the ability of the soap
molecules to achieve close molecular "packing" (which will be
expected to decrease with increasing cis bonds) and the efficiency

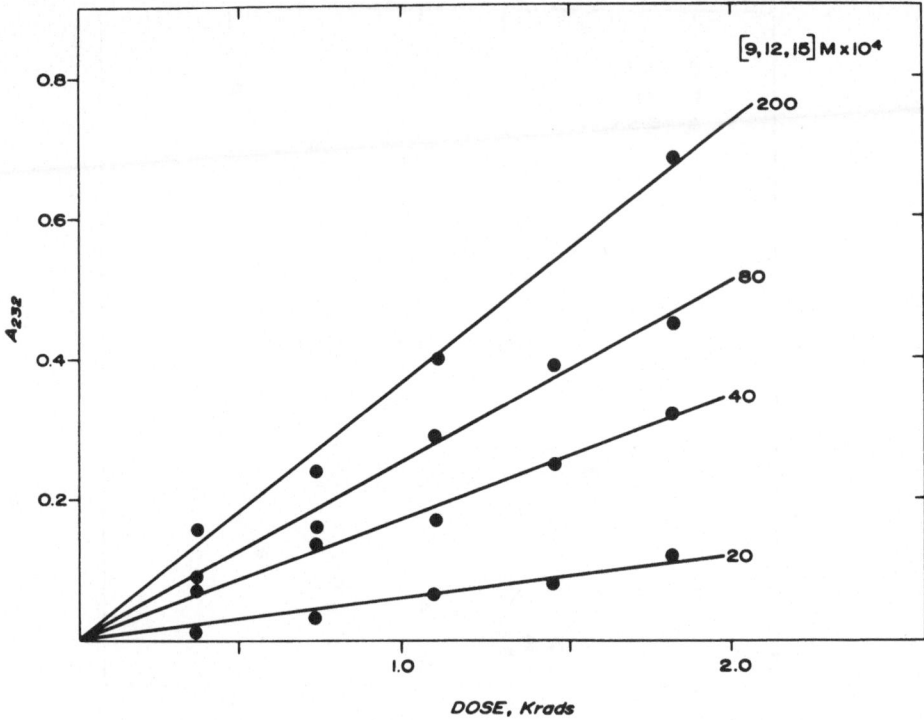

Figure 1.   Conjugated diene absorbance at 232 nm as a function of radiation dose in buffer solutions of linolenate at several con-centrations.

of the hydrogen transfer to $RO_2 \cdot$ upon which the extent of peroxidation has been shown to depend.

Role of OH· as a Peroxidation Initiating Radical

The identity of the primary radical which initiates peroxidation has been open to question. Irradiation of oxygenated aqueous solutions will produce OH·, H· and O_2^- radicals if solutes with high reactivity toward e_{aq}^- are absent. Of these, the O_2^- radical is thought by several workers to be a principal source for peroxidative damage in biological lipid systems.[18,19]   It is well established that $S_2O_8^=$ reacts readily with e_{aq}^- to produce the ·SO_4^- radical which may react in part with OH⁻, giving OH·. This process occurs at the expense of O_2^- formation. When quantities of $S_2O_8^=$ adequate to completely scavenge e_{aq}^- were introduced into an oxygenated linoleate solution at pH 11, the hydroperoxide yield was

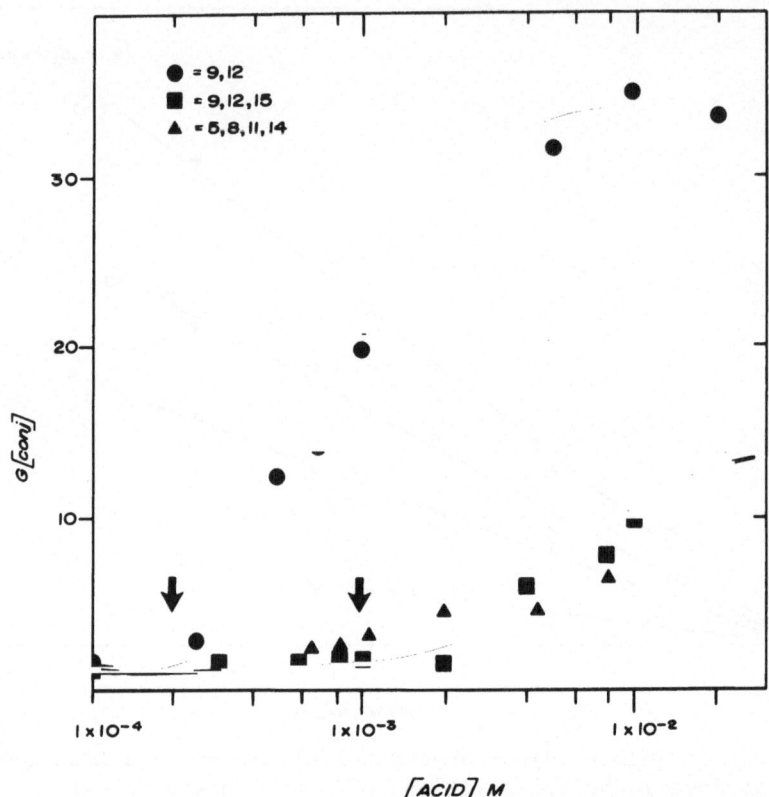

Figure 2.   Dependence of G-values for conjugated diene yield on
soap concentration in buffer solutions; linoleate = (9,12),
linolenate = (9,12,15), arachidonate = (5,8,11,14).   Vertical
arrows mark the c.m.c.'s for linoleate and linolenate.

effectively doubled.   This behavior indicates a dominant contribution by OH· to the peroxidation process in these systems and lead
to competition studies with known OH· scavengers.   The role of H·
in peroxidation was not studied here.

If a compound, X, which competes for the initiating radical
and which does not perturb other steps in the peroxidation mechanism is introduced into fatty acid soap solution, the reduction in
diene absorbances at a fixed dose level will be expected to obey
the relation:

$$A_o/A_m \ = \ 1 + \frac{k_x(X)}{k_{sp}(SP)} \tag{1}$$

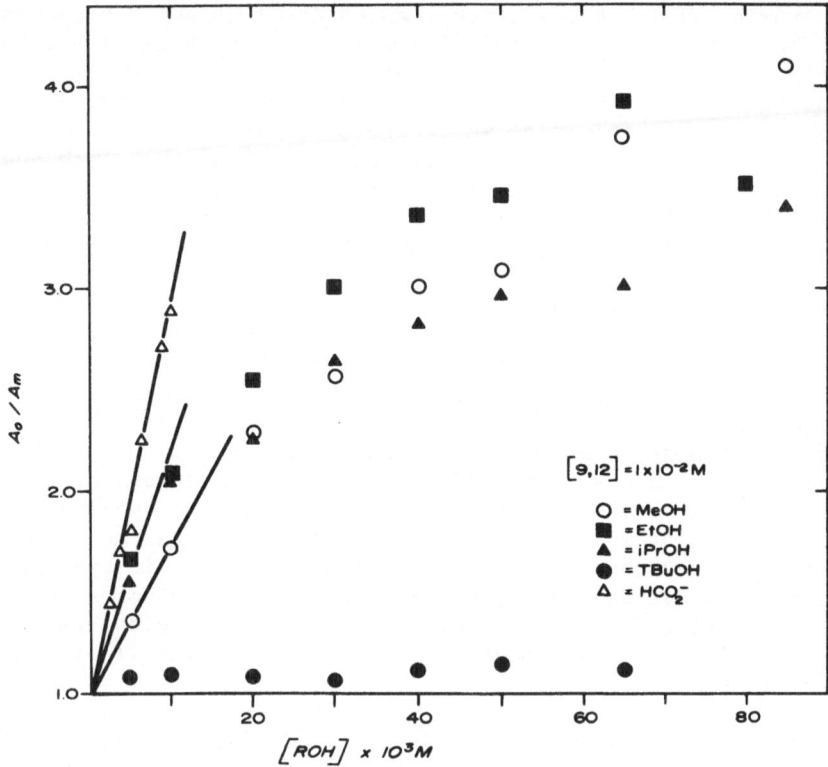

Figure 3. Effects of several OH· scavengers on conjugated diene absorbance plotted in terms of competition parameters from equation (1); MeOH = methanol, EtOH = ethanol, i-PrOH = isopropanol, t-BuOH = tert-butanol, HCO_2^- = formate ion.

where A_o is the absorbance of conjugate for soap concentration (SP) in the absence of X and A_m is the absorbance of conjugate for soap concentration (SP) at some concentration of scavenger, (X). Rate constants for reaction of the initiating radical with soap and competitor are represented by k_{sp} and k_x respectively. Results of such studies using several alcohols and HCO_2^- as competitors in linoleate solution are given in Figure 3. Save for the case of t-butanol, the initial slopes as given in the figure are all consistent with a competition between soap and scavenger for the OH· radical. Analysis of these slopes using established first order rate constants $k_{(x+OH·)}$ gives $k_{(linoleate + OH·)} = 1.3 \times 10^9$ $M^{-1}sec^{-1}$. With increasing alcohol concentration, these slopes diminish. This behavior would indicate that additional competitor is not decreasing

hydroperoxide yield as effectively as at lower competitor concentration. The extent of this deviation shows some dependence on the
character of the alcohol alkyl groups; the effect increases with
increasing alkyl group size. Methanol data show the least decrease
in slope over the whole plot while t-butanol exhibits little effect
on diene absorption throughout. It has been shown that free energies
for transfer of alcohol molecules from pure alcohol to water is positive and become larger with increasing aliphatic chain length.[20] A
pronounced relationship is indicated between hydrophobicity and
alkyl group size even for these low molecular weight alcohols. It
appears, then, that changes of slope observed in Figure 3 are related
to solubilization of alcohols by linoleate soap micelles. At lower
concentrations alcohol partitioning favors the aqueous phase where
the probability of radical recombination will be relatively high.
With increasing concentration, the alcohol distribution shifts
toward the soap hydrocarbon pseudophase bringing greater proportions of alcohol and alcohol radical into the region of the linoleate unsaturate moiety. The results suggest that solubilized
alcohol may be shielded from OH· attack and/or that alcohol radicals
at such sites may themselves abstract soap α-methylenic hydrogens to
initiate peroxidation processes. However, further studies to characterize reactivity of various radicals toward fatty acids are warranted. Analogous measurements were carried out with linolenate
soap and gave similar results.

The high negative surface charge on soap aggregates, repelling
the negative formate charge to prevent interaction of the two
species, would appear to make formate the best choice for determining rate constants by competition. Such studies with 1×10^{-2} M
linoleate and linolenate gave rate constants of 1.3×10^9 and 2.5×10^9 $M^{-1}sec^{-1}$ respectively. However, at soap concentrations of
2.5×10^{-4} M, both gave rate constants of $1 \times 10^{10} M^{-1}sec^{-1}$, effec-
tively the diffusion controlled rate constant. It might be sug-
gested that the greater number of labile hydrogens on the linolenate ion accounts for its relatively higher rate constant observed
in micellar solutions. However, the presence of some linolenate
monomers may contribute to the higher rate constant also. The differences observed between soaps in monomeric compared to micellar
form are similar to those reported for OH· reactions with detergent
surfactants[21] and would appear to reflect the relative accessibility
of soap molecules to attack by radicals generated in the aqueous
phase.

While the above observations demonstrate initiation of peroxide
by OH·, they do not alone exclude the possibility of initiation by
O_2^- in the biological lipids. In addition to repelling formate ions,

the negative surface potential of the soap micelle should prevent entry of O_2^- into the hydrocarbon core. One would expect any reactions between O_2^- and soap molecules to be limited to residual monomers in solution. However, membrane systems do not exhibit these high surface potentials and contact between O_2^- and the lipid bilayer may occur.

Mixed Micelle Competition Studies

Competition studies were also carried out with mixed linoleate-oleate, linolenate-oleate and linoleate-laurate micelles of varying concentration ratios. In these cases the total soap concentrations were maintained at 2×10^{-2} M. The results are plotted in Figure 4. In contrast to formate competition the decrease in diene formation as a function of the concentration ratio is more than twice as great for linolenate as for linoleate. This behavior cannot be due

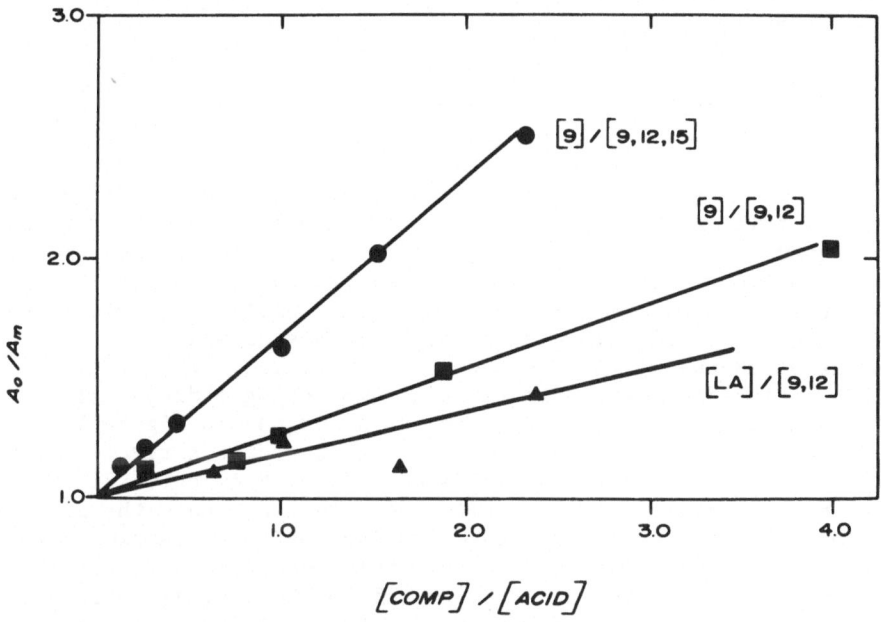

Figure 4. Competition plots for mixed micelles in which ratios of competitor soaps, oleate, (9), or laurate, (LA), to linoleate or linolenate are varied. Total soap concentration is maintained at 2×10^{-2} M. A_0 used here was that from unmixed 2×10^{-2} M linoleate or linolenate.

simply to relative reactions of OH· at unsaturate sites as linolenate would compete more effectively and produce a lower slope for
the plot. Oleate itself has been reported to undergo peroxidative
chain reaction, although no significant amount of diene is produced
to contribute to the absorption at 232 nm.[22] The possibility of a
mixed chain reaction cannot be dismissed without further study. If
such processes do indeed occur, these data may again suggest a
steric influence on the extent of the reaction. Packing of oleate
moieties with linolenate groups for reaction chain propagation
might – due to the greater number of cis double bonds – be more
difficult than with linoleate. It may be noted that the difference
in slopes between the two cases is approximately the same as between
conjugate yields in the two separate soap solutions at 2 x 10^{-2} M
(Figure 1). However, these results depend only on relative optical
densities and not the assignment of an extinction coefficient. The
laurate ion has a shorter hydrocarbon group and no double bonds for
possible participation in a mixed reaction with linoleate. From
data with oleate and the alcohols one would expect, as was observed,
a lower effect on conjugated diene production in the laurage containing system.

Effects of α-tocopherol on Peroxidation
in Soap Micelles

It is well established that α-tocopherol (Vitamin E) plays an
important role in the protection of biomolecules against peroxidative degradation.[23] The efficiency of this compound as an antioxidant in linoleate soap aggregates was examined by solubilizing
α-tocopherol in the micellar systems. Insolubility of this molecule in aqueous solution assures favorable distribution in the
micellar pseudophase. It was found that very low concentrations
of this antioxidant were effective in restricting hydroperoxide
yield in irradiated solutions. Results are shown in Figure 5.
Even when relative concentrations of α-tocopherol constituted about
one molecule in 10^4 of soap, yields were dropped by half (Figure
5a). The high antioxidant activity seems to accentuate the point
that not only is α-tocopherol an effective chain termination agent
but that the propagation step in the chain reaction must be indeed
slow to permit diffusion of α-tocopherol to the radical site to
compete effectively with hydrogen transfer to RO_2·. Polyster and
Mead using methyl linoleate emulsions, found that d-γ-tocopherol
provided protection against radiation induced peroxidation but only
at somewhat higher concentrations than found here.[24] A study was
also conducted to determine a relationship between dose and α-
tocopherol inactivation (Figure 5b). It was found that 2.5 x 10^{-6}
M α-tocopherol in 1 x 10^{-2} M linoleate was sufficient to suppress
most peroxidation until 2 kilorads had been delivered. Further
irradiation is accompanied by a sharp rise in conjugate yield with

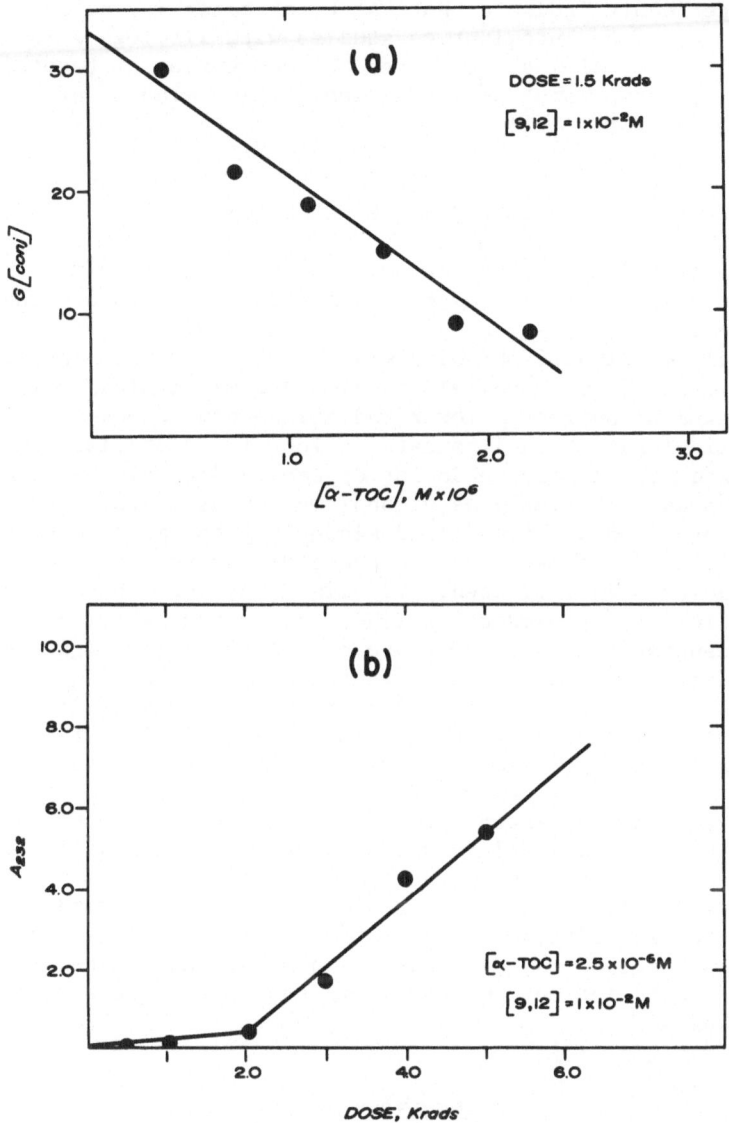

Figure 5. Effect of α-tocopherol on peroxidation in linoleate micelles as a function of: (a) α-tocopherol concentration at fixed dose and (b) dose at fixed α-tocopherol concentration.

a slope almost double that observed in the absence of α-tocopherol. The break point occurs in a region where approximately two hydroxyl radicals have been produced per molecule of α-tocopherol present. Recent experiments indicate that the break point may occur at even higher doses. From autoxidation studies with linoleic acid dispersions, Cillard et al. have reported the conversion of tocopherol from anti- to prooxidant as a function of time elapsed after exposure of the acid to air.[25]

SUMMARY AND CONCLUSIONS

There is evidence that conformation of unsaturate moieties in fatty acid hydrocarbon chains may significantly influence the course of peroxidative degradation induced by ionizing radiation. Though the initial reaction of OH· with soap in micellar form appears to be rapid, evidence from dose-yield studies and mixed micelle measurements appears to suggest a lower radiation-induced peroxidation yield in linolenate than linoleate. Alcohols solubilized in or comicellized with linoleate do not appear to decrease peroxidative degradation as efficiently as alcohol molecules in the aqueous phase. Protection of solubilized alcohol by the micelle structure and/or initiation of peroxidative processes by the intermediate alcohol radicals is indicated. Vitamin E has been shown to be a highly efficient antioxidant in linoleate aggregates even at molecular concentrations relative to soap of $1:10^4$. Its higher efficiency under these conditions indicates that interaction must take place with a component of the chain propagation step. At levels of radical production corresponding to about 2 radicals per molecule of Vitamin E, the antioxidant is transformed into a prooxidant.

In these studies fatty acid soap micelles have provided an elementary, but useful, model in which to study effects of ionizing radiation on peroxidative degradation. Such aggregates provide an extensive water-lipid interface and an associated hydrocarbon pseudophase for the localization of radiation modifiers. Findings from these investigations may well bear on problems associated with the effects of radiation on membrane lipid systems and other biological macromolecules.

ACKNOWLEDGMENTS

The authors wish to thank Ms. Mary Lou Tortorello and Mr. Raymond David of our laboratory for their assistance in the experimental work reported here.

REFERENCES

1. C. Tanford, "The Hydrophobic Effect: Formation of Micelles and Biological Membranes", Chapter 13, John Wiley and Sons, New York, 1973.

2. "Handbook of Chemistry and Physics", 45th ed., R. C. Weast, Editor, The Chemical Rubber Co., Cleveland, 1964.

3. J. Seelig, F. Axel and H. Limacher, Ann. N.Y. Acad. Sci., 222, 558 (1973) and references therein.

4. T. de Saussure, Ann. Chim. et Phys.[2], 13, 337 (1820).

5. D. Swern in "Autoxidation and Antioxidants", W. O. Lundberg, Editor, Chapter 1 and references therein, John Wiley and Sons, New York, 1961.

6. A. L. Tappel, Geriatrics, 23, 97 (1965).

7. B. D. Goldstein, C. Lodi, C. Collinson and O. J. Balchum, Arch. Environ. Health, 18, 631 (1969).

8. H. B. Demopoulos, Fed. Proceedings, 32, 1859 (1973) and references therein.

9. E. D. Wills and A. E. Wilkinson, Radiation Res., 31, 732 (1967).

10. A. Petkau, W. S. Chelack, S. D. Pleskach, B. E. Meeker and C. M. Brady, Biochem. Biophys. Res. Commun., 65, 885 (1975) and references therein.

11. J. F. Mead in "Autoxidation and Antioxidants", W. O. Lundberg, Editor, Chapter 8, John Wiley and Sons, New York, 1961.

12. L. Bateman, Quart. Rev. (London), 8, 147 (1954).

13. W. E. Link and M. W. Formo in "Autoxidation and Autoxidants", W. O. Lundberg, Editor, Chapter 10 and references therein, John Wiley and Sons, New York (1961).

14. J. F. Mead, Science, 115, 470 (1952).

15. J. M. Gebicki and A. O. Allen, J. Phys. Chem., 73, 2443 (1969).

16. E. D. Wills and J. Rotblat, Int. J. Rad. Biol., 8, 551 (1965).

17. B. A. Brice, M. L. Swain, S. F. Herb, P. L. Nichols, Jr. and R. W. Riemen-Schneider, J. Amer. Oil Chem. Soc., 29, 279 (1952).

18. A. Petkau and W. S. Chelack, Red. Proc., 33, 1505 (1974).

19. D. D. Tyler, FEBS, 51, 180 (1975).

20. K. H. Kinoshita, H. Ishikawa, and K. Shinoda, Bull. Chem. Soc., Japan, 31, 1081 (1958).

21. L. K. Patterson, K. M. Bansal and J. H. Fendler, Chem. Commun., 152 (1971).

22. S. M. Hyde and D. Verdin, Trans. Faraday Soc., 64, 155 (1968).

23. E. Aaes-Jørgensen in "Autoxidation and Antioxidants", W. O. Lundberg, Editor, Chapter 21, John Wiley and Sons, New York, 1961.

24. B. H. Polister and J. F. Mead, J. Agricul. Food Chem., 2, 199 (1954).

25. J. Cillard, M. Cormier and R. Girre, C. R. Acad. Sci. Paris, [D], 455 (1975).

BIFUNCTIONAL MICELLAR CATALYSIS

R. A. Moss, R. C. Nahas, and S. Ramaswami

Wright & Rieman Laboratories, Rutgers
The State University of New Jersey
New Brunswick, New Jersey 08903

Surfactants $\underline{16}$, $\underline{16}$-OH, $\underline{16}$-Im, and $\underline{16}$-OH,Im were examined as micellar catalysts for the hydrolysis of p-nitrophenyl acetate (PNPA) and hexanoate (PNPH) in 0.01 $\underline{M}$ and 0.4 $\underline{M}$ aqueous phosphate buffers at pH 8.0. The kinetic data is presented and discussed.

$$\underline{n}\text{-}C_{16}H_{33}\overset{+}{N}(CH_3)R_1R_2, \ Cl^-$$

($\underline{16}$, $R_1=R_2=CH_3$; $\underline{16}$-OH, $R_1=CH_3$, $R_2=CH_2CH_2OH$; $\underline{16}$-Im, $R_1=CH_3$, $R_2=CH_2$-4-imidazolyl; $\underline{16}$-OH,Im, $R_1=CH_2CH_2OH$, $R_2=CH_2$-4-imidazolyl)

Whereas the hydrolysis of PNPA in micellar $\underline{16}$-Im involved nucleophilic attack of an (anionic) imidazole moiety on PNPA, accompanied by the spectroscopically detectable formation of an $\underline{N}$-acetylimidazole intermediate, the hydrolysis of PNPA in $\underline{16}$-OH,Im did not provide an observable $\underline{N}$-acetylimidazole intermediate. The mechanistic implications of these contrasting observations are presented, and then discussed in the light of additional experimental data. It is concluded that the most likely course of the PNPA + $\underline{16}$-OH,Im reaction is "slow" acetylation of the imidazolyl moiety followed by rapid transfer of the acetyl fragment to the hydroxyl moiety. The transfer step is most probably a composite of intermolecular (intramicellar) and intramolecular processes.

INTRODUCTION

Much has been made of the analogy between micelles and enzymes; the analogy itself is now commonplace.[1-7] However, the rational design of functionalized surfactants to provide increasingly exact enzyme analogs is a relatively recent innovation.[8] A key feature of the chymotrypsin-catalyzed cleavage of esters and amides is basic activation, by the imidazole group of His-57, of the Ser-195 hydroxyl moiety; the latter's oxygen is the nucleophile which attacks the substrate's carbonyl group.[9-11] It has also been postulated that the carboxylate group of Asp-102 interacts with the imidazole of His-57, so as to strengthen the basicity of the latter.[12]

Many "model enzymes" have been designed to mimic the mechanism of chymotrypsin catalysis.[13] Micellar chymotrypsin models have included hydroxyl[14-22] and imidazole-functionalized surfactants. The imidazole moiety has been supplied as an hydrophobic acylhistidine or benzimidazole, solubilized by a "carrier" micelle.[23-28] Alternatively, the imidazole functionality was part of the surfactant itself.[29-33] In contrast to this appreciable body of work, rather fewer reports of bifunctional micellar catalysts have appeared,[34-36] although several multifunctional polymeric[37-39] and cyclodextrin[40-42] catalysts have been examined.

We therefore began a comparative study of several simple mono- and bifunctional surfactant catalysts. The initial stages of this investigation were concerned with comparative catalytic effectiveness as it related to surfactant functionalization,[43] and with delineation of the catalytic mechanisms. Later stages will focus on the construction of more sophisticated micellar enzyme analogs.

The structures of the surfactants chosen for initial study are shown below, together with their abbreviations.

Only 16-OH,Im is a new construction.[44] Surfactant 16-Im, with an octadecyl hydrophobic group, was first prepared by Tagaki,[29] whereas hydroxyl-functionalized surfactants are older still.[14-17] In our hands, cmc values for 16-Im and 16-OH,Im were 7.9×10^{-5} and 6.8×10^{-5} M, respectively, in 0.01 M phosphate buffer at pH 8, 25°. We also prepared an analogous set of model compounds in which the cetyl moieties were replaced by methyl groups.

$$\underline{n}\text{-}C_{16}H_{33}\text{-}\overset{\overset{\displaystyle CH_3}{+|}}{\underset{\underset{\displaystyle CH_3}{|}}{N}}\text{-}CH_3, \; Cl^- \qquad\qquad \underline{16}$$

(structures continued)

$$\underline{n}\text{-}C_{16}H_{33}\text{-}\overset{\overset{\displaystyle CH_3}{\overset{+}{|}}}{\underset{\underset{\displaystyle CH_3}{|}}{N}}\text{-}CH_2CH_2OH,\ Cl^-$$ $\underline{16}\text{-OH}$

$$\underline{n}\text{-}C_{16}H_{33}\text{-}\overset{\overset{\displaystyle CH_3}{\overset{+}{|}}}{\underset{\underset{\displaystyle CH_3}{|}}{N}}\text{-}CH_2\text{-}\text{(imidazole)},\ Cl^-$$ $\underline{16}\text{-Im}$

$$\underline{n}\text{-}C_{16}H_{33}\text{-}\overset{\overset{\displaystyle CH_3}{\overset{+}{|}}}{\underset{\underset{\displaystyle CH_2CH_2OH}{|}}{N}}\text{-}CH_2\text{-}\text{(imidazole)},\ Cl^-$$ $\underline{16}\text{-OH,Im}$

RESULTS AND DISCUSSION

In initial studies, the hydrolyses of p-nitrophenyl acetate (PNPA) or of p-nitrophenyl hexanoate (PNPH) were studied in phosphate buffers at pH 8.0, 25°, as a function of the concentrations of added surfactants $\underline{16}$, $\underline{16}$-OH, and $\underline{16}$-Im. The substrates were present in low concentration (2.0×10^{-5} M), and the pseudo-first-order hydrolytic rate constants were determined by monitoring the release of p-nitrophenolate ion at 400 (or 440) nm. From profiles of k_{ψ} versus [surfactant], values of k_{ψ}^{max} were obtained for each surfactant. Typical data are shown (rounded to 2 significant figures) in Table I.

Table I. $10^5\ \underline{k}_{\psi}^{max}(\text{sec}^{-1})$ for Catalyzed Hydrolyses of PNPA or PNPH

| Buffer | Subst. | Surfactant | | | |
		None	$\underline{16}$	$\underline{16}$-OH	$\underline{16}$-Im
0.01 M	PNPA	1.8	16.	190.	20,000
	PNPH	2.1	27.		43,000
0.4 M	PNPA	8.1		200.	11,000
	PNPH	2.7		130.	17,000

It is clear from these results that the imidazole-functionalized surfactant, 16-Im, is a much more effective catalyst for the cleavage of p-nitrophenyl esters than is the hydroxyl-functionalized surfactant, 16-OH. The latter, in turn, is more effective than the unfunctionalized surfactant, 16. Although there are differences in the concentration of each surfactant at which k_ψ^{max} is observed, the differences are relatively small; e.g., for PNPA cleavage in 0.01 M buffer, the concentrations required for k_ψ^{max} are (x 100): 16, 1.8; 16-OH, 1.4; and 16-Im, 4.0 M.

A more direct comparison of surfactant effectiveness is shown in Table II, where the k_ψ^{max} values have been adjusted to a relative scale on which the uncatalyzed (buffer) rate constant has been assigned a value of unity. In 0.01 M buffer, the effectiveness order of the surfactants toward PNPA, in terms of k_ψ^{max}, is 16-Im> 16-OH>16 ~ 1200: 12: 1. Toward the more hydrophobic PNPH, 16-Im is 1500 times more effective than non-functionalized 16. These observations parallel previous findings, but quantify the very large rate enhancements obtainable with 16-Im at relatively low pH, and the marked superiority of 16-Im over the choline-surfactant, 16-OH. Similar trends are observed in the more concentrated buffer, although k_ψ^{max} values are generally lower (salt inhibition of micellar catalysis), and the relative rate scales are compressed.

Table II. Relative Catalytic Abilities

		Catalyst			
Buffer	Subst.	None	16	16-OH	16-Im
0.01 M	PNPA	1.0	8.9	110	11,000
	PNPH	1.0	13.		20,000
0.4 M	PNPA	1.0		25	1,400
	PNPH	1.0		48	6,300

In Table III, we compare the catalytic abilities of the monofunctional surfactants, 16-OH and 16-Im, to those of model compounds which, although identical in functionalization, are unable to micellize. The superiority of 16-OH, relative to its choline chloride model, is, of course, largely due to its ability to micellize and to bind the substrate. However, studies of hydroxyl-functionalized surfactants have indicated that the effective catalyst is the alkoxide form of the surfactant,[20,21] and, in this regard, micellization of 16-OH confers an added benefit. The pK_a

Table III.   Surfactants Versus Models[a]

	Substrates			
Catalyst	PNPA	k_{rel}	PNPH	k_{rel}
16-OH	200	14	130	4.3
$Me_3\overset{+}{N}CH_2CH_2OH$	14	1.0	30	1.0
16-Im	11,000	300	17,000	740
$Me_3\overset{+}{N}CH_2Im$	37	1.0	23	1.0
Imidazole	1,700	46	290	13

[a] $10^5 \, k_\psi^{max}$ vs. $10^5 \, k_\psi^{model}$ at identical concentrations, in 0.4 M
phosphate buffer, pH 8.0.

of the micellized surfactant's hydroxyl is lowered (apparently as
far as ~10.5[20]) because its conjugate alkoxide base is stabilized
in the micelle's positive field. This added stabilization of the
alkoxide form means that, at a given pH, the conversion of micellar
16-OH to the catalytically active alkoxide form will exceed that of
the choline chloride model (pK_a ~ 12.8[20]), leading to a further
enhancement of its catalytic advantage.

Does a similar effect hold for imidazole-surfactant 16-Im?
Recent work of the Russian group indicates that the anion is the
catalytic form of an imidazole moiety in cationic surfactant
micelles.[26-28] This conclusion was anticipated by Tagaki,[29] and
accords with both our work[43] and that of Tonellato.[33]

Although imidazole itself is a good catalyst for the hydrolysis
of PNPA or PNPH (Table III), acylation involves a dipolar transition state, with positive charge development on the imidazole:

This step would be disfavored if the imidazole group were bonded
to a cation. For example, $Me_3\overset{+}{N}CH_2Im$ is a poorer catalyst than
imidazole by factors of 46 or 13 toward PNPA or PNPH in 0.4 M
buffer. Were the imidazole group solely in its neutral form in
micelles of 16-Im, then its catalytic effectiveness would be no
more, and perhaps less, than that of the model compound. The contribution of substrate binding would be offset by the interaction
of the additional micellar positive charge with the positive charge

formed on the neutral imidazole moiety in the acylation transition
state. Indeed, the acylation of neutral benzimidazole or of N-
methylbenzimidazole has been shown to be inhibited in a cationic
micelle.[28]

However, we observe that 16-Im is a much better catalyst than
the model compound (by factors of 300 and 740 toward PNPA or PNPH
in 0.4 M buffer) and also better than imidazole (factors of 6.5
and 59). The great effectiveness of 16-Im micelles therefore
requires that the imidazole anion, and not the neutral imidazole
moiety, be the catalytic center. With the anionic form as the

$$\underline{n}\text{-}C_{16}H_{33}\overset{+}{N}(CH_3)_2CH_2\text{-}\quad N\text{:}^-$$

catalyst, the transition state for acylation by PNPA or PNPH will
be charge-diffuse, relative to the ground state, and will be stabilized by the micellar environment.

In Table IV, we turn to a comparison of monofunctional surfactant 16-Im with bifunctional surfactant 16-OH,Im, and with a 1:1
mixture of 16-Im and 16-OH. The pseudo-first-order rate constants
are for the cleavage of PNPA or PNPH (acylation step) in either 0.01
or 0.4 M phosphate buffers. Judged strictly from the k_{ψ}^{max} values,
16-OH,Im would appear to be a less effective catalyst for p-nitrophenyl ester cleavage than monofunctional 16-Im. No evidence for
synergism or cooperation between the two functional groups of 16-
OH,Im (or of the 1:1 mixture of 16-OH and 16-Im) is apparent on these
strictly kinetic grounds. The apparent kinetic advantage of 16-Im

Table IV. Bifunctional Catalysts[a]

Catalyst	PNPA		PNPH	
	0.01 M	0.4 M	0.01 M	0.4 M
16-Im	20,000	11,000	43,000	17,000
16-OH, Im	13,000	7,500	21,000	15,000
16-Im + 16-OH[b]		5,600		7,000

[a] $10^5 \; k_{\psi}^{max}$ sec^{-1}, pH 8.0, 25°, phosphate buffer.   [b] 2.5×10^{-2} M in
each surfactant.

over 16-OH,Im largely vanishes if the k_ψ^{max} if the k_ψ^{max} values are divided by the corresponding [surfactant] at which they were determined, so as to afford second order catalytic constants, k_c; see Table V. In this comparison, the effectiveness of 16-Im is very similar to that of 16-OH,Im, except that the bifunctional catalyst is somewhat superior toward PNPH in 0.4 M buffer.

Table V. Catalytic Rate Constants, k_c [a]

Catalyst	PNPA		PNPH	
	0.01 M	0.4 M	0.01 M	0.4 M
16-Im	5.0	3.7	172.	17.
16-OH,Im	4.6	2.5	161.	30.

[a] In l/mol-sec, derived from the data of Table IV as described in the text.

The data of Tables IV and V give no reason to suspect any significant difference in mechanistic behavior between 16-Im and 16-Im,OH. A closer look at the reactions of these catalysts with PNPA, however, provides a surprising observation. Whereas the reaction of 16-Im with PNPA leads to the clear build-up and decay of an acylimidazole intermediate, readily observable at 245 nm, no such intermediate can be observed during the comparable reaction of bifunctional catalyst 16-OH,Im with PNPA. The conditions under which these reactions were carried out will be designated as standard conditions for the remainder of this discussion, and include: [PNPA] = 2.0×10^{-4} M, [surfactant] = 5.0×10^{-3} M, pH 8.0, 25°, 0.4 M phosphate buffer.

From a preparative scale reaction of PNPA and 16-OH,Im, we quantitatively isolated O-acetylated-16-OH,Im. Thus, PNPA is not acetylating water under the influence of the bifunctional catalyst. The failure of this catalyst to display an acetylimidazole intermediate in its reaction with PNPA must therefore be attributed to either of two mechanistic situations.

(a) No intermediate is visible at 245 nm because none is ever formed; bifunctional catalyst 16-OH,Im behaves as a chymotrypsin analog and undergoes direct O-acetylation:

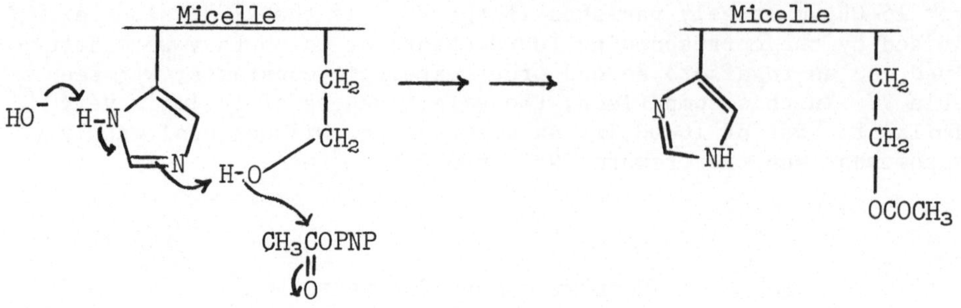

(b) Alternatively, no intermediate is observed at 245 nm because the intermediate is not permitted to accumulate to a suffi- cient concentration to be visible. In a two-step mechanism, a rela- tively slow N-acetylation of 16-OH,Im by PNPA is followed by a very rapid, hydroxyl-catalyzed deacylation:

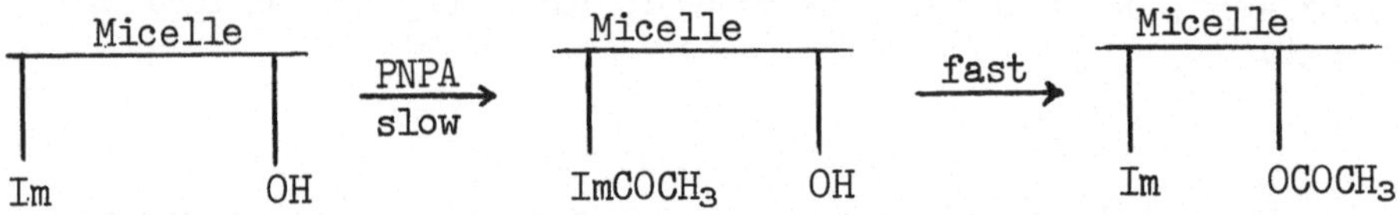

In neither case need we now specify whether the involved imidazole and hydroxyl moieties are on the same surfactant molecule or on adjacent surfactant molecules within a single micelle.

What evidence is available to decide between alternatives (a) and (b)? In preliminary experiments, we have observed the formation and decay of what is apparently hexanoylimidazole-16-OH,Im, when 16-OH,Im reacts with PNPH under "standard conditions". The inter- mediate's absorption was monitored at 245 nm, and the pseudo-first- order rate constant for dehexanoylation was extracted from the time dependence of the 245 nm absorption.[45] The data is shown in Table VI.

Although deacylation of hexanoylimidazole-16-OH,Im is very rapid, the acylation step is also rapid, and the intermediate can be observed. The reaction of 16-OH,Im with PNPH appears, therefore, to follow the two-step mechanism. It is simpler to believe that the reaction of 16-OH,Im with PNPA follows the same mechanism, than that the mechanism changes from chymotrypsin-like to two-step as the substrate changes (although this is not intrinsically impossible).

Table VI. Acylimidazole Intermediates[a]

Catalyst	Substrate	k_{acyl}[b]	k_{deacyl}[c]
16-OH,Im	PNPH	0.14	"0.32"
16-OH,Im	PNPA	0.038	0.6 (est)[d]

[a]Standard conditions, pseudo-first-order rate constants (sec^{-1}).
[b]From the release of p-nitrophenolate ion at 440 nm. [c]See text and ref. 45. [d]Estimated value; see text.

If, indeed, the two-step mechanism holds in each case, then it is simply an unfavorable ratio of the rates of acylation (slow) and deacylation (rapid) which prevents the observation of acetylimidazole-16-OH,Im in the reaction of 16-OH,Im and PNPA. Based on its non-observation with PNPA, where k_{acyl} = 0.038 sec^{-1} (Table VI), we estimate[45] that k_{deacyl} for acetylimidazole-16-OH,Im must be at least 0.6 sec^{-1}, under our standard conditions. A priori, this is not unreasonable, because k_{deacyl} for hexanoylimidazole-16-OH,Im is ~0.3 sec^{-1}, and removal of an acetyl group might well be a more rapid process than removal of an hexanoyl group.

In another preliminary experiment, 16-OH,Im under otherwise standard conditions, but diluted with a five-fold excess of surfactant 16, was treated with acetic anhydride (or PNPA). The formation and decay of an intermediate was now observable at 245 nm. We suspect that the intermediate is acetylimidazole-16-OH,Im, surrounded by unfunctionalized surfactant 16:

$$(\underline{16} \big< ^{ImCOCH_3} _{OH} \quad) \; \underline{16}$$

These results can be interpreted on the basis of either the two-step or chymotrypsin mechanisms for 16-OH,Im catalysis. Presently, we feel that the two-step mechanism is more likely (see above). On this basis, we interpret the dilution experiment as follows. Deacetylation of initially formed acetylimidazole-16-OH,Im probably involves acetyl transfers to hydroxyl groups on the same (intramolecular) and adjacent (intermolecular) surfactant molecules. The overall, normal deacetylation rate is thus a composite rate. Dilution of 16-OH,Im with excess surfactant 16, however, interferes with the intermolecular acetyl transfer, reduces the overall deacetylation rate, and permits the accumulation of sufficient acetylimidazole-16-OH,Im to render the intermediate observable.

If the normal catalytic mode of 16-OH,Im with PNPA were chymo-trypsin-like, then one would have to maintain that the cooperativity of hydroxyl and imidazolyl moieties was mainly intermolecular. Dilution with surfactant 16 would then destroy the cooperativity, alter the catalytic mechanism to nucleophilic imidazole attack, and thus lead to an observable acetylimidazole intermediate. Whether the mechanism of the PNPA + 16-OH,Im reaction is two-step or chymo-trypsin-like, the 16-dilution experiment points to the importance of intermolecular hydroxyl-imidazolyl cooperativity.

Finally, it should be emphasized that there is most definitely hydroxyl catalysis of the deacetylation of acetylimidazole surfac-tants. The data in Table VII speak to this point. The formation and decay of an acetylimidazole intermediate is readily observed when 16-Im is treated with either PNPA or acetic anhydride, and the rate constant for deacetylation, 0.015 sec^{-1}, can be easily deter-mined from the decay of the 245 nm absorption. When a 1:1 comicelle of 16-Im and 16-OH is acetylated, an acetylimidazole intermediate can also be observed at 245 nm, although the deacetylation of the acetyl-16-Im is now considerably faster than it was previously. Indeed, based on a preliminary value[45] of 0.16 sec^{-1} for k_{deacyl}, we find the deacetylation of acetylimidazole-16-Im to be accelerated ~11-fold by comicellized 16-OH.

If the bifunctional catalyst 16-OH,Im cleaves PNPA by the two-step mechanism, then catalysis of the deacetylation step must be even greater than in the 1:1 16-OH, 16-Im comicelle. k_{acyl} for 16-OH,Im with PNPA is 0.038 sec^{-1} (Tables VI, VII), leading to a minimum estimate of 0.6 sec^{-1} for k_{deacyl} of (unobserved) acetyl-imidazole-16-OH,Im under our standard conditions (see above). As shown in Table VII, this requires a minimum 40-fold enhancement of

Table VII. Comicellar vs Bifunctional Micellar Catalysis with PNPA[a]

Catalyst	k_{acyl}[b]	k_{deacyl}	k_{rel}^{deacyl}
16-Im	0.051	0.015[c]	1.0
16-Im + 16-OH[d]	0.036	"0.16"[e]	11.
16-OH,Im	0.038	0.60[e,f]	40.[f]

[a]Standard conditions, rate constants in sec^{-1}. [b]From the appear-ance of p-nitrophenolate ion at 440 nm. [c]Monitored at 245 nm, analysis by simple first order kinetic treatment. [d]1:1 comicelle, 5×10^{-3} M in each surfactant. [e]Monitored at 245 nm, consecutive first order reactions analysis.[45] [f]Estimated[45] minimum value.

the deacetylation rate constant, relative to 16-Im, and about a 4-fold enhancement of the deacetylation rate constant relative to the 16-Im, 16-OH comicelle. Part of the additional catalysis is probably due to intramolecular hydroxyl-mediated deacetylation of acetyl-imidazole-16-OH,Im. We are not yet certain, however, that this pathway can quantitatively account for all of the additional catalysis.

We are continuing our studies of bifunctional surfactant 16-OH,Im, and refining some of the preliminary experiments described here.

ACKNOWLEDGMENTS

We are grateful to the National Cancer Institute (National Institutes of Health) and to the National Science Foundation for their generous support of our research. R.A.M. wishes to acknowledge a fellowship from the A. P. Sloan Foundation. We also wish to recognize the experimental contributions of Mr. Winfred J. Sanders, who carried out several studies involving surfactant 16-OH. Thanks are due to Pergamon Press for permission to quote small segments from Tetrahedron Letters.

REFERENCES

1. J. H. Fendler and E. J. Fendler, "Catalysis in Micellar and Macromolecular Systems," Academic Press, New York, N.Y., 1975.
2. E. H. Cordes, Editor, "Reaction Kinetics in Micelles," Plenum Press, New York, N.Y., 1973.
3. C. Tanford, "The Hydrophobic Effect: Formation of Micelles and Biological Membranes," Wiley-Interscience, New York, N.Y., 1973.
4. C. A. Bunton, Prog. Solid State Chem., 8, 239 (1973).
5. E. H. Cordes and C. Gitler, Prog. Biorg. Chem., 2, 1 (1973).
6. I. V. Berezin, K. Martinek, and A. K. Yatsimirski, Russ. Chem. Rev., 42, 787 (1973).
7. For an early, critical view, see H. Morawetz, Advan. Catal., 20, 341 (1969).
8. For a review, see reference 1, pp. 169-189.
9. E. Zeffren and P. L. Hall, "The Study of Enzyme Mechanisms," Wiley-Interscience, New York, N.Y., 1973, pp. 167-193.
10. W. P. Jencks, in "Chemical Reactivity and Biological Role of Functional Groups in Enzymes," R. M. S. Smellie, Editor, Academic Press, New York, 1970, pp. 59 ff.
11. A. D. B. Malcolm and J. R. Coggins, Ann. Rep. Chem. Soc., 71B, 540-544 (1974).

12. D. M. Blow, Acc. Chem. Res., 9, 145 (1976), and references cited therein.
13. T. H. Fife, Advan. Phys. Org. Chem., 11, 1 (1975).
14. C. A. Bunton, L. Robinson, and M. Stam, J. Amer. Chem. Soc., 92, 7393 (1970).
15. M. Chevion, J. Katzhendler, and S. Sarel, Israel J. Chem., 10, 975 (1972).
16. G. Meyer, Tetrahedron Lett., 4581 (1972).
17. G. Meyer, Compt. Rend. Acad. Sci. (Paris), 276, Ser. C, 1599 (1973).
18. C. A. Bunton and L. G. Ionescu, J. Amer. Chem. Soc., 95, 2912 (1973).
19. V. Gani, C. Lapinte, and P. Viout, Tetrahedron Lett., 4435 (1973).
20. K. Martinek, A. A. Levashov, and I. V. Berezin, Tetrahedron Lett., 1275 (1975).
21. C. A. Bunton and M. McAneny, J. Org. Chem., 41, 36 (1976).
22. C. A. Bunton and S. Diaz, J. Org. Chem., 41, 33 (1976).
23. A. Ochoa-Solano, G. Romero, and C. Gitler, Science, 156, 1243 (1967).
24. C. Gitler and A. Ochoa-Solano, J. Amer. Chem. Soc., 90, 5004 (1968).
25. P. Heitmann, R. Husung-Bublitz, and H. J. Zunft, Tetrahedron, 30, 4137 (1974).
26. A. P. Osipov, K. Martinek, A. K. Yatsimirski, and I. V. Berezin, Dok. Akad. Nauk SSSR, 215, 914 (1974).
27. K. Martinek, A. P. Osipov, A. K. Yatsimirski, V. A. Dadali, and I. V. Berezin, Tetrahedron Lett., 1279 (1975).
28. K. Martinek, A. P. Osipov, A. K. Yatsimirski, and I. V. Berezin, Tetrahedron 31, 709 (1975).
29. W. Tagaki, M. Chigira, T. Ameda, and Y. Yano, Chem. Commun., 219 (1972).
30. J. M. Brown and C. A. Bunton, Chem. Commun., 969 (1974).
31. J. M. Brown, C. A. Bunton, and S. Diaz, Chem. Commun., 971 (1974).
32. D. G. Oakenfull and D. E. Fenwick, Aust. J. Chem., 27, 2149 (1974).
33. U. Tonellato, J. C. S. Perkin II, 771 (1976).
34. U. Tonellato, private communication.
35. J. Sunamoto, H. Okamoto, H. Kondo, and Y. Murakami, Tetrahedron Lett., 2761 (1975).
36. T. Kunitake, Y. Okahata, and T. Sakamoto, Chem. Lett., 459 (1975).
37. N. Ise, T. Okubo, H. Kitano, and S. Kunugi, J. Amer. Chem. Soc., 97, 2882 (1975).
38. C. G. Overberger and Y. Okamoto, Macromolecules, 5, 363 (1972).
39. C. G. Overberger and J. C. Salamone, Accounts Chem. Res., 2, 217 (1969).

40. Y. Iwakura, K. Uno, F. Toda, S. Onozuka, K. Hattori, and M. L.
     Bender, J. Amer. Chem. Soc., 97, 4432 (1975).
41. D. W. Griffiths and M. L. Bender, Advan. Catal., 23, 209 (1973).
42. F. Cramer and G. Mackensen, Angew. Chem. Int. Ed. (Engl.), 5,
     601 (1966); Chem. Ber., 103, 2138 (1970).
43. R. A. Moss, R. C. Nahas, S. Ramaswami, and W. J. Sanders,
     Tetrahedron Lett., 3379 (1975).
44. This surfactant was first prepared, in our laboratory, by Dr.
     W. L. Sunshine.
45. The acylation and deacylation steps were treated as consecutive
     first order reactions; the analysis is described in A. A. Frost
     and R. G. Pearson, "Kinetics and Mechanism," 2cd ed., pp. 166-
     169, John Wiley and Sons, New York, N.Y., 1961.

THE USE OF PHASE TRANSFER CATALYSTS WITH EMULSION AND MICELLE SYSTEMS IN ELECTRO-ORGANIC SYNTHESIS

Thomas C. Franklin and Tadatoshi Honda

Chemistry Department, Baylor University

Waco, Texas 76703

It was shown that in systems where diphenyl-acetonitrile were solubilized in organic-aqueous sodium hydroxide emulsions and micelles that one could obtain appreciable yields of the dehydro-dimer by anodic oxidation at platinum electrodes if one used a cationic surfactant. It was also shown that it was possible anodically to dimerize diethyl malonate solubilized in micelles in a 2M sodium carbonate solution using a cationic surfact-ant if bromide ion was present. Best results in the anodic dimerization of diethyl malonate were obtained in emulsions prepared by adding tetra-butylammonium bromide to an aqueous 2M sodium carbonate solution.

INTRODUCTION

Organic electrochemistry, having in the past gone through several waves of interest, is now the focus of more diversified investigations than at anytime in the history of the field. Work is moving beyond academic problems towards industrial synthesis.[1-10]

One of the biggest hindrances to the development of practical organic electrosyntheses has been the lack of suitable methods of solubilizing the reactants and products. The solvent system must

dissolve the organic reactants and products and also ionize an
electrolyte to carry the current. On the one hand water is not
capable of dissolving most organic molecules while nonaqueous
systems are not suitable solvents for dissolving and ionizing
electrolytes.

One approach has been to solubilize the organic compounds
in aqueous solutions by the use of hydrotropic salts such as
"McKee's salts" (arylsulfonate salts). This has been used in a
number of organic syntheses including the commercial synthesis
of adiponitrile.[11-13]

Another approach has been to prepare emulsions of liquid
organic reactants in aqueous electrolyte solutions with and
without emulsifying agents. These emulsions solubilized
reasonable amounts of the organic compounds while retaining
the high conductivity of the aqueous solution.[14-20]

Recently, Eberson and co-workers [21-22] introduced the use of
dichloromethane-aqueous sodium cyanide emulsions in the anodic
cyanation of several aromatic compounds. They observed that,
similar to results obtained in other organic reactions performed
in emulsion systems,[23-24] the introduction of catalytic amounts
of a phase transfer catalyst (the tetrabutylammonium ion)
markedly increased the yield of product.

There have also been some recent polarographic studies in
which the organic compound was solubilized as micelles and sus-
pended in an aqueous medium.[25-29] The results with both emulsions
and micelles show that it is possible to solubilize the organic
compounds by these procedures, and still maintain the high
conductivity of the aqueous solution.

Recently a report was made on the electrooxidation of benzhydrol to benzophenone in systems in which the organic compounds were
solubilized in both emulsions and micelles.[30-31] The results were
similar to those of Eberson, since the use of cationic (quaternary
ammonium) surfactants to form the micelles or act as emulsifying
agents caused marked increases in the yields of product whereas
the addition of anionic or neutral surfactants caused only small
increases in the yields. These cationic surfactants are recognized
to be phase transfer catalysts.

This paper is a description of some further studies using
emulsions and micelles to solubilize organic compounds in electrochemical oxidations. The reactions studied were the anodic dehydrodimerization of diphenylacetonitrile and diethyl malonate.

EXPERIMENTAL

Synthesis Experiments

The apparatus used in the synthesis studies consisted of a 150 ml porous porcelain cup in a 400 ml beaker. Inside the cup was placed a magnetic stirrer, a bright platinum foil anode with a 16 cm^2 surface area and a Luggin capillary connected to a saturated calomel reference electrode. The cup was surrounded with a platinum foil which was used as a counter electrode. The electrode potential was controlled in the diethyl malonate experiments by an Anotrol Model 4100 Potentiostat. The diphenylacetonitrile experiments were performed at a constant current.

In the beaker outside the porcelain cup was placed approximately 100 ml of aqueous 2M sodium carbonate; this served as the catholyte. In the electrolysis of diethyl malonate, 65 ml of sodium carbonate or 32.5 ml of 2M sodium carbonate plus 32.5 ml of the organic solvent was placed in the cup. To this system was added the desired amount of the various salts and 5 ml of diethyl-malonate. The electrolyses were performed potentiostatically while passing approximately 31.5 milli-Faradays of electricity, which is the amount of electricity needed to oxidize the 5 ml of diethyl malonate completely. The diphenylacetonitrile experiments were the same except the electrolyte was 2M sodium hydroxide and the total volume was 50 ml. In all experiments the solutions were stirred vigorously enough to maintain the emulsion or suspension formed prior to the experiment.

Analytical Procedures

After electrolysis the anolyte mixture was placed in a separatory funnel. In the micelle experiments the anolyte was extracted with 50 ml of benzene. The organic and aqueous phases were then separated and the analysis was performed using the organic phase.

The concentration of the dehydrodimer of diphenylacetonitrile was determined by liquid chromatography using a Waters ALC 202 instrument.

The concentration of the dehydrodimer of diethyl malonate was determined by gas-liquid chromatography using a 6 ft Apiezon grease column in an Aerograph 200 Gas Chromatograph manufactured by Wilkens Instrument and Research, Inc.

The conversion percentages are calculated based on the amount of product per amount of added reactant. The conversions would be higher if they were calculated on the basis of amount of material consumed.

RESULTS AND DISCUSSION

Electrochemical Dehydrodimerization of Diphenylacetonitrile

In the study of the anodic oxidation of benzhydrol[30-31] it had been concluded that the free radical intermediate [(A) in Equation (1)] was unstable on the electrode surface so that it dissociated rapidly giving H atoms and benzophenone with very little dimerization to form the pinacol.

$$Ph_2COH \xrightarrow{-e} Ph_2\overset{.}{C}OH \longrightarrow Ph_2C=O + H.$$ (1)
$$H \quad OH-$$
$$(A)$$

For this reason the compound selected for dimerization studies was

$$Ph_2C \ CN$$
$$\overset{|}{H}$$

since the free radical formed in this oxidation ($Ph_2 \overset{.}{C} CN$) should be relatively stable and therefore should give the dimer,

$$\overset{CN}{\overset{|}{Ph_2 \ C-C \ Ph_2}}$$
$$\underset{NC}{}$$

as the primary product. Diphenylacetonitrile has been reported to be dehydrodimerized quantitatively in benzene by addition of a stoichiometric amount of nickel peroxide.[32]

Table I shows the results obtained from a series of experiments containing 13 millimoles of the nitrile in 50 mls total volume.

It can be seen in Exp. 1 that when the diphenylacetonitrile was dissolved in a benzene-aqueous 2 N sodium hydroxide emulsion, even though it was stirred vigorously, no measurable amount of product was obtained. However, when a cationic emulsifying agent (which also can act as a phase transfer catalyst) was added, the amount obtained in one hour rose to 19.7%. In the benzhydrol study [31] it had been concluded that at least part of the effect was due to the formation by the surfactant on the surface of the electrode of a conducting hydrophobic film which excluded water but did not exclude the organic compound. That this explanation is involved in this system also is indicated by the fact that an acetone-sodium hydroxide emulsion produces no product (Exp.3). In the presence of acetone this film is destroyed.

Table I. The Conditions and Results of Anodic Dimerization of Diphenyl Acetonitrile at
Constant Current (815 mA) in Organic-2 M Sodium Hydroxide Emulsions and in Micelles

Exp. No.	Concn. of (MDB)Me$_3$NCl[1] % by wt.	Organic Solvent	Time (hrs)	Conversion %
1	0.2	Benzene	1.0	0
2	0.2	Benzene	1.0	19.7
3	0.2	Acetone	1.0	0
4	0.2	Acetonitrile	0.5	34.8
5	6.0	None	0.5	13.9

[1](MDB)Me$_3$NCl--refers to the commercial surfactant Hyamine 2389 manufactured by Rohm and Haas.

The use of acetonitrile as the organic phase gave much higher conversions than when benzene was used. This is consistent with the results obtained in the oxidation of benzhydrol, and is probably caused by the fact that the acetonitrile is partially soluble in water and this probably increases the solubility of the organic compound in the water.[33-34]

Exp. 5 in the table also shows that appreciable amounts of product can be obtained by solubilizing the organic reactant as micelles using larger amounts of the cationic surfactant as a dispersing agent without using any organic solvent.

Anodic Dimerization of Diethyl Malonate

The next reaction that was studied was the anodic dimerization of diethyl malonate:

$$2 \ OH- + H_2C(COOEt)_2 \longrightarrow (EtOOC)_2 \overset{H}{C}-\overset{H}{C}(COOEt)_2 + 2 \ H_2O + 2 \ e-$$

Various quaternary ammonium salts were added to the aqueous sodium carbonate solution to solubilize the ester. The results are summarized in Table II. From the table it can be seen that to obtain any yield one must not only solubilize the ester but also have bromide ion present. The product was obtained only with tetrabutylammonium bromide (28%), tetraethylammonium bromide (1%), and a mixture of the commercial surfactant Hyamine 2389 ((MDB)Me$_3$NCl) and potassium bromide (13%).

It is well known that in nonaqueous solvents the addition of bromine to diethyl malonate produces the dimer. It was observed in these studies that the addition of bromine to the water–tetrabutylammonium bromide–diethyl malonate mixture also readily produced the dimer. Since a brown film, presumably bromine was observed around the anode, it was concluded that the dimerization in the presence of bromide probably proceeds by the same mechanism as the chemical reaction, with the bromine being produced electrolytically. The chemical reaction has been shown to proceed by a substitution mechanism.[35]

$$OH- + (EtOOC)_2CH_2 \longrightarrow (EtCOOC)_2CH- + H_2O$$

$$(EtOOC)_2CH- + Br-Br \longrightarrow (EtOOC)_2CHBr + Br-$$

$$(EtOOC)_2CHBr + -HC(COOEt)_2 \longrightarrow (EtOOC)_2\overset{H}{C}-\overset{H}{C}(COOEt)_2 + Br-$$

The tetrabutylammonium bromide–sodium carbonate–water system was somewhat unusual in that it separated into two liquid layers. The corresponding compound tetrabutylammonium iodide is insoluble

Table II. The Conditions and Results of Anodic Dimerization of Diethyl Malonate
in Aqueous 2 M Sodium Carbonate Solutions in the Presence of Various Electrolytes

Exp. No.	Electrolytes	Concentration of Electrolyte	Potential (vs. SCE)	Electrolysis Time (hr)	Current mA	Conversion %
1	$(n-Bu)_4NBr$	0.52 M	1.6	3.3	300	28
2	Et_4Nbr	0.52 M	1.6	3.3	350	1
3	KBr	0.52 M	1.5	2.0	500	0
4	$(n-Bu)_4NOH$	0.52 M	1.7	3.0	330	0
5	$(n-Bu)_4NC1$	0.52 M	1.7	3.0	300	0
6	$(MDB)Me_3NC1$	7 wt %	2.0	3.0	360	0
7	$(MDB)Me_3NC1$	2.5 wt %	2.1	3.0	250	0
8	$(MDB)Me_3NBr^1$	2.5 wt %	1.9	3.3	300	13
9	$(n-Bu)_4NBr^2$	0.52 M	1.6	3.3	320	0

$^1(MDB)Me_3NC1 + KBr$

$^2(n-Bu)_4NBr$ was added to water instead of aqueous 2 M Na_2CO_3 solution. This solution was homogenous.

Table III. The Conditions and Results of Macro Electrolysis of Diethyl

Malonate in Tetrabutyl Ammonium Bromide-2 M Sodium Carbonate Two Phase System

Exp. No.	Conc. (eq./1) of $(n-Bu)_4N^+Br^-$	Current	Potential volts (vs. SCE)	Milli-Faradays	Conversion %
1	0.13	300	1.7	36.9	52
2	0.52	310	1.6	38.1	28
3	0.52	460	1.7	30.4	42
4	0.52	640	1.8	26.2	64
5	0.52	835	2.0	26.1	69
6	0.52	1050	2.2	31.3	32

in aqueous potassium carbonate while tetrabutylammonium chloride at this concentration is completely soluble in the sodium carbonate solution. The ion paired compound, $n-Bu_4N^+Br^-$, apparently carries enough water to maintain it in the liquid state, and is apparently salted out by the addition of the sodium carbonate. This layer is nonpolar in its behavior and diethyl malonate and the dimer dissolve primarily in this layer. The tetrabutylammonium bromide layer behaves like "McKee's salts"[29-30] dissolving the diethyl malonate and its dimer, compounds that would normally be insoluble in water.

One can alter the yield obtained in these experiments by changing the experimental conditions (Table III). It can be seen (Exp. Nos. 1 and 4) that, although the quaternary ammonium salt is necessary to obtain a yield, one can increase the yield by decreasing the concentration of the salt. One can also see that increasing the potential causes an increase in yield, but this goes through a maximum at about 2.0 volts.

In summary it has been shown that it is possible to solubilize diphenylacetonitrile in emulsions and micelles, and that with the aid of a cationic surfactant, one can obtain appreciable yields of the dehydrodimer by anodic oxidation at platinum electrodes.

It has also been shown that the dehydrodimer of diethyl malonate can be obtained anodically by solubilizing the compound in micelles using a cationic surfactant if bromide ions are present in the solution. Best results were obtained in an emulsion formed by mixing concentrated solutions of tetrabutylammonium bromide and sodium carbonate.

ACKNOWLEDGEMENT

We thank the Robert A. Welch Foundation of Houston, Texas for their financial support of this work.

REFERENCES

1. R. N. Adams, "Electrochemistry at Solid Electrodes," Marcel Dekker Inc., New York, N.Y., 1969.
2. C. K. Mann and K. K. Barnes, "Electrochemical Reactions in Nonaqueous Systems," M. Dekker, Inc., New York, N.Y., 1970.
3. A. J. Fry, "Synthetic Organic Electrochemistry," Harper and Row, New York, N.Y., 1972.
4. M. M. Baizer, Editor, "Organic Electrochemistry, "Marcel Dekker, Inc., New York, N.Y., 1973.
5. M. Rifi and F. H. Covitz, "Introduction to Organic Electrochemistry," M. Dekker, Inc., New York, N.Y., 1974.

6.   N. L. Weinberg and A. Weissberger, Editors, "Techniques of
     Electroorganic Synthesis," Vol. 5, Pt. 1, Wiley – Interscience,
     New York, N.Y., 1974.
7.   N. L. Weinberg, Editor, "Techniques of Electroorganic Synthesis,"
     Vol. 5, Pt. 2, Wiley – Interscience, New York, N.Y., 1975.
8.   F. D. Popp and H. P. Schultz, Chem. Rev. 62, 19 (1962).
9.   N. L. Weinberg and H. R. Weinberg, Chem. Rev. 68, 449 (1968).
10. G. Popp, Org. Chem. Bull. 45 (3), (1973). From Eastman Kodak
     Co., Rochester, N.Y.
11. R. H. McKee and C. J. Brockmann, Trans. Electrochem. Soc. 62,
     203 (1932).
12. R. H. McKee and B. G. Gerapostolou, Trans. Electrochem. Soc.
     68, 329 (1935).
13. M. M. Baizer, J. Electrochem. Soc. 111, 215 (1964).
14. Fr. Fichter, Z. Electrochem. 19, 781 (1913).
15. Fr. Fichter and R. Stocker, Ber. 47, 2007 (1919).
16. P. N. Anantharam and H. V. K. Udupa, Extended Abstracts of the
     Meeting of the Electrochemical Soc. Washington, D.C., May 1976,
     Abs. No. 283.
17. J. P. Millington and J. Trotman, Extended Abstracts of the
     Meeting of the Electrochemical Soc. Washington, D. C., May
     1976, Abs. No. 279.
18. Fr. Fichter and O. Muller, Helvet, Chim. Acta 18, 831 (1935).
19. Fr. Fichter and G. Schetty, Helvet. Chim. Acta 20, 150 (1937).
20. C. W. Proudfit and W. G. France, J. Phys. Chem. 46, 42 (1942).
21. L. Eberson and B. Helgee, B. Chem. Scri. 5, 47 (1974).
22. L. Eberson and B. Helgee, Acta Chem. Scand. B29, 451 (1975).
23. C. M. Stark, J. Am. Chem. Soc. 93, 195 (1971).
24. E. V. Dehmlow, Chemtech 1975, 210.
25. S. Hayano and N. Shinozuka, Bull. Chem. Soc. Japan 42, 1469
     (1969).
26. S. Hayano and N. Shinozuka, Bull. Chem. Soc. Japan 43, 2083
     (1970).
27. S. Hayano and N. Shinozuka, Bull. Chem. Soc. Japan 44, 1503
     (1971).
28. P. G. Westmoreland, R. A. Day, Jr., and A. L. Underwood, Anal.
     Chem. 44, 737 (1972).
29. T. Erabi, H. Huira, and M. Tanaka, Bull. Chem. Soc. Japan 48,
     1354 (1975).
30. T. C. Franklin and L. Sidarous, Chem. Comm. 1975, 741.
31. T. C. Franklin and L. Sidarous, Extended Abstracts of the
     Meeting of the Electrochemical Society in Washington, D.C.,
     May 1976, Abstract No. 305.
32. S. Terabe and R. Konaka. J. Am. Chem. Soc. 91, 5655 (1969).
33. F. W. Steuber and K. Dimroth, Chem. Ber. 99, 258 (1966).
34. E. Juday, J. Org. Chem. 22, 532 (1957).
35. C. R. Noller "Chemistry of Organic Compounds" p. 879, W. B.
     Saunders Co., Philadelphia, PA., 1965.

THE CATALYTIC ROLE OF MICELLE-BISULFITE COMPLEXATION IN VINYL POLYMERIZATION

Oh-Kil Kim

Naval Research Laboratory

Washington, D. C. 20375

The present paper is concerned with the polymerization of vinyl monomers such as acrylamide and methyl acrylate with sodium bisulfite as catalyst in the presence of cetyltrimethylammonium bromide (CTAB). Comparative studies are made with systems containing non-micellar salts or polycationics.

Evidence for complexation was obtained from changes in pH, UV absorption spectra and the rates of autoxidation of bisulfite in the presence of CTAB. An enhanced rate of polymerization and a high molecular weight were obtained in acrylamide polymerization in the presence of CTAB and sodium bisulfite in the absence of air. Methyl acrylate was readily polymerized in micellar systems in air, whereas no polymerization took place in non-micellar systems.

Although non-micellar cationic polyelectrolytes showed a somewhat similar effectiveness to that of CTAB on the oxidation of bisulfite in the absence of monomer, their catalytic effect on the polymerization of methyl acrylate was negligible as compared to that of CTAB or Polysoap. This discrepancy is considered to be attributable to the difference in the configuration between micellar and non-micellar cations in solution, which suggests the importance of the site of monomer adsorption and of the complex stability.

INTRODUCTION

In recent years, a growing research effort has been directed to the area of micellar catalysis in organic reactions. Micellar particles with a very high density of surface charges are expected to exhibit some phenomena analogous to those observed in polyelectrolyte solutions[1,2]. Nevertheless, there are distinctive differences between the two systems[3], since micelles form only above a critical concentration of detergent ions, and the microenvironment of the surface charges is different from that of macromolecular ions. We are interested in exploring the catalytic role of cationic micelles and non-micellar polycationics in vinyl polymerization catalyzed by sodium bisulfite.

Emulsion polymerization is an example of a micellar system in which monomer is solubilized, and accelerated reaction rates result. However, polymerization in the present study is carried out in a homogeneous system to avoid the complexities of emulsion polymerization; no excess of the monomer is present as a separate phase. A feature of the present work is that both micellar and non-micellar cationics are used as a part of the catalyst system rather than as an emulsifying agent.

Sodium bisulfite is an effective reducing agent and it can function as a radical initiator for certain vinyl monomers[4]. For acrylamide polymerization[5] in the absence of oxygen or other oxidizing agents, the monomer is considered to act as the oxidizing component that forms a redox pair with bisulfite ion. Interest in this catalyst system has been extended to the polymerization of methyl acrylate and acrylonitrile, which have high electron affinities, to gain further insight into the monomer dependency of the micellar effect and redox mechanism.

In this report some unique features of micellar and non-micellar catalysis are discussed in terms of complexation of bisulfite, monomer selectivity and bisulfite redox mechanisms.

EXPERIMENTAL

Materials

Acrylamide was recrystallized twice from acetone and dried in vacuo at room temperature. Methyl acrylate and acrylonitrile were purified by distillation. Sodium bisulfite was reagent grade and used without further purification. The following additives were used in this work:

Cetyltrimethylammonium bromide (CTAB) (micelle-forming)
Tetramethylammonium bromide (TMAB)
Poly(p-vinyl-N-ethylpyridinium bromide) (NEPVP)
Poly(p-vinyl-N-ethylpyridinium-co-p-vinyl-N-dodecylpyridinium
   bromide) (68:32) (Polysoap) (micelle-forming)
Poly(N,N-diallyldimethylammonium chloride) (PDADM)
Poly [(p-vinylbenzyl)trimethylammonium chloride] (PVBTM)
Sodium lauryl sulfate (micelle-forming)
Poly(oxyethylene) (20) sorbitan monolaurate (micelle-
   forming)

Among these additives, CTAB was recrystallized from ethanol-
acetone, and NEPVP and Polysoap were prepared by alkylation of
the polymer bases[6]. Poly(p-vinylpyridine) was prepared by bulk
polymerization of the monomer with benzoyl peroxide; its
molecular weight was estimated to be 3.5 x 10^5 by viscometry[7].
Poly(N,N-diallyldimethylammonium chloride), PDADM, (Cat-Floc T, a
product of the Calgon Corporation) and PVBTM (Polyscience, Inc.)
were purified by reprecipitation from water solution with
acetone. Water used for polymerizations had a resistivity
1.5 x 10^6 ohm cm.

Polymerization Procedures

Polymerizations were carried out in Pyrex tubes containing
magnetic stirrers.

For acrylamide polymerizations, nitrogen gas was bubbled
vigorously for 12 minutes through 20 ml. of a 10% aqueous
monomer solution held at 30° or 40°C. The gas inlet tube was
then raised above the solution to maintain a nitrogen atmosphere.
With stirring, a weighed amount of detergent or other additive
was added, the pH adjusted to 4.5 - 6.5 with hydrochloric acid
or ammonium hydroxide, and, finally, sodium bisulfite solution
was added with a microsyringe. The final pH of the reaction
system was measured soon after the addition of the catalyst.
The polymerization was initiated almost instantaneously upon
adding the sodium bisulfite. After a specified reaction time
(10 or 20 minutes) a weighed amount of the reaction mixture
was treated with acetone to precipitate the polymer which was
then dried in vacuo.

Polymerization of methyl acrylate was carried out in air
with a solution of 0.8 g of monomer in 20 ml water or in
aqueous CTAB or polycation solutions with stirrring at 45°C; the
solutions were homogeneous at this temperature. Sodium
bisulfite solution was added with a micro-syringe and the
reaction tube was sealed. After the desired reaction time, the
reaction mixture was diluted with alcohol to clear any haze, and
a large amount of water was added to precipitate the polymer.

In some cases where an emulsion-like solution was produced, sodium sulfate was added to break the emulsion. The precipitated polymer was washed repeatedly with water.

Measurements of pH Shift

The shift in pH of bisulfite in the acrylamide solution and the acrylamide-blank solution upon the addition of the detergent was followed over a 5-minute period (in air or in a nitrogen atmosphere, 25°C) using a Corning Expanded Scale pH Meter (Digital 110). All pH measurements were made at least in triplicate and they agreed within ± 0.02.

Oxidation Rates

Oxidation of bisulfite in the presence of or absence of methyl acrylate was measured by means of an oxygen analyzer (YSI Model 54ARC). The rate of oxidation was determined from the decrease in the concentration of dissolved oxygen in the initially air-saturated solution (at 23°C) containing bisulfite and CTAB or polycationics.

Ultraviolet Spectra

A Cary 118 spectrophotometer was used. Water used in preparing the solutions was nitrogen-purged to reduce oxidation during spectral measurements. Spectra of solutions of sodium bisulfite with CTAB or other additives were determined at pH 4.3-4.4 in air at room temperature.

Molecular Weight Determination

Polyacrylamide molecular weights were determined by viscosity measurements and application of the known relationship[8] of molecular weight to intrinsic viscosity: $[\eta] = 6.31 \times 10^{-5} \overline{M}^{0.80}$.

RESULTS AND DISCUSSION

The discussion is focused on the polymerization behavior of acrylamide and methyl acrylate in relation to the interactions occurring between bisulfite and cationic micelles and between cationic micelles and monomer. Acrylamide is a neutral, extremely hydrophilic monomer, and as a consequence its interaction with micelles is minimal. With micelle-monomer interaction minimized, the interaction between bisulfite and micelles can be evaluated in this polymerization system. On the other hand, methyl acrylate is a hydrophobic monomer with greater electron affinity than that of acrylamide, so that it is expected that the monomer would have some degree of interaction with micellar hydrocarbons or with the backbone of cationic polymers. Also, since methyl acrylate is known to readily undergo

addition of bisulfite[9] the influence of such interactions on the
polymerization can be evaluated.

Micellar Complexation of Bisulfite Ions

The addition of cationic micelle-forming CTAB to an aqueous
sodium bisulfite solution in air induced a shift in pH to the
acidic side[5,10]. This shift in pH occurred with CTAB concentrations at or above the cmc (9.2 x 10^{-4} M), as shown in Figure 1.

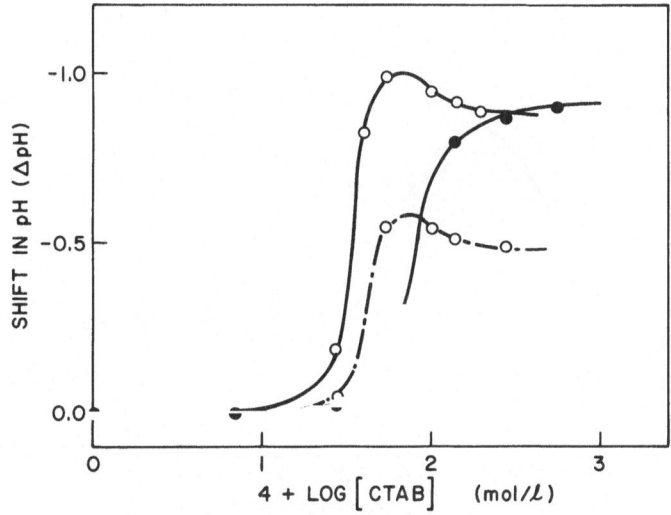

Figure 1. Shift in pH induced by CTAB in sodium bisulfite
(1.92 x 10^{-4} M) solutions at room temperature: —●— in
air, acrylamide (1.40 M), initial pH 6.28; —o— in air,
no monomer, initial pH 5.69; - —o— - in nitrogen, no monomer,
initial pH 6.90. (From Ref. 5)

In the presence of acrylamide, somewhat higher CTAB concentrations
were required. Smaller pH shifts took place under nitrogen,
suggesting the changes in equilibrium of sodium bisulfite by
CTAB, and these shifts were increased by the introduction of air.
No pH change took place in air when anionic (sodium lauryl
sulfate) or nonionic (polyoxyethylene (20) sorbitan monolaurate)
micelle-forming substances, TMAB, or NaCl were added to sodium
bisulfite solutions. Clearly, CTAB interacts in some way with

bisulfite ion to increase the rate of its reaction with oxygen.
It is proposed that a reactive complex is involved.

Evidence for complexation in bisulfite-containing systems
was obtained from oxidation rate and from ultraviolet spectral
studies. With the mole-ratio method, oxidation rate determinations in the bisulfite-CTAB system indicated the presence of
1:1 and 2:1 bisulfite:CTAB complexes[11]. In Figure 2 are shown
ultraviolet spectra of solutions obtained by adding CTAB to a
constant concentration of bisulfite; the limited solubility of
CTAB precluded the use of higher concentrations than those shown.

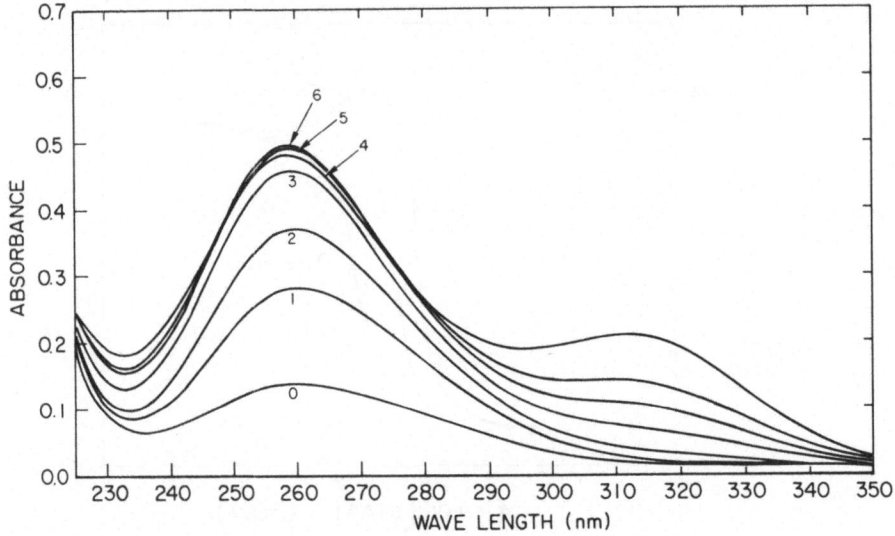

Figure 2. Effect of CTAB on the UV absorption spectra of 2.50 x
10^{-2} M sodium bisulfite solutions. CTAB :(0) 0; (1) 6.80 x 10^{-3}
M; (2) 1.35 x 10^{-2} M; (3) 2.70 x 10^{-2} M; (4) 4.05 x 10^{-2} M; (5)
5.40 x 10^{-2} M; and (6) 8.10 x 10^{-2} M; pH 4.3-4.4.

Similar spectra were obtained when bisulfite was added to CTAB
solutions. In both cases, new absorption bands having λ_{max} 259
and 314 nm were formed, and an isosbestic point was observed at
226 nm. It is evident that two or more species produce these
bands and that the species are related. This can be seen in
Figure 3, in which the changes in absorbance with the addition

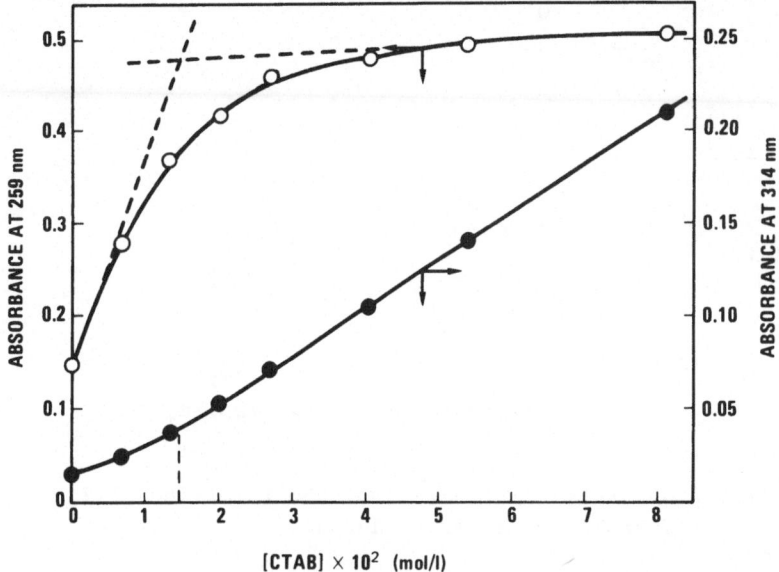

Figure 3.   Absorbance of 2.50×10^{-2} M sodium bisulfite
solutions at 259 nm and 314 nm as a function of CTAB concentration.

of CTAB are plotted. A break in the plot of absorbance at
259 nm determined from the tangents at the extremes of the
continuous curve suggests the presence of complex(es) having a
bisulfite to CTAB ratio in the region of 1.7.

    Preliminary experiments indicated that similar changes in
absorbance at 259 nm took place when PDADM was added to
bisulfite solution, but no band corresponding to the 314 nm
for CTAB was observed. The addition of TMAB, sodium lauryl
sulfate, or sodium chloride to bisulfite solutions produced
no new absorption bands in the 200–400 nm region. With NEPVP or
Polysoap, interfering absorption by the pyridine chromophore
precluded the observation of possible complex band formation
in this region.

    According to Golding[12], four species are in equilibrium in
bisulfite solutions; their absorption bands were assigned as
indicated:

$$
\underset{\substack{\text{(I)}\\\text{(205 nm)}}}{HO-S\!\!\!\diagup^{O^-}_{\diagdown O}} \;\rightleftharpoons\; \underset{\text{(II)}}{H-\underset{\underset{O}{\|}}{\overset{O^-}{S}}=O}
$$

$$
I + II \;\rightleftharpoons\; \underset{\substack{\text{(III)}\\\text{(215 nm)}}}{\overset{^-O}{\underset{O}{\diagdown}}\underset{\diagup}{\overset{\diagup}{S}}\cdots\overset{OH}{\underset{H-\underset{\underset{O}{\|}}{S}-O^-}{O}}} \;\rightleftharpoons\; \underset{\substack{\text{(IV)}\\\text{(255 nm)}}}{\overset{^-O}{\underset{O}{\diagdown}}\underset{\diagup}{\overset{\diagup}{S}}-\underset{\underset{O}{\|}}{\overset{O^-}{S}}=O} \;+\; H_2O
$$

As bisulfite concentration increases, the formation of III and IV are favored and I and II are reduced.

In bisulfite-CTAB solutions, the results presented here suggest the following equilibria, where $RR_3'N^+$ represents the CTAB cation:

$$
\underset{\text{(V)}}{RR_3'N^+ \;+\; HSO_3^- \;\rightleftharpoons\; [RR_3'N^+\cdots^-O_3SH]} \tag{1}
$$

$$
\underset{\text{(VI)}}{V \;+\; HSO_3^- \;\rightleftharpoons\; [RR_3'N^+\cdots^=O_5S_2]^-} \tag{2}
$$

$$
VI \;\rightleftharpoons\; RR_3'N^+ \;+\; S_2O_5^= \tag{3}
$$

$$
\underset{\text{(VII)}}{V \;+\; RR_3'N^+ \;\rightleftharpoons\; [(RR_3'N)_2^{++}\cdots^=O_3S] \;+\; H^+} \tag{4}
$$

$$
\underset{\text{(VIII)}}{VI \;+\; RR_3'N^+ \;\rightleftharpoons\; [(RR_3'N)_2^{++}\cdots^=O_5S_2]} \tag{5}
$$

As a consequence of the high local concentration of bound bisulfite ions that results from reaction (1), the formation of complex VI by reaction (2) is favored. The absorption band at

259 nm is therefore likely to be due to complex VI with a possible contribution from the dissociation (reaction 3) of VI to form $S_2O_5^=$. As CTAB is added, reaction (5) becomes important; complex VIII may be responsible for the 314 nm band, although other complexes may also be involved. Concomitant oxidation with a resulting decrease in pH will tend to suppress reaction (4). With PDADM, similar equilibria may be set up; the lack of a band corresponding to the 314 nm band with CTAB may reflect the magnitude of the equilibrium constants or a low extinction coefficient for the species analogous to VIII.

Acrylamide Polymerization

The kinetic behavior of acrylamide polymerization under nitrogen in micellar CTAB-containing systems and non-micellar systems is illustrated in Figure 4. The rate of polymerization, R_p, of acrylamide catalyzed by bisulfite alone or by TMAB (a non-micellar analog of CTAB) is strictly dependent upon the square root of the bisulfite concentration, whereas the R_p of the bisulfite/CTAB catalyst system showed an increasing deviation from linearity with increasing bisulfite concentration

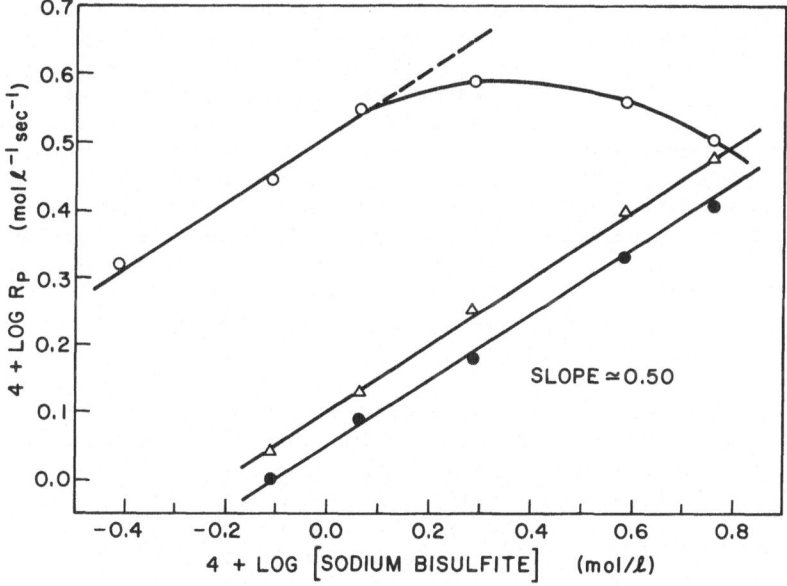

Figure 4. Double logarithmic plot of the rate, R_p, of polymerization, of acrylamide against sodium bisulfite concentration: [CTAB] = [TMAB] = 1.37×10^{-2} M, pH = 5.3, 30°C, nitrogen atmosphere; ——•—— (no additive), ——Δ—— (with TMAB), and ——o—— (with CTAB). (From Ref. 5)

$\geq 1.1 \times 10^{-4}$ M; below this bisulfite concentration, CTAB is an
activator for the bisulfite catalysis. The reason for the
decreased catalytic efficiency of bisulfite ion in the high
concentration region in the CTAB-containing system is unknown,
but it may be a result of an increase in ionic strength in the
system. At the lower bisulfite concentrations, the R_p of the
CTAB-containing system is nearly three times that of the
nonadditive system.

Although the addition of TMAB to the bisulfite/acrylamide
system increased the R_p about 15%, the rate enhancement was even
smaller than the 30% increase obtained by the addition of sodium
chloride at the same ionic strength (I = 0.05). Thus, there were
salt effects[10] in the polymerization process. The R_p increased
gradually with increasing concentration of salts, suggesting
that the oxidation of bisulfite is facilitated by an increase of
ionic strength. The addition of salts to the CTAB-containing
system, however, resulted in only a slight decrease in the R_p,
even though salts are known to retard or even eliminate ionic
micellar catalysis in most cases[14].

Dependences of R_p and $\overline{M}_v$ on CTAB concentration are
illustrated in Figure 5. At concentrations of CTAB below the
cmc (9.2×10^{-4} M), the R_p of the CTAB-containing system is nearly
the same as that of the nondetergent system. The R_p increase
was small in the cmc region, but there was a sharp increase in
R_p at CTAB $\sim 8 \times 10^{-3}$ M, after which R_p approached a limiting
value with a further increase in CTAB concentration. A similar
tendency was observed with the molecular weight of the polymer
formed. The $\overline{M}_v$ at CTAB $\sim 2 \times 10^{-2}$ M was approximately double
that obtained with the nonmicellar system. There is also a
good correspondence between the pH shift in bisulfite solutions
(Figure 1) and increases in R_p and $\overline{M}_v$ (Figure 5) as a function of
added CTAB. This suggests a common cause for the pH shift, the
high R_p and the high $\overline{M}_v$, namely, an interaction of bisulfite
ions with CTAB micelles.

As stated earlier, in the presence of CTAB, bisulfite
becomes highly susceptible to oxidation through the complexation
between the two reagents. In air, the reactive complex is
strongly subject to autoxidation (discussed in the following
section), and this is accompanied by a decrease in pH. In
nitrogen, the complex would not be oxidized unless another
oxidizing agent was present. However, a significant decrease
in the pH of bisulfite,did take place, and this may be ascribed
to the change in the dissociation constant by the micelles.
Micellar effects on the apparent dissociation constants of weak
acids have been observed in micellar reactions[15].

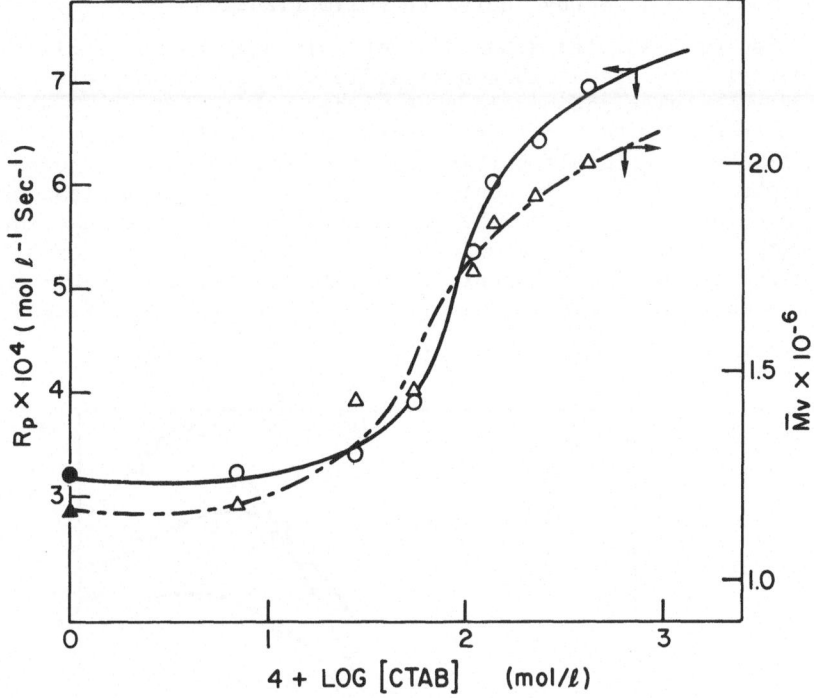

Figure 5. Plot of the initial rate of acrylamide polymerization, R_p, and molecular weight, $\bar{M}_v$, as a function of the logarithm of CTAB concentration: $[NaHSO_3] = 1.92 \times 10^{-4}$ M, pH = 5.4, 40°C nitrogen atmosphere; solid points represent the CTAB-free system. (From Ref. 5)

Questions arise concerning the species involved in the initiation process and how the initiating radicals are generated. There is a noticeable effect of pH[10] on the rate of polymerization, especially in a CTAB-containing system. The rate maximum was observed at pH $\sim$ 5.3 in the nonadditive, CTAB- and TMAB-containing systems. This finding strongly suggests that bisulfite ion is the most active species involved in the radical-generating process. Therefore, in the CTAB-containing system, a bisulfite complex ought to be the most active species. According to the mechanism proposed by Palit et al.[4], bisulfite radicals are generated by a redox reaction between bisulfite ion and monomer; this will be discussed in a later section.

Methyl Acrylate Polymerization

Sodium bisulfite alone did not initiate the polymerization of methyl acrylate at pH 4-6 regardless of the atmosphere. However, polymerization did take place by bisulfite catalysis in the presence of cationic activators such as CTAB or polycationics in air, but negligible polymerization occurred in a nitrogen atmosphere. This is in contrast to acrylamide polymerization in which oxygen inhibits the polymerization with the same catalyst system. Thus, it can be said that the oxygen molecule can act as an effective oxidizing agent for bisulfite in the presence but not in the absence of the activator. Studies of the bisulfite

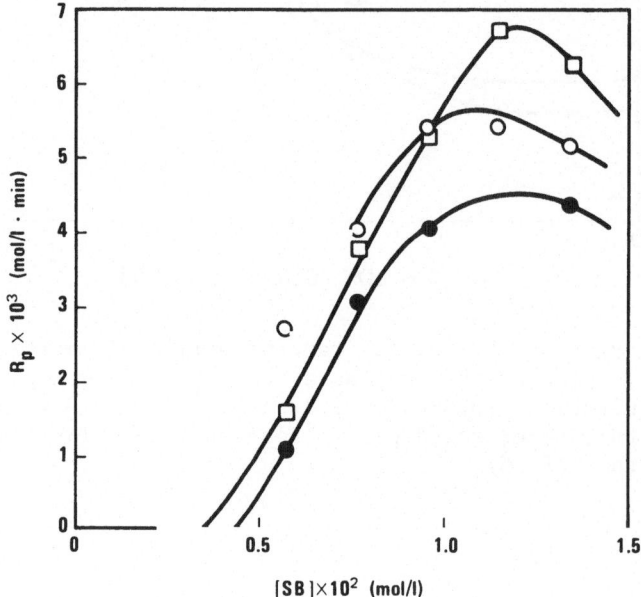

Figure 6. Dependence of the polymerization rate, R_p, of methyl acrylate on sodium bisulfite (SB) concentration in air at 45°C; [methyl acrylate] = 4.65 x 10^{-1} M; [CTAB]: —●— , 1.37 x 10^{-2}M; —□— , 2.74 x 10^{-2} M and —o—, 4.12 x 10^{-2} M. (From Ref. 13)

oxidation reaction in the absence of monomer were made to relate the oxidation rate to the bisulfite catalysis in the polymerization, and this will be discussed below.

As illustrated in Figure 6, the polymerization rate of methyl acrylate in the presence of CTAB is initially increased

with increasing bisulfite concentration but shows a downward
tendency after attaining a maximum. A similar feature was
observed in acrylamide polymerization (Figure 4) where an
initially linearly increasing rate of polymerization with
increasing bisulfite concentration showed a gradual downward
curvature. This is further evidence for micellar involvement in
the bisulfite catalysis. Figure 7 shows the distinct catalytic
effect of micelles on the bisulfite-catalyzed polymerization.
No polymerization occurred in the absence of a cationic activator.
A sharp rate acceleration was observed in the CTAB concentration
region (0.5 - 1.0 x 10^{-2} M) where a similar rate acceleration
was also attained in acrylamide polymerization using the same
catalyst system in nitrogen. The leveling-off or slowly
declining tendency of the rate with further increase in CTAB
concentration has been often observed in organic reactions
catalyzed by CTAB micelles[16].

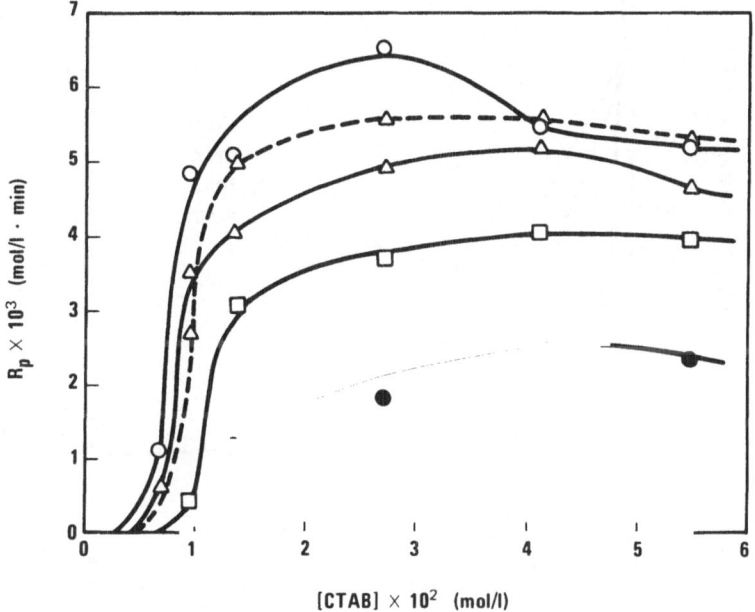

Figure 7.   Effect of micellar cationics on the polymerization
rate, R_p, of methyl acrylate; [methyl acrylate] = 4.65 x 10^{-1} M;
[NaHSO$_3$] : -●-, 5.77 x 10^{-3} M; -□-, 7.68 x 10^{-3} M; -Δ-, 9.61 x
10^{-3} M (solid line and dotted line refer to CTAB and Polysoap,
respectively); and -o-, 1.15 x 10^{-2} M.   (From Ref. 13)

Micellar Cations and Non-micellar Polycations

It was found that monomers, especially hydrophobic ones like methyl acrylate, interacted strongly with cationic activators, and as a consequence, the catalytic efficiency of the activator showed a marked decrease; such interactions are operative in the polymerization systems under study. A comparison in the catalytic effect on bisulfite oxidation was made between micellar cations and non-micellar polycations in order to evaluate the structural

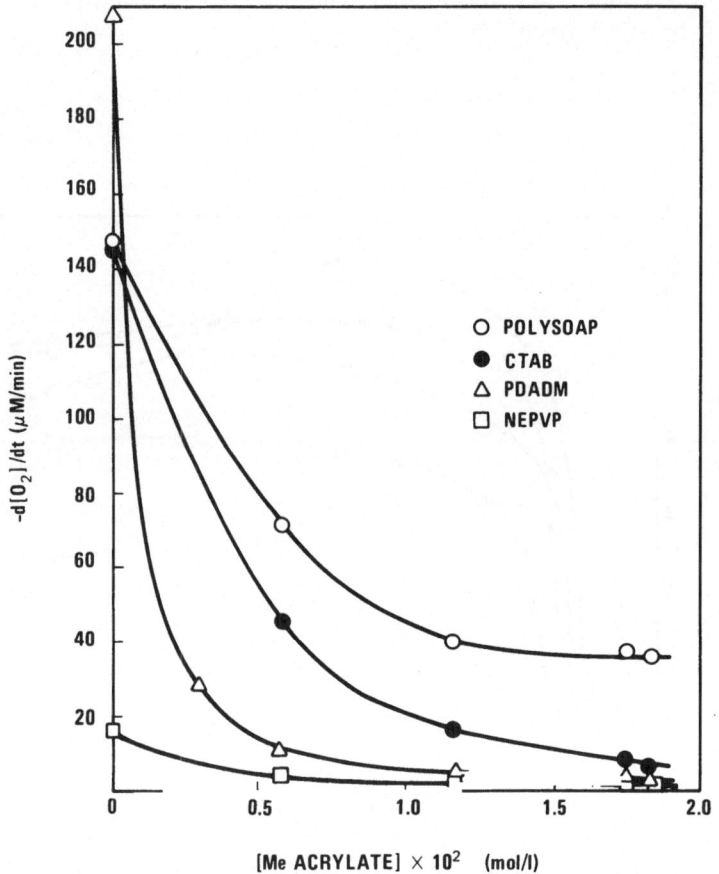

Figure 8. Oxidation of bisulfite catalyzed by micellar or non-micellar cationics as a function of methyl acrylate concentrations; $[NaHSO_3]$ = [cationics] = 6.0 x 10^{-3} M, initial $[O_2]$ = 2.53 x 10^{-4} M at 23.0°C.   (From Ref. 13)

influence of the cationic substance on the rate. As can be seen
in Figure 8, in the absence of methyl acrylate the oxidation rate
of bisulfite is strongly affected by CTAB or polycationics,
especially PDADM. Without such activators, however, the oxidation
rate was only 0.078 µM/min. The addition of the CTAB, Polysoap,
or PDADM activators increased the rate of bisulfite oxidation
about 2000 times[11]. The reason for the effect of PDADM is unclear,
but is likely that the rather rigid chain of closely inter-
connected rings directs the charged groups favorably toward
interaction with bisulfite ion. Polysoap adopts an organized
configuration by intramolecular micellization[18], while NEPVP is in
a random coil structure that does not form micelles. The high
oxidation rates of bisulfite are reduced by the addition of methyl
acrylate to the system; the reduction is greatest with PDADM.
These results may be explained in terms of the monomer adsorption
on the hydrophobic backbones of non-micellar PDADM (or NEPVP) and
subsequently ionic addition reaction of bisulfite to the monomer
is highly facilitated, owing to the locally high monomer concentration in the neighborhood of polycations. This interference
in the oxidation is less in the case of the micelle-forming
CTAB and Polysoap because the monomer is more likely to be
solubilized in the core of the micelle.

   In the presence of methyl acrylate, the rate of bisulfite
oxidation controls the polymerization rate of the monomer. The

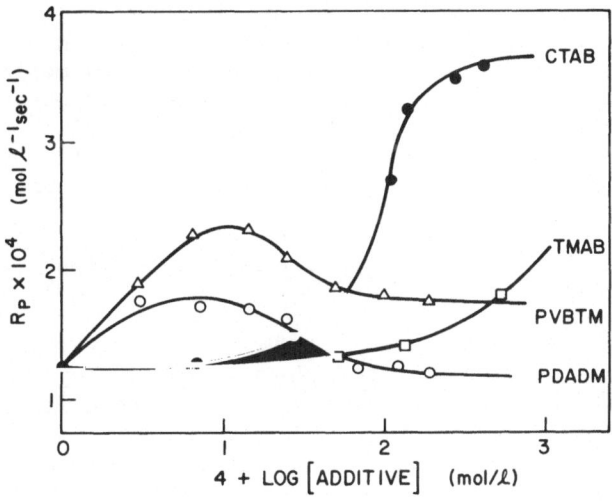

Figure 9. The rate of polymerization, R_p, of acrylamide (1.40 M)
as a function of cationic additive concentrations; $[NaHSO_3]$ =
1.15×10^{-4} M, pH = 5.3, 30°C.

non-micellar polycationics, PDADM and NEPVP, showed a negligible
effect on the polymerization as compared with CTAB and Polysoap.
In the polymerization of acrylamide (Figure 9), however, the
polymerization was significantly affected by non-micelle-forming
polycations (PVBTM and PDADM) at a very low concentration
($\lesssim 1 \times 10^{-3}$ mol/l). Such a catalytic behavior of polyions at a
concentration below a detergent cmc has been previously observed
in organic reactions[3]. This catalytic activity of non-micelle-
forming polycations in acrylamide but not in methyl acrylate
polymerization is thought to be due to acrylamide/bisulfite redox
reaction as shown in reaction (6). In agreement with the
assumption, oxidation of bisulfite in the present polycationic
systems was enhanced in the presence of acrylamide as well as
methyl methacrylate[11].

Redox Polymerization Processes in Bisulfite Systems

     In the preceding Sections, micellar participation in
bisulfite-initiated polymerizations and micellar complexation of
bisulfite ions have been demonstrated. With acrylamide under
nitrogen, micellar participation strongly increased the rate of
polymerization. With methyl acrylate in air, micellar
participation allowed polymerization that in the absence of
micelles would not have occurred. Acrylonitrile behaved in a
manner similar to methyl acrylate. It is evident that a high
degree of monomer selectivity is involved.

     A degree of selectivity can also be seen where micellar
participation is not involved. The polymerization of acrylamide
was initiated by bisulfite under nitrogen but not in air.
Neither methyl acrylate nor acrylonitrile were polymerized by
bisulfite under nitrogen or in air. Palit et al.[4] observed that
the polymerization of methyl methacrylate by bisulfite was
retarded by oxygen. They proposed that the initiating radical
species were generated in a redox reaction between bisulfite ion
and monomer:

$$2\ HSO_3^- \ + \quad \underset{/}{\overset{\backslash}{C}}{=}\underset{\backslash}{\overset{/}{C}} \quad + \quad 2H_2O \longrightarrow 2HSO_3 \ + \quad \overset{\bullet}{\underset{/}{\overset{\backslash}{C}}}HCH\underset{\backslash}{\overset{/}{{}}} \ + \ 2OH^- \quad (6)$$

The bisulfite radical is known[17] to be a chain-carrying species
in the addition of bisulfite to α,β-unsaturated compounds.
Autoxidation of bisulfite by oxygen is a slow reaction under the
conditions of polymerization. The major process competing with
reaction (6) is ionic addition of bisulfite to monomer:

$$HSO_3^- \ + \quad \underset{/}{\overset{\backslash}{C}}{=}\underset{\backslash}{\overset{/}{C}} \quad \longrightarrow \quad \underset{\underset{SO_3^-}{|}}{\overset{\backslash}{C}} {-}\!\!-\!\! CH\underset{\backslash}{\overset{/}{{}}} \qquad (7)$$

It would be expected that the electron affinity of the monomer would more strongly influence the rate of reaction (7) than (6). The electron affinity of methyl acrylate is higher than that of acrylamide. A consequence is that addition of bisulfite to methyl acrylate occurs to the exclusion of reaction (6), and no polymerization takes place; it has been reported that the rate of bisulfite addition to methyl acrylate is some 200 times that of addition to methyl methacrylate[9]. In the case of acrylamide, reaction (7) is less important and polymerization can occur in the absence of oxygen.

Bisulfite reactivity is strongly altered by complexation with micelle-forming substances such as CTAB. The complexes (V, VI, and VIII) themselves appear to be stronger reducing agents than the corresponding sulfur-containing anions, perhaps as a result of high local concentrations of anions bound to the micelle. They can thus participate in reactions analogous to (6) and (7). One result is that the rate of polymerization of acrylamide in nitrogen is accelerated, at least at low concentrations of bisulfite. Methyl acrylate is not polymerized in this system under nitrogen, probably because of competition from the ionic addition reaction.

In the presence of oxygen, interaction of the complexes with either oxygen or monomer will tend to yield polymerization-initiating species. Methyl acrylate polymerization does indeed occur under these conditions, in part as a result of monomer solubilization within the micelle. Acrylamide polymerization does not take place, however. The inhibitory action of oxygen in this system is probably a result of stable peroxide formation resulting from interaction between the propagating acrylamide species and oxygen[19].

ACKNOWLEDGEMENT

The author is indebted to Dr. Robert B. Fox for assistance and suggestions on preparation of the manuscript.

REFERENCES

1. C. G. Overberger and J. C. Salamone, Acct. Chem. Res., 2, 217 (1969).
2. H. Morawetz, Acct. Chem. Res., 3, 354 (1970).
3. T. Okubo and N. Ise, J. Amer. Chem. Soc., 95, 2293 (1973).
4. A. R. Mukherjee, P. Ghosh, S. C. Chadha and S. R. Palit, Makromol. Chem., 80, 208 (1964).
5. O.-K. Kim and J. R. Griffith, J. Polym. Sci. (Polym. Letters Ed.), 13, 525 (1975).
6. U. P. Strauss and N. L. Gershfeld, J. Phys. Chem., 58, 747 (1954).

7. J. B. Berkowitz, M. Yamin and R. M. Fuoss, J. Polym. Sci., $\underline{28}$, 69 (1958).
8. W. Sholtan, Makromol. Chem., $\underline{14}$, 169 (1954).
9. M. Morton and H. Landfield, J. Amer. Chem. Soc., $\underline{74}$, 3523 (1952).
10. O.-K. Kim and J. R. Griffith, J. Colloid Interface Sci., $\underline{55}$, 191 (1976).
11. O.-K. Kim (1976), to be published.
12. R. M. Golding, J. Chem. Soc., 3711 (1960).
13. O.-K. Kim (1976), J. Polym. Sci. (Polym. Letters Ed.) (submitted for publication).
14. C. A. Bunton, in "Reaction Kinetics in Micelles", E. H. Cordes, Editor, pp. 73-95, Plenum Press, New York, 1973.
15. C. A. Bunton and M. J. Minch, J. Phys. Chem., $\underline{78}$, 1490 (1974).
16. C. A. Bunton and L. Robinson, J. Org. Chem., $\underline{34}$, 773 (1969).
17. C. Walling, "Free Radicals in Solution", pp. 326-328, John Wiley & Sons, Inc., New York, 1957.
18. U. P. Strauss, N. L. Gershfeld and E. H. Crook, J. Phys. Chem., $\underline{60}$, 577 (1956).
19. R. Schulz, G. Renner, A. Henglein, and W. Kern, Makromol. Chem., $\underline{12}$, 20 (1954).

DISCUSSION

On the paper by L. S. Romsted

S. Goldwasser, *Consultant*: What is your picture of the role of the reaction products: (a) if they are compatible with the micelle? (b) if they are not compatible and rapidly "desorb"?

L. S. Romsted: No systematic effort has yet been made to study the effect of reaction products on the rates of micelle catalyzed reactions. In the systems studied to date, the catalyst concentration is usually in large excess relative to substrate concentrations. In fact, the interpretation of the effect of micelles on reaction rates are considered most reliable only when a large micelle/substrate ratio is obtained, thus minimizing substrate perturbation of micelle structure. Consequently, unlike enzyme catalyzed reactions the buildup of product would not have a significant effect on the observed rate of reaction.

Under experimental conditions which would make product accumulation important, hydrophobic products, which are the most compatible with the micelle, would probably have the most pronounced effect on micelle size and shape, the c.m.c., and substrate binding; and thus the greatest effect on reaction rates. Hydrophilic products would probably desorb rapidly, and therefore have only minimal effects on micelle stability and reaction rates.

On the paper by A. J. Frank

J. H. Fendler, *Texas A & M University*: I hope that you did not really mean to imply that pyrene is solubilized in the CTAB micellar interior. First of all, it is known that nothing is solubilized in the aqueous micellar core. Secondly, aromatic molecules (benzene and naphthalene) are solubilized at the CTAB-water interface (abstract presented by Dr. Mukerjee at this conference and J.A.C.S., 97, 89-95, 1975).

A. J. Frank: There is evidence that indicates that pyrene is solubilized in the CTAB micellar interior. Drs. Grätzel and Thomas have shown by H^1-NMR and C^{13}-NMR spectroscopy that pyrene is dissolved

mainly near the C^{16} to C^{10} carbon atoms in a CTAB micelle. (M.
Grätzel and J. K. Thomas, in "Modern Fluorescence Spectroscopy",
E. L. Wehry, Editor, Plenum Press, New York, to appear). Even the
paper in J.A.C.S., which you cited, suggests that acetophenone and
benzophenone have their C=O group pointing towards the surface –
but below it – and the rest of the molecule oriented towards the
center of the micelle.

On the paper by R. A. Moss, R. C. Nahas,
and S. Ramaswami

J. H. Fendler, *Texas A & M University*: Effects of imidazole bound
to macromolecular backbones on reaction rates are for smaller than
that on your micellar systems (Fendler and Fendler, 1975).

R. Moss: The reactions referred to by Dr. Fendler involve hydrolyses
of PNPA and PNPHeptanoate catalyzed by poly-4(5)-vinylimidazole
(PVIm), cf. C. G. Overberger and M. Morimoto, J. Amer. Chem. Soc.
93, 3222 (1971); C. G. Overberger and J. C. Salamone, Acc. Chem.
Res., 2, 217 (1969). With PVIm, k_{cat}. (1/mol-sec) for hydrolysis
of PNPA = 1.01 (10% ethanol in water, pH 7.71, μ = 0.02, 26°), and
for hydrolysis of PNPHeptanoate, k_{cat} = 0.575 (20% ethanol in water,
pH 8.03, μ = 0.02, 26°). Comparative k_{cat} values for micellar 16-Im
(pH 8.0, 0.01M phosphate buffer, 25°) appear in Table V, and are 5.0
and 172, respectively, for hydrolyses of PNPA and PNPHexanoate.
Taken as ratios of the k_{cat}'s, the kinetic advantages of the micellar
over the polymeric imidazole catalyst here range from 5-300, under
roughly comparable conditions.
      The main cause of these advantages is the greater contribution
made by anionic imidazole residues to the 16-Im-mediated hydrolyses.
It has been estimated that pK_{a2} for the imidazole of PVIm is >
14.5 (cf. Overberger and Salamone, p. 218, note 24); whereas the
corresponding value for micellar 16-Im must be much lower due to
the electrostatic stabilization of anionic imidazole groups by both
indwelling ammonium ion centers and by the aggregate positive charge
on the micelle. In this regard, pK_{a2} for $(C_2H_5)_3N^+CH_2Im$, which is
not micellized, is 11.2 (reference 33), so that a pK_{a2} value of
~10 seems reasonable for micellar 16-Im.
      Therefore, under comparable conditions at pH 8, micellar 16-Im
will be more extensively converted to its anionic imidazole form
than PVIm. Because an anionic imidazole residue is more nucleophilic than the neutral residue, micellar 16-Im will be the more
effective catalyst.
      We note that a related discussion, with important emphasis on
polarity effects, has been given by Martinek et al. (reference 28).

Part V

Reactions in Micelles and Micellar Catalysis
in Nonaqueous Media

SOME KINETIC STUDIES IN THE REVERSED MICELLAR SYSTEM-AEROSOL OT

(DIISOOCTYL SULFOSUCCINATE)/H_2O/HEPTANE SOLUTION

M. Wong and J.K. Thomas

Radiation Laboratory and Department of Chemistry

University of Notre Dame, Notre Dame, Indiana 46556

From ^{1}H and ^{23}Na NMR studies, the water molecules partitioning into the hydration shell of the counter ions are found to be highly immobilized in the reversed micelles Aerosol OT/H_2O/Heptane. Upon the completion of hydration shell, the mobility of water is enhanced and approaches that of the ordinary water. Fluorescent probes Pyrene (Py) and Pyrene Sulfonic Acid (PSA) are incubated into the reversed micelles and excited by the pulsed ruby Laser 347.1 nm light. The subsequent decay of the excited singlet states of the probes were followed to study the dynamic motions of the probe and the added quenchers. For the ionic quenchers, the motion is very restricted at low water content. On the other hand, O_2 and CH_2I_2 can undergo unrestricted diffusion. The quenching rate constants are lower for the hydrophobic probe Py than the amphiphillic probe PSA in the case of ionic quenchers which is attributed to the larger encountered probability between the quencher and PSA. Nanosecond pulsed radiolysis of biphenyl in the reversed micelles leads to the formation of biphenyl anion and triplet. The subsequent electron and energy transfer from these two transients to acceptors located at different sites in the micelles is investigated. The charge on the donor, the availability of the acceptor as well as the microenvironment around the acceptor are found to affect the efficiency of these transfer processes significantly.

INTRODUCTION

Water occupies an unique role amongst solvents due to its
abundance and status in organic life systems. One feature of
water to which attention is often drawn is the so called "bulk"
water phase as distinct from monomeric water.[2] It is the properties of the "bulk" phase or polymerized hydrogen bonded water
with which we are so familiar. The opportunity rarely arises to
observe the chemical reactions in some other form of condensed
water. Certain reversed micelles provide experimental conditions
for initiating reactions in water clusters or bubbles of varying
size and of different aggregation states of water. In this study
the surfactant Di-isooctyl sulfosuccinate, Aerosol OT, is used to
form reversed micelles in heptane. Varying amounts of water are
then accomodated in the micellar interior. Some of the properties
of these systems have already been characterized.[3] The micelles
may be briefly described as roughly spherical structures with the
polar sulfosuccinate group and sodium ions in the micellar interior,
and the iso-octyl chains on the exterior of the micelle in contact
with the solvent heptane. The radii of the micelles and degree of
aggregation under the several experimental conditions used in
this work are known.

In earlier work[4] several fluorescence aromatic molecules
were used to gain information on the nature of water in the
micellar interior. It was noted that an abrupt change in the
physical properties of the water bubble occurred at 1% water in
heptane (AOT 3%), a condition where the number of water molecules
per sodium ion exceeded 6. Recent laser Raman and N.M.R. data
have confirmed this observation.[5] In the present work we have
attempted to initiate reactions which require reactant penetration
of the micelle, and study the effect of micellar size, water content
and water phase, on these reactions.

EXPERIMENTAL

Laser Photolysis

Aromatic probe molecules such as pyrene sulfonic acid or
pyrene which were located in the micelle or the heptane respectively, were excited by 10^{-8} sec pulses of 3471Å light from a Q
switch frequency doubled ruby laser. Excitation produces reactive
species of these probes. The reactions of the short lived
entities thus produced were observed by fast kinetic emission or
absorption spectro-photometry.[6]

Pulse Radiolysis

Micellar systems containing biphenyl in the heptane phase were irradiated with 10^{-8} sec pulses of 10 Mev electrons Anions and excited triplet states of biphenyl were produced via electron capture and energy transfer from heptane.[7] The kinetics of these species as they entered the micelle to react with micellar components, were observed by fast spectro-photometry.

Fluorescence spectra and fluorescence polarization were measured on an Aminco-Bowman spectrofluorometer.[4]

Di-(2 ethyl hexyl) sodium sulfosuccinate (Aerosol OT) was obtained from American Cyanamid Corp. The soap was purified further by solubilizing in methanol and filtering. The solvent was evaporated under vacuum to yield a white precipitate, which in turn was dried under vacuum at 40°C. Pyrene sulfonic acid, PSA, pyrene tetra solfonic acid, PTSA, (Pfaltz and Bauer) and heptane (Phillips 66 research grade) were used as supplied. Other chemicals were analytical grade.

Micelles were prepared by adding AOT to heptane and stirring the solution with added water for ca. 20 minutes at room temperature. Oxygen was removed from the solutions by flushing with nitrogen.

RESULTS

Reactions of Biphenyl Anion

Pulse radiolysis of heptane solutions of biphenyl gives rise to the anion, excited singlet and triplet state of biphenyl. The energy is initially absorbed in the heptane leading to excited states $C_7H_{16}^{*}$[7] and ionization.

$$C_7H_{16} \rightarrow C_7H_{16}^{+} + e^{-}$$
$$\rightarrow C_7H_{16}^{*}$$

Electron capture by a solute such as biphenyl, ϕ_2 leads to solute anions:

$$e^{-} + \phi_2 \rightarrow \phi_2^{-}.$$

Energy transfer from excited heptane leads to solute excited states

$$C_7H_{16}^{*} + \phi_2 \rightarrow \phi_2^{*} + C_7H_{16}.$$

The singlet excited state of ϕ_2 is short lived ($\sim$11 nsec)[8] and converts to the longer lived triplet excited state well before the present observations are made. The anion of biphenyl shows unusual decay kinetics which have been described as geminate ion neutralization of ϕ_2^- with the solvent cation.[6] The inserts in Fig. 1 show ϕ_2^- decay in heptane and in heptane/6% H_2O/3% AOT solutions. This also leads to a small yield of excited states. Fig. 1. shows the spectrum at the end of, and 100 nsec following

the irradiation of deaerated 3 x 10^{-2} $\underline{M}$ ϕ_2 in heptane with a 10 nsec pulse of 10 Mev electrons. The spectrum at the end of the 10 nsec pulse is typical of the U.V. band of ϕ_2^- at 408 nm. The spectrum rapidly decays and leaves a smaller residual absorption of the triplet spectrum of ϕ_2 which has a maximum absorption

towards the U.V. In the presence of 6 x 10^{-6} M micellar, i.e. a solution of 6% H_2O/3% AOT, the ϕ_2^- decays more rapidly. The initial spectrum at 10 nsec is blue shifted with respect to that in the absence of micelles. The spectrum at 100 nsec is quite different to that observed in heptane alone. A shoulder is noted at 395-400 nm which is absent in the pure heptane solution. This spectrum decays with $\tau_{1/2}$ = 350 nsec. The enhanced decay of this latter spectrum with O_2, and SF_6, two solutes which rapidly react with ϕ_2^- indicate that the spectrum is due to ϕ_2^-.

In aqueous solution ϕ_2^- has an absorption maximum at 398 nm.[9] Hence, it is suggested that ϕ_2^- migrates to the micellar water bubble where it is stabilized. This is illustrated by the rapid decay of the 408 nm peak in the hydrocarbon environment, to that in the aqueous environment at λ_{max} = 395 nm. This effect becomes less prominent in micelles of smaller water content. The inserts in Fig. 1 show the enhanced rate of decay of ϕ_2^- in heptane in the presence of reversed micelles. The enhancement is not large compared to the natural rate of decay of ϕ_2^- in heptane alone. Hence, only a fraction of ϕ_2^- becomes associated with micelles.

Reactions of Biphenyl Anion ϕ_2^- with Species in Micelles

Various species such as (H_3O^+), Cu^{++}, and pyrene sulphonic acid PSA, which are solubilized in reversed micelles, react efficiently with ϕ_2^-. The reaction of ϕ_2 and PSA leads to the anion of PSA which is observed at 493 nm. It is suggested that the reaction of ϕ_2^- with PSA and probably with Cu^{2+} are via electron transfer from ϕ_2^- to the acceptor. A typical oscilloscope trace of the growth of the PSA anion is shown in Fig. 2. The body of the figure shows a first order plot of the growth from which a rate constant is calculated for electron transfer from ϕ_2^- to PSA. Table I shows rate constants for reactions of ϕ_2^- with PSA, H_3O^{++}, and Cu^{++} in water bubbles of various size.

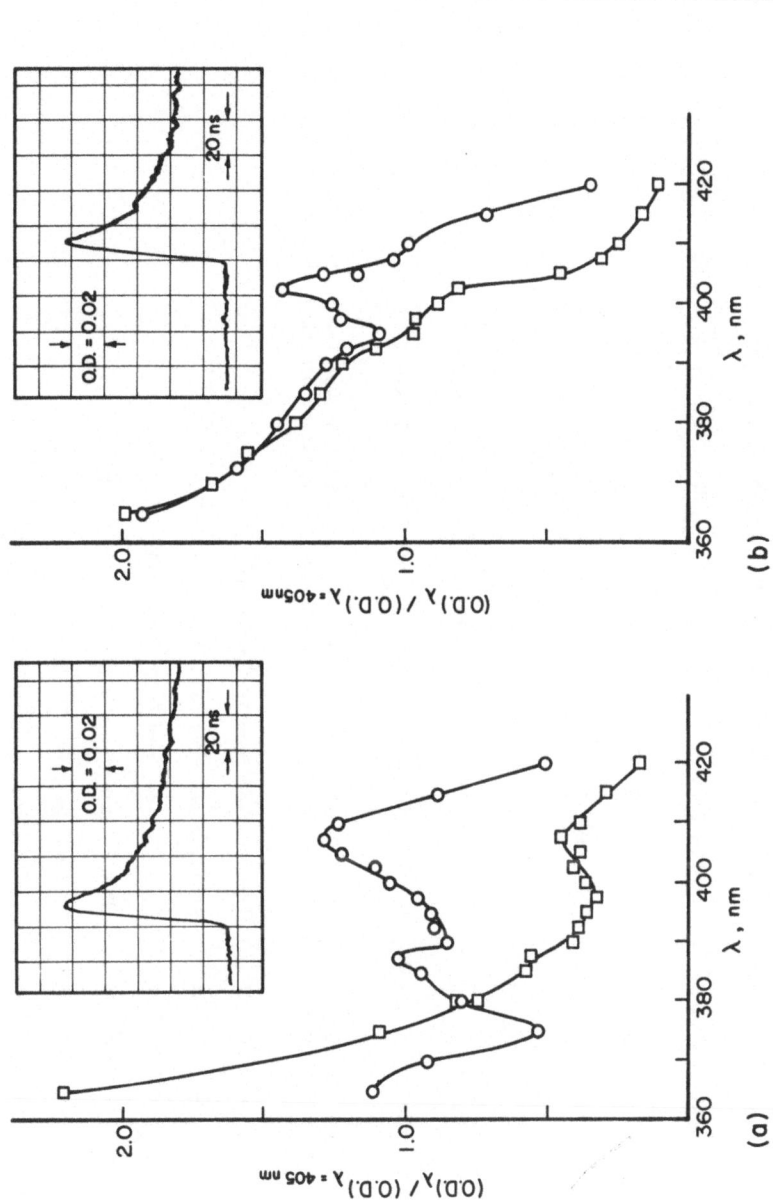

Figure 1. Transient spectra of ϕ_2^- observed in the irradiation of 2×10^{-2} M ϕ_2 in heptane (a), and 3% AOT/6% H_2O/heptane micelles (b); o taken at the end of the pulse; □ taken 100 ns after the pulse. The inserted pictures are the oscilloscope traces of the decay of ϕ_2^- (λ = 410 nm) in the corresponding systems.

Table I. Bimolecular Rate Constants[a] of Some Reactions Taking Place in the Reversed Micellar System, 3% AOT/H_2O/Heptane.

Reactions H_2O % (v/v)	\multicolumn{4}{c}{K ($M^{-1}sec^{-1}$)}			
	6	3	2	1
[b] $\phi_2^- + PSA \to \phi_2 + PSA^-$	$1.05 \times 10^{10}(6.3 \times 10^8)$	$1.84 \times 10^{10}(5.5 \times 10^8)$	$3.7 \times 10^{10}(7.4 \times 10^8)$	
[b] $\phi_2^T + PSA \to \phi_2 + PSA^T$	$6 \times 10^9(3.6 \times 10^8)$	$7.5 \times 10^9(2.25 \times 10^8)$		$6.5 \times 10^9(6.5 \times 10^7)$
[c] $\phi_2^T + PBA \to \phi_2 + PBA^T$	$6.4 \times 10^9(3.8 \times 10^8)$			
[d] $\phi_2^T + Py \to \phi_2 + Py^T$	2.7×10^{10}			
[e] $\phi_2^T + PTSA \to \phi_2 + PTSA^T < 2 \times 10^9$				
[f] $H_3O^+ + \phi_2^- \to \phi_2 H + H_2O$	$5 \times 10^9(3 \times 10^8)$	$5.8 \times 10^9(1.78 \times 10^8)$	$4.6 \times 10^9(9.2 \times 10^7)$	$4.7 \times 10^9(4.7 \times 10^7)$
[g] $Cu^{++} + \phi_2^- \to Cu^+ + \phi_2$	$6.3 \times 10^9(3.78 \times 10^8)$	$3.0 \times 10^9(1.0 \times 10^8)$	$3.4 \times 10^9(6.8 \times 10^7)$	$1.7 \times 10^9(1.7 \times 10^7)$

a: rate constants are calculated from the bulk solute concentration over the whole solution. The rate constants in the parenthesis are calculated from the local solute concentration in the micelle.

b: $[\phi_2] = 2 \times 10^{-2}$ M   [PSA] varies from 10^{-4} M to 10^{-3} M.

c: $[\phi_2] = 2 \times 10^{-2}$ M   [PSA] varies from 5×10^{-5} M to 5×10^{-4}M.

d: $[\phi_2] = 2 \times 10^{-2}$ M   [Py] varies from 5×10^{-5} M to 10^{-4} M.

e: $[\phi_2] = 2 \times 10^{-2}$ M   [PTSA] varies from 5×10^{-5} M to 2×10^{-3} M.

f: $[\phi_2] = 2 \times 10^{-2}$ M   [H$^+$] varies from 10^{-3} M to 10^{-2} M.

g: $[\phi_2] = 2 \times 10^{-2}$ M   [Cu^{++}] varies from 10^{-4} M to 2×10^{-3} M.

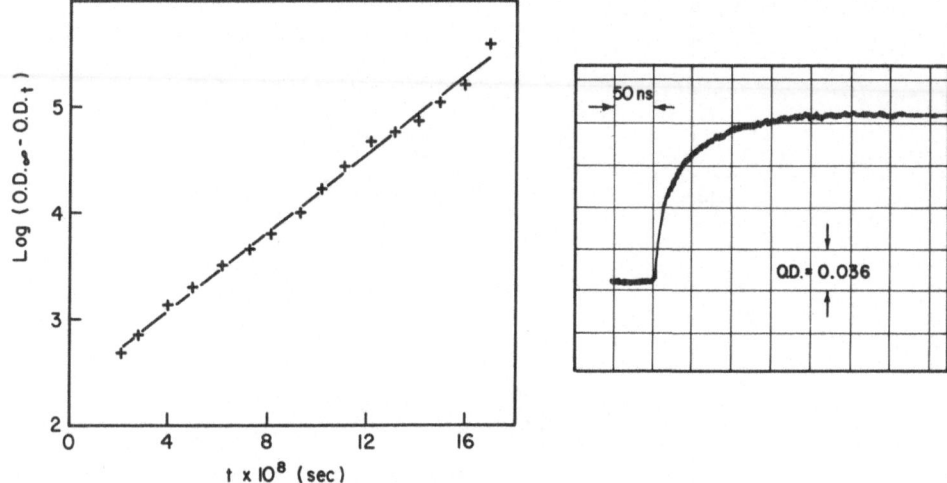

Figure 2.   Oscilloscope trace and computer plotted growth rate of
PSA⁻ (λ = 493 nm) in 2 x 10^{-2} M ϕ_2/3% AOT/6% H_2O/heptane micelles.

The rate constants are calculated from the quotient of the rate
of reaction and the concentration of solute.   Two effective
solute concentrations are used, the bulk solute concentration
over the whole solution, and the local or micellar concentration
of the solution.   In most cases both rate constants are indicated.
The rate constant for electron transfer from ϕ_2^- to PSA was
independent of PSA concentration over the range 2 x 10^{-4} to 10^{-3}
M.   (Figure 3).   Similar data are also shown in Figure 3 for the
reaction of ϕ_2^- with Cu^{++}.   The reaction rate versus Cu^{++} concentration is only linear at lower Cu^{++} concentrations, a plateau in
reaction rate is indicated at higher concentrations.   The rate
constants in Table I were calculated from the slopes of rates of
reaction versus solute concentration.   This procedure corrects
for any natural decay of the reacting species viz. ϕ_2^- reacting
with countercations.

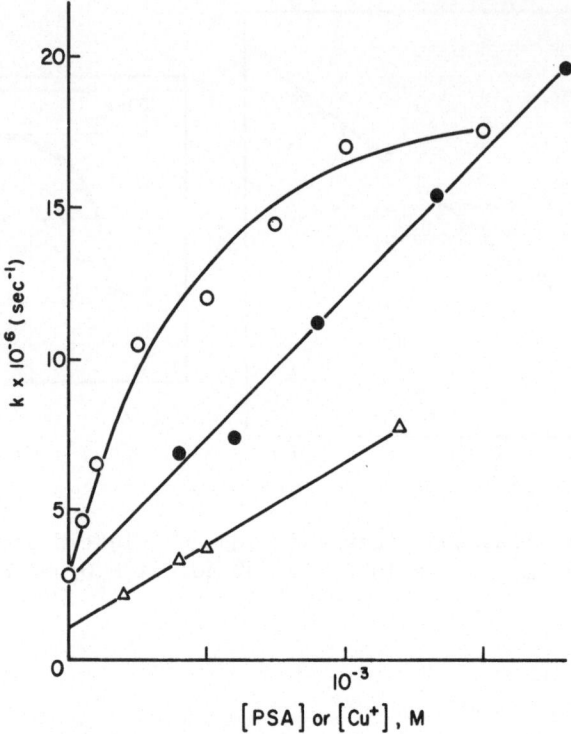

Figure 3.   Growth rate of PSA$^-$ via ϕ_2^- + PSA → PSA$^-$ ●
Reaction of ϕ_2^- + Cu$_T^{++}$ → ^{2}Cu$_T^+$ O
and ϕ_2^- + PSA^2 → PSA$^-$ △

The rates of electron transfer from ϕ_2^- to PSA versus micelle concentration for a constant concentration of PSA/micelle are plotted in Figure 4. At high PSA concentrations viz. 250 PSA/micelle, the rate of electron transfer from ϕ_2^- to PSA varies non-linearly with micellar concentration. Similar data are also obtained at lower PSA concentration viz. 40 PSA/micelle, although the linearity of the data is better here. The curve at lower [PSA] extrapolates at [micelle] = 0 to the decay observed in the absence of micelles. The data for higher [PSA] are also drawn through this point.

The rate constant for reaction of H_3O^+ with ϕ_2^- increases with increasing water content, provided the local or micelle

concentration of H_3O^+ is considered.   The increased rate corresponds to the increased mobility of species in the micelle as the
water content increases.

Triplet Energy Transfer

The pulse radiolysis of ϕ_2 in heptane both in presence and
absence of reversed micelles gives rise to the triplet excited
state of biphenyl, ϕ_2^T.   The mechanisms for formation of ϕ_2^T
have been established elsewhere.[7]   The triplet was monitored by
its characteristic absorption spectrum with spectral maximum at
360 nm and extinction coefficient ε = 37000.[10]   The rate of decay
$\tau_{1/2}$ = $\sim$ 1 μsec of the triplet is not affected by the presence of
micelles.   However, certain solutes associated with micelles lead
to an enhanced rate of decay of ϕ_2^T.   Data for triplet energy
transfer to pyrene butyric acid PBA, pyrene, pyrene tetrasolfonic
acid and PSA giving rise to the excited triplets of these molecules,
are given in Table I.   The rate constants of these reactions are

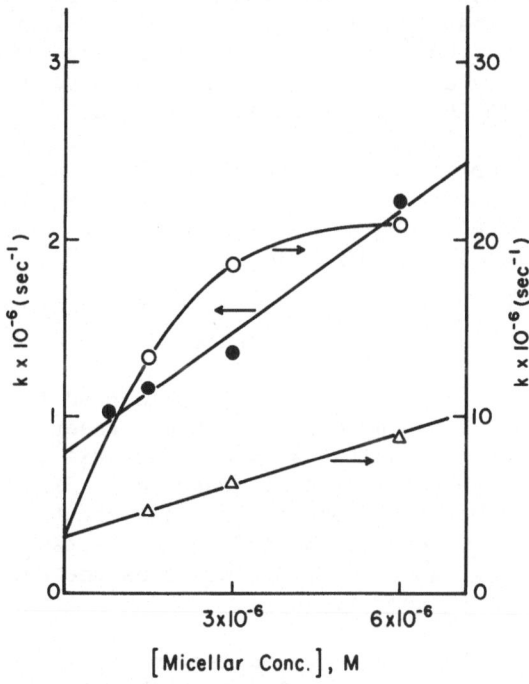

Figure 4.   The effect of micelle concentration on the rate of
reaction of ϕ_2^- + PSA $\rightarrow$ PSA$^-$ (O$\triangle$); ϕ_2^T + PSA $\rightarrow$ PSA^T ($\bullet$); in
2 x 10^{-2} M ϕ_2/3% AOT/6% H_2O/heptane micelles.   (0) [PSA]/[M] =
250;   ($\triangle$) [PSA]/[M] = 40.

independent of solute concentration over the range $3 \times 10^{-4}M$ to
3×10^{-3} $\underline{M}$ (Figure 3). The rate of triplet energy transfer from
ϕ_2^T to PSA varies with the micellar concentration at a constant
PSA/micelle ratio. Typical data are shown in Figure 4.

Quenching of Pyrene Singlet Excited State

Pyrene is not associated with reversed micelles but is
dissolved in the heptane phase. Laser excitation λ_{max} = 3471A
of pyrene leads to the singlet excited state P^S which is observed
by its characteristic fluorescence. The rate of decay of P^S is
independent of the presence of micelles. Several solutes dissolved
either in the micelle or in the heptane enhance the rate of
decay of the pyrene excited singlet state. Typical data for
various conditions are shown in Figures 5 and 6. Bulk solute
concentrations are used in the calculations.

The effect of water content on the rate constant for quenching
of P^S by I^- and Cu^{++} ions is shown in Figure 5. A substantial
drop in rate constant k, is observed as the water content increases.
The relative decrease in k from 1% to 6% water is much larger for
I^- (x 6.5) than for Cu^{++} (x 1.5). The quenching rate constants
in micellar systems are all lower than those observed in homogeneous polar solution such as water or alcohol.

Increasing the ionic strength of the water bubble by direct
addition of NaCl and $MgCl_2$ also reduces the rate of reaction of
P* and Ca^{++} in micelle containing 1% H_2O. $MgCl_2$ has a significantly
larger effect than NaCl, Figure 6.

Quenching of Singlet Excited States of PSA

In reversed micellar systems PSA is located in the water
bubble and removed from the heptane phase. Laser excitation, λ =
3471A of PSA leads the excited singlet state PSA^*, which is
monitored by its characteristic fluorescence. Figure 7 shows the
effect of micelle water content on the rate constant for quenching
of PSA^* by various molecules located in the heptane and micelle
phases. Increasing water constant increases the rate of quenching
by all solutes apart from oxygen.

Local or micelle solute concentrations were used in calculating the rate constants. For solutes which are soluble in heptane
and micelles it is probably better to calculate rate constants
from bulk concentrations. These are shown for CH_3NO_2, and CH_2I_2
in Figure 8. The quenching rate with CH_3NO_2 still increases with
increasing water content, but a decrease is observed for CH_2I_2.

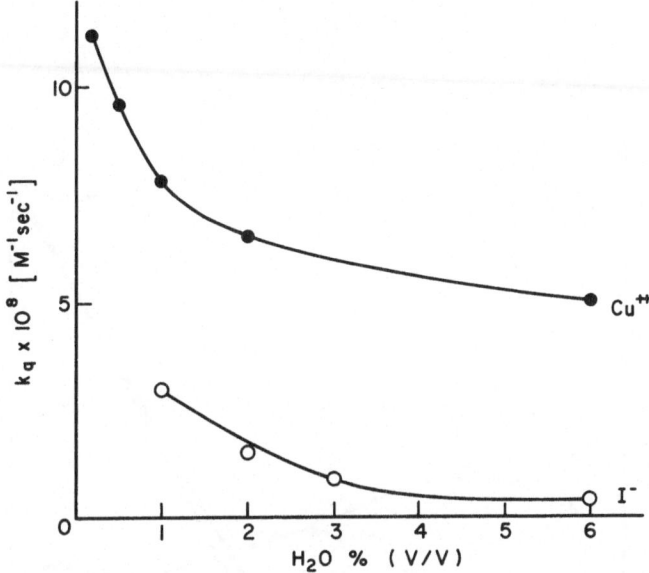

Figure 5. The effect of water content on the rate of quenching of excited pyrene by Cu^{++} (•) and I^- (o) in the 3% AOT/H_2O/heptane micelles.

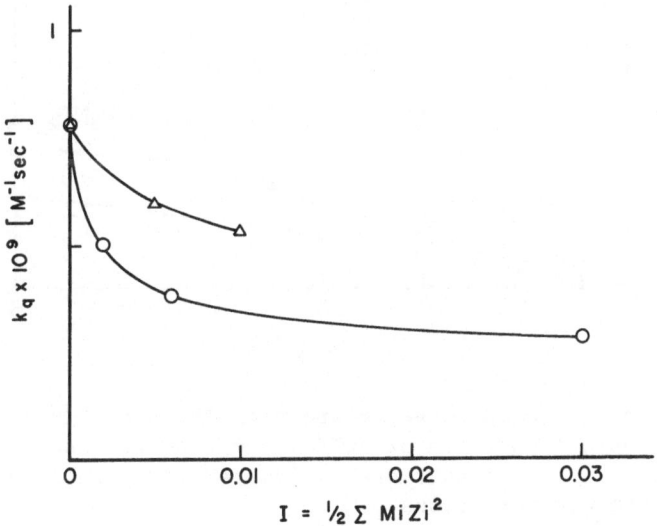

Figure 6. The effect of NaCl (Δ) and $MgCl_2$ (o) on the quenching of excited pyrene by Cu^{++} in 3% AOT/1% H_2O/heptane micelles.

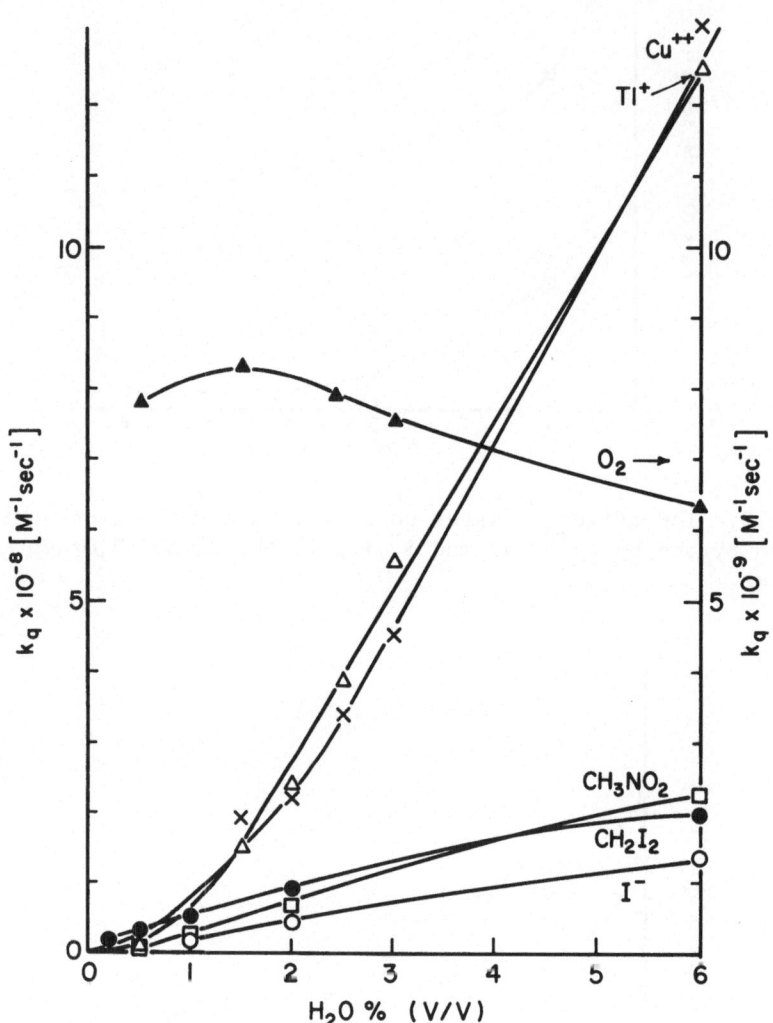

Figure 7. The effect of water content on the quenching of excited PSA by various quenchers in 3% AOT/H$_2$O/heptane reversed micelles. The quenching rate constants are calculated from the local or micelle solute concentrations.

DISCUSSION

The studies outlined earlier indicate the influence of reversed micelles on the rates of selected chemical reactions. Micelles influence the reactions only if at least one reactant is confined to the micelle. The reactant in the heptane phase is randomly dispersed, while that in the micelle has the possible choices of the micellar interior or surface. A reactive species such as ϕ_2^- produced in the heptane could migrate to the micelle where subsequent reaction of ϕ_2^- and a micelle-located solute could occur. A reactive species such as ϕ_2^T may have no specific affinity for the micelle, and it will only carry out diffusion controlled encounters or collisions with the micelle. Reactions of ϕ_2^- with a micelle located solute will then depend on the possible encounters of ϕ_2^T and the solute during the ϕ_2^T micelle encounter. This will differ in detail from that experienced in homogeneous solution where ϕ_2^T solute encounters are random throughout the solution. In the micellar systems ϕ_2^T tends to encounter regions viz micelle, of high local solute concentration. Finally the internal viscosity of the micelle decreases with increasing water content.[5] This leads to an increase in collisional encounters of micellar reactants with a corresponding increase in reaction rate. These features will be considered in discussing the foregoing data.

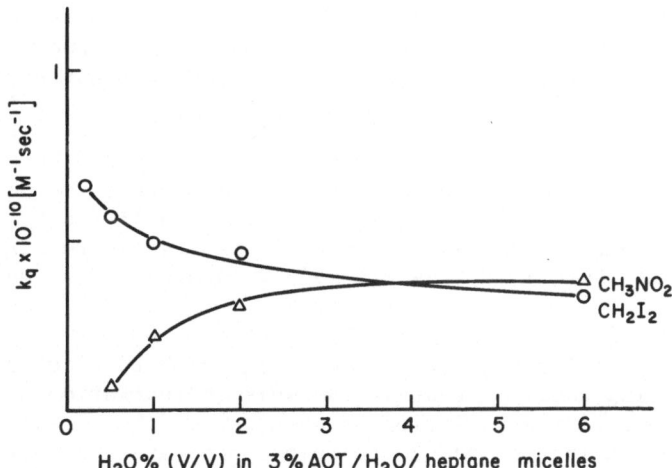

Figure 8. The effect of H_2O on the quenching of excited PSA by CH_3NO_2 and CH_2I_2. The quenching rate constants are calculated from the bulk solute concentration.

Reactions Involving ϕ_2^-

Data in Figure 1 show that ϕ_2^- produced in the radiolysis of
heptane solutions of ϕ_2 are captured by reversed micelles.
Similar data have been noted previously for the capture of
electrons by reversed micelles.[4] During the lifetime of ϕ_2^-
(~200 ns) some of the anions are captured and stabilized by the
micelle. Under the conditions of the experiments the natural
rate of decay of ϕ_2^- via geminate recombination with the cation
is comparable to or greater than the capture by the micelle. The
absorption maximum of the ϕ_2 in the heptane is at 407.5 nm. The
ϕ_2^- captured by the micelle with λ_{max} at 395 nm similar to that
in water,[9] has an extended lifetime so that it may transfer an
electron to PSA which is located in the micelle. The rate of the
electron transfer reaction must be comparable to that of the
capture of ϕ_2^- by the micelle ($\tau_{1/2} >$ 70 nsec). This follows from
the observations that the rate of the reaction of ϕ_2^- with PSA
increases with increasing micellar concentration at constant
[PSA]/[micelle] ratio. Increasing PSA concentration causes a
concommittant increase in the rate of the electron transfer reaction.
This indicates that ϕ_2^- once captured by the micelle diffuses in
the micelle to react with PSA. Encounters of ϕ_2^- and PSA lead to
the formation of PSA$^-$. The exact locations of PSA and ϕ_2^- in
reversed micelles need some consideration. The similar chemical
nature of ϕ_2^- and PSA suggest similar micellar sites of solubilization to a first approximation. More data are available for PSA
as it is a stable species. Earlier H'N.M.R. data[11] showed that
PSA is located in the water/micelle interface in cetyltrimethyl
ammonium bromide, CTAB, micelles. Presumably the positive charge
of the CTAB binds the acidic sulfate group of PSA. However, no
binding is visualized for a negatively charged surface such as
found in Aerosol OT micelles. However,[1]H N.M.R. data indicate
interaction of PSA with the surface of the micelles at high
concentrations of PSA. Most probably, PSA is located both in the
water bubble and in the micellar surface. The relative degree to
which both sites occur is unknown. However, earlier data[4]
with anilino-naphthalene sulfonate which is similar to PSA,
supports that these molecules are mainly located in the water
bubble.

The rate constant for reaction of ϕ_2^- with PSA increases
with decreasing water content if the bulk concentration of PSA is
used. However, little variation is observed if the micellar or
local concentration is used to calculate the rate constant. This
indicates the large factor played by the diffusion of PSA and ϕ_2^-
within the micelle. However, the increased rigidity of the
micelle with decreasing water content[4] is not apparent in these
data.

The reaction of Cu^{++} with ϕ_2^- is slower in reversed micelle systems compared to homogeneous solution where $k = 3.6 \times 10^9 M^{-1} s^{-1}$. The Cu^{++} is located in the micelle and probably distributed between the aqueous phase and the micelle head group region where it is partially bound to the sulfosuccinate groups. Binding of Cu^{++} to negative surfaces reduces its rate of reaction with negative species.[12] Decreasing water content leads to an increased binding of Cu^{++} to the head groups as the solvation of the cation by H_2O decreases. The rate of reaction of ϕ_2^- and Cu^{++} thus decreases with increased binding of Cu^{++} to the head groups viz. by lowering the micellar water content. The rate of reaction of ϕ_2^- with Cu^{++} initially increases linearly with Cu^{++} concentration, but the rate tends to plateau at higher Cu^{++} concentrations (Figure 3). This is due to an increasing binding of Cu^{++} to the head groups with increasing Cu^{++} concentration leading to a decrease in reaction rate.

The reactivity of ϕ_2^- with H_3O^+ does not show a large increase with changing micellar water content. The main point of interest is the very low rate constant observed compared to homogeneous solution[9] ($\sim 1/1000$). Due to the hydrophillic nature of H_3O^+ it is expected that it will reside in the water bubble itself, with little binding to the head group region.

Reactions of Biphenyl Triplet ϕ_2^T

The rate constant of transfer of triplet energy from ϕ_2^T originating in heptane to PSA located in the micelle is independent of the micellar size and water content, Table I. The rate constant for transfer is of the same order as that observed for PBA attached to the micelle ($k \sim 6 \times 10^9 M^{-1} s^{-1}$ and slower than the transfer to pyrene in the solution bulk $k = 2.7 \times 10^{10} M^{-1} s^{-1}$. A hydrophobic species such as ϕ_2^T is not expected to enter the hydrophillic interior of the micelle. However, species such as PBA and to some extent PSA which are located in the vicinity of the micellar surface, do encounter ϕ_2^T. The decrease in rate constant compared to homogeneous solution may reflect on the aggregating of PSA or PBA by the micelle as compared to random dispersion in homogeneous solutions. The micelle itself may also impose restrictions on the movement of both PBA, PSA and ϕ_2^T which will also lead to a reduction in rate constant. The rate of transfer of triplet energy from ϕ_2^T to pyrene tetrasulfonic and PTSA is much slower ($< 2 \times 10^9 M^{-1} sec^{-1}$) than the corresponding reaction of ϕ_2^T and PSA. This is to be expected if PTSA resides in the more hydrophillic portions of the micelle, near the bubble center. Chance encountered with the hydrophobic ϕ_2^T is then decreased.

Reactions of Pyrene Excited State

Both I^- and Cu^{++} rapidly quench excited states of pyrene P^s, in homogeneous solution.[4,5] The presence of reversed micelles decreases the rates of these reactions, the effect increasing with increasing water content, Figure 5. As with ϕ_2^I the hydrophobic nature of P^s prohibits its ready access to the hydrophillic core of the micelle. Quenching of P^s by Cu^{++} and I^- then depends on chance encounters close to the micellar surface. This is increased in the case of Cu^{++}, as it binds to the anionic surface head groups. Evidence to support this latter point is forthcoming in Figure 6. If inert electrolytes such as NaCl and $MgCl_2$ are added to the Cu^{++}/micelle/P^s, then the quenching rate of P^s by Cu^{++} decreases with increasing ionic strength. Figure 6 illustrates this effect for a 1% H_2O micelle using NaCl and $MgCl_2$. The effect of $MgCl_2$ is bigger than NaCl. This is in accord with a mechanism whereby Cu^{++} adsorbed on the micellar surface as replaced by Na^+ or Mg^{++}. The latter ion, due to its double charge will be bound more strongly to the anionic surface than Na^+. Iodide ion is expected to reside in the aqueous micelle interior. Increasing water content increases the size of the micelle and as P^s does not penetrate the aqueous region of the micelle, then it progressively sees a lower effective concentration of I^-. The rate of quenching of P^s by I^- should then decrease with increasing water content, as observed.

Quenching of PSA Excited Singlet

It was indicated earlier that PSA and consequently, the excited singlet state of PSA are principally located in reversed micelles. The precise effect of micellar water content on the quenching of excited PSA depends on the quencher as shown in Figure 7. Increasing water content causes an increase in the quenching rates for all quenchers apart from oxygen, Figure 7. The increased quenching with increasing water content is more pronounced for Cu^{++} and Tl^+.

Increasing rates of reaction with excited PSA indicate an increase in the fluidity of the micelle with increasing water content. This effect has been alluded to previously. The very marked increase in the rates of Cu^{++} and Tl^+ with excited PSA with increasing water content may be associated with the decreased extent of binding of these cations to the head groups over this range. As the water content increases, the cations Tl^+ and Cu^{++} become progressively located in the water bubble while bound cations decrease. The effective concentration of quencher in the bubble then increases. This coupled with the increased mobility of PSA and cation in the bubble leads to an increase in the rate of quenching of excited PSA. These data are in accord with the P^s and Cu^{++} quenching data. Here only the bound Cu^{++} contributed to the quenching which decreased with increasing water content. As CH_3NO_2 and CH_2I_2 may reside both in the aqueous and heptane phase it is pertinent to reconsider the data of Figure 8 where the reaction rate constants are calculated from bulk solute concentrations. The rate of CH_3NO_2 quenching still increases with water content, which indicates a high solubility of this hydrophillic molecule in the micelle. However, the quenching rate by CH_2I_2 now decreases with water content. This species is hydrophobic and of low solubility in the aqueous micelle. Reaction occurs by diffusion of CH_2I_2 from the heptane phase into the micelle. As the water content increases and the micelle becomes progressively more hydrophillic, the presence of CH_2I_2 within micelle deteriorates, andthe reaction rate of excited PSA and CH_2I_2 decreases. This is analogous to the quenching of P^s in the heptane by Cu^{++} in the micelle. If one species is located in the micelle and one in the heptane, then the reaction rate drops with increasing water content.

CONCLUSION

     Several features significantly affect the courses of selected reactions in reversed micelles. Increasing micelle size and water content tend to increase the rate of reactions where two reactants are hydrophillic and in the micelle. This is clearly seen for the quenching of excited PSA by I^-, Cu^{++}, Tl^+ and nitromethane, and the electron transfer from ϕ_2^- to Cu^{++}. The effect is attributed to enhanced mobility of solutes in the micelle with increasing water content. This is borne out by other physical measurements. Additional effects occur with cations which may bind to the anionic micelle surface or reside in the water bubble. Greatly enhanced rates are observed with increasing water content if both reactants are in the micelle viz., excited PSA and Tl^+ or Cu^{++}.

     If one reactant is hydrophobic and located in the heptane and one hydrophillic in the micelle, then increasing water content decreases the rate of reaction. This is due to a decrease in the probability of reactant encounters with increasing water content. This is solely due to the lack of penetration of the hydrophillic reactants into the micelle. This is seen with reactions of excited pyrene with I^-, and Cu^{++}, and excited PSA with CH_2I_2.

     The rates of reaction of non-charged reactants are slightly decreased by reversed micelles if one reactant is located in the micelle. This is due to a clustering of this reactant in one region, which leads to a decrease in its mobility.

Reversed micellar systems containing water provide a useful method of investigating certain reactions at the molecular level.

REFERENCES

1. The Radiation Laboratory of the University of Notre Dame is operated under contract with the U.S. Energy Research and Development Administration. This is ERDA Document No. COO-38-1054.

2. D. Eisenberg and W. Kaufman, "The Structure and Properties of Water", Oxford (1969).

3. (a) P. A. Winsor, Chem. Rev. $\underline{68}$, 1 (1968); C. Singelterry, J. Amer. Oil Chem. Soc. $\underline{32}$, 446 (1955); P. H. Elworthy, A. T. Florence, and C. B. MacFarlane, "Solubilization by Surface Active Agents", Chapman and Hall Ltd., London (1968); J. H. Fendler and E. J. Fendler, "Catalysis in Micellar and Macro-molecular Systems", Academic Press, New York, N.Y. (1975); J. H. Fendler, Accounts Chem. Res. $\underline{9}$, 153 (1976).
    (b) J. B. Peri, J. Amer. Oil Chem. Soc. $\underline{35}$, 110 (1958); P. Ekwall, L. Mandell and K. Fontell, J. Colloid Interface Sci. $\underline{33}$, 215 (1970); W. J. Knox and T. O. Parshall, ibid. $\underline{40}$, 290 (1972); A. Kitahara, T. Kobayashi and T. Tochibana, J. Phys. Chem. $\underline{66}$, 363 (1962).

4. (a) M. Wong, M. Grätzel and J. K. Thomas, Chem. Phys. Lett. $\underline{30}$, 329 (1975).
    (b) M. Wong, M. Grätzel and J. K. Thomas, J. Amer. Chem. Soc. $\underline{98}$, 2391 (1976).

5. M. Wong, T. Nowak and J. T. Thomas, unpublished work.

6. J. K. Thomas, K. Johnson, T. Klippert and R. Lowers, J. Chem. Phys. $\underline{48}$, 1608 (1968).

7. J. K. Thomas, Int. J. Radiat. Phys. and Chem. $\underline{8}$, 1 (1976).

8. I. S. Berlman, "Handbook of Fluorescence of Aromatic Molecules, New York, Academic Press (1965).

9. (a) S. C. Wallace and J. K. Thomas, Rad. Res. $\underline{54}$, 49 (1973).
    (b) I. A. Taub, D. A. Harter, M. C. Sauer and L. M. Dorfman, J. Chem. Phys. $\underline{41}$, 979 (1964).
    (c) L. M. Dorfman, Acc. Chem. Res. $\underline{7}$, 224 (1970).
    (d) I. A. Taub, M. C. Sauer and L. M. Dorfman, Disc. Faraday Soc. $\underline{36}$, 30 (1963).
    (e) P. Chang, R. V. Slates and M. Szware, J. Phys. Chem. $\underline{70}$, 3180 (1966).

10. (a) E. J. Land, Proc. Roy. Soc. Ser. $\underline{A}$, $\underline{305}$, 1457 (1968).
    (b) R. Benasson and E. J. Land, Trans. Faraday Soc. $\underline{67}$, 1904 (1971).

11. K. Kalyanasundaram, M. Grätzel and J. K. Thomas, J. Amer. Chem. Soc. $\underline{97}$, 3915 (1975).

12. M. Grätzel and J. K. Thomas, J. Phys. Chem. $\underline{78}$, 2248 (1974).

CATALYSIS BY CATIONS IN CORES OF NON-AQUEOUS MICELLES

F. M. Fowkes, D. Z. Becher, M. Marmo,
C. Silebi and C. C. Chao

Department of Chemistry
Lehigh University
Bethlehem, Pennsylvania 18015

The cationic core of inverse micelles of metal sulfonates in hydrocarbon solvents has a catalytic microenvironment like a molten salt but it is easily available to hydrocarbon-soluble reactants. Not only can a wide variety of cations be used as catalysts, but mixed cation micelles provide the intimate contact between unlike cations desirable for many special catalysts. The acidity of cations is not appreciably masked by the weak basicity of the sulfonate anions, and acid-catalyzed reactions are shown to depend strongly on which cations are used.

INTRODUCTION

The metal salts of oil-soluble surfactants such as high molecular weight sulfonates readily form "invert" micelles in organic
solvents. These micelles have a highly concentrated inorganic
core, a microenvironment which can catalyze reactions of organic
solutes.

Micelle formation by oil-soluble metal sulfonates has been
studied by several investigators.[1-4] Aggregation numbers and critical concentrations have been studied as a function of organic
structure of the sulfonate, and of different cations or solvents.
Solubilization of water and of highly polar molecules by such
micelles has also been studied; this phenomenon appears related to
acid-base interactions between the acidic cations in the micelle
core and basic polar molecules.[5-7]

Catalysis by such micelles involves the same considerations
that govern catalysis by any other catalyst or enzyme: 1, the
reactant must concentrate at the catalytic sites; 2, the catalyst
must tend to form a reaction intermediate with the reactant; and
3, the product must be less strongly bound to the catalytic site
than the reactant.

AN ACID-CATALYZED ESTER DECARBOXYLATION

A model reaction involving acid-catalysis was chosen to illustrate the above principles. Acid catalysis was chosen because in
a sulfonate micelle the very weak Lewis basicity of the sulfonate
anions allows the acidity of the cations to dominate the core microenvironment, and the acidity of micelles with different cations can
be compared by using Hammett indicators to determine the Hammett
acidity function H_O of the micellar cores.[8]

For an acid catalyst to attract a reactant and reject a product, we need basic reactants and less basic products. Possible
reactions include hydrolysis of esters where both water and ester
are basic, and the freed acid would be indeed less basic. However
we must concern ourselves with the solubilization of four species
(ester, water, acid, and alcohol). For a model system it appeared
better to concentrate on unimolecular decompositions so as to
reduce the number of components. We chose the decomposition of
benzylchloroformate:

$$C_6H_5CH_2-O\overset{\overset{O}{\|}}{C}Cl \;\rightarrow\; C_6H_5CH_2Cl + CO_2\uparrow$$

In this reaction the reactant is basic, both products have no basicity, the temperature required for the reaction lies in a suitable range, and the progress of the reaction can easily be followed by measuring CO_2 evolution.

At first we did not know that this reaction was acid-catalyzed. According to the literature[9,10] the reaction has an ionic intermediate and the reaction rate depends strongly on dielectric constant. However, as shown in Table I, reaction rates measured in a variety of solvents indicate that the reaction is catalyzed more strongly by acidic or basic solvents than by solvents of high dielectric constant.

Catalysis of the above reaction by various metal sulfonates in n-decane was studied with Aerosol OT (sodium di-2-ethylhexyl sulfosuccinate) and various metal salts derived from it by aqueous metathesis as devised by Kitahara.[11] The reaction was studied in the temperature range of 95°C to 126°C. Experimental details have been published elsewhere.[12] At 95°C the salts catalyzed the reaction in the order $Cr^{3+} > Al^{3+} > Ca^{2+} > Na^+$, but even the sodium salt catalyzed the reaction appreciably; 0.056M Aerosol OT increased the rate thirty-fold. At 126°C the first order reaction rate constant was found to increase with catalyst concentration as shown in Figures 1 and 2. Extrapolation toward zero concentration shows that the reaction rate becomes negligibly small at a finite sulfonate concentration (1.5 g/liter for the aluminum salt and 6.6 g/liter for the sodium salt), presumably the CMC.

Table I. Reaction Rate Constants k_1 vs. Dielectric Constant ε of Solvents for Decarboxylation of Benzylchoroformate.

Solvent	k_1 (hr^{-1})	Temperature (°C)	ε (at Temperature)
n-decane	2.3×10^{-4}	92-97	1.89
dioxane	4.0×10^{-2}	90-95	2.09
toluene	2.2×10^{-3}	93-95	2.21
tri-n-butylamine	50	95	2.9
trichloroacetic acid	10	98	4.5
tetrahydrofuran	1.0×10^{-2}	65-73	7.3
nitrobenzene	1.1×10^{-2}	96-97	24.3

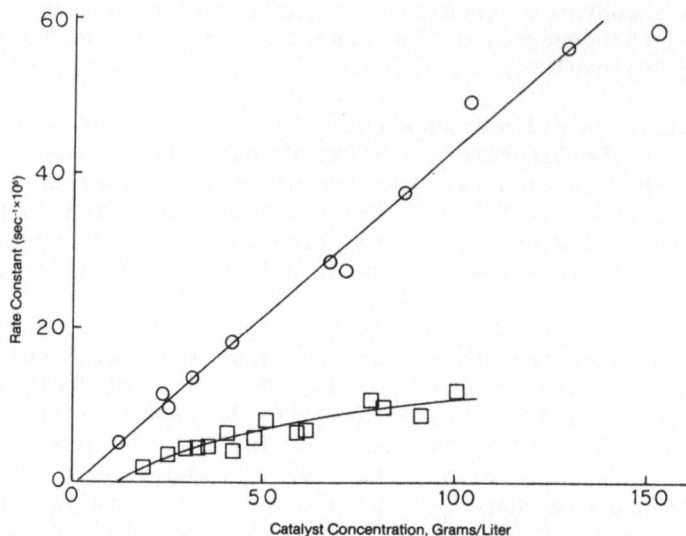

Figure 1. The rate of decomposition of benzyl chloroformate as a function of concentration of sodium Aerosol OT (O) and of aluminum Aerosol OT (☐) in n-decane at 126°C.

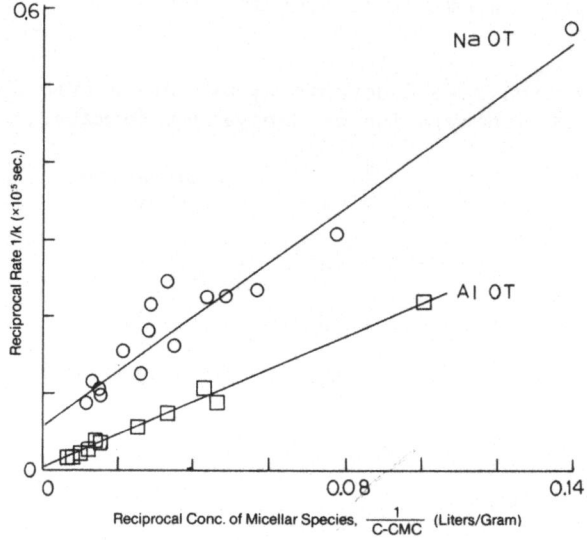

Figure 2. Lineweaver-Burk plot of the data in Figure 1 to illustrate saturation kinetics.

The reaction rates with Zn^{2+}, Al^{3+}, Ce^{3+}, and Na^{+} salts at different temperatures are shown as an Arrhenius plot in Figure 3. The curious crossing-over of these lines indicates that the pre-exponential terms vary appreciably and that any comparison of reaction rates vs. different cations will hold only in a limited temperature range. The activation energy, E_a, a pre-exponential factor A, activation enthalpy $\Delta H^{\neq}$, and activation entropy $\Delta S^{\neq}$ are shown in Table II. The very large positive $\Delta S^{\neq}$ values for the more active catalysts suggests the name "entropic catalysis".

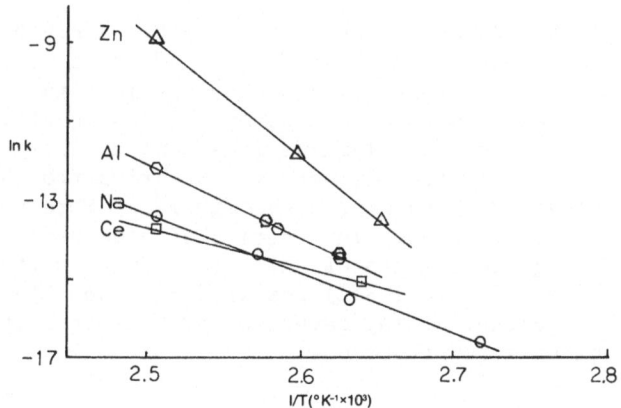

Figure 3. Arrhenius plots for the catalyzed decompositions of benzyl chloroformate in n-decane solutions of Aerosol OT salts.

Table II. Kinetic Parameters for Temperature-Dependence of Benzyl-chloroformate Decomposition in n-Decane Solutions of Aerosol OT Salts.

Cations in Micelle	E_a	lnA	$\Delta H^{\neq}$	$\Delta S^{\neq}$
Na^{+}	129±4	25.3±1.2	124	−15.5
Al^{+3}	148±7	32.2±2.0	147	50.3
Ce^{+3}	84±4	11.7±1.2	83	−122
Zn^{+2}	267±4	71.4±1.4	266	375

$\Delta H^{\neq}$, E_a units − kJ/mole $\Delta S^{\neq}$ units − kJ/mole-deg

Table III compares the catalytic activity at 126°C with the Hammett acidity function H_O determined at room temperature in benzene solutions with 20 g/liter of sulfonate, using standard Hammett indicators.[6] The correlation is good except for the Zn^{2+} sulfonate which is exceptionally catalytic for its moderate acidity. There are two possible reasons for the lack of correlation. The Hammett acidity function H_O is a single-valued system of acidity and cannot account for "hard" and "soft" acids or bases.[5,13] It is possible that the higher activity of Zn^{2+} sulfonates results from some degree of "softness" as compared to Al^{3+}. Another reason for the lack of correlation could be that traces of water can change acidity and the comparison at two very different temperatures may involve some difference in water content.

In general the effect of water on catalytic activity of metal sulfonates is very marked and this effect leads to some experimental irreproducibility. The preparation of the metal sulfonate catalysts included many precautions to remove water, but especially with the aluminum sulfonates, for the aluminum ions tend to retain adjacent water molecules. All sulfonates were freeze-dried from benzene and pumped out at elevated temperatures through large bore tubing with a large liquid nitrogen trap. In a series of aluminum sulfonate samples, increasing temperature of the freeze-dried sulfonate during final pumping (from 85°C to 113°C) resulted in a doubling of catalytic activity. Further drying gave even greater activity, sometimes catalyzing unwanted reactions.

Table III. Comparison of Reaction Rates at 126°C of Benzylchloroformate Decomposition with Hammett Acidity of Micelles.

Cation	Acidity (H_O)	Rate Constant (liters-gr^{-1}-sec^{-1}x10^6)
Zn^{+2}	0.7	131
Cr^{+3}	-2.3	6.6
Al^{+3} (from nitrate)	0.1	5.6
Al^{+3} (from chloride)	0.7	4.3
Na^+	5.3	1.4
Ce^{+3}	3.1	1.0

AN ACID-CATALYZED DIELS-ALDER CONDENSATION

A model bimolecular reaction was also chosen for study with the same micellar catalysts, the Diels-Alder condensation of benzoquinone with cyclopentadiene in benzene at 20-40°C. Wasserman[14,15] had shown this reaction to be catalyzed by trichloroacetic acid to about 300 times faster than the uncatalyzed reaction. The reaction was followed by measuring UV absorption of benzoquinone at 454 nm. Our findings for the Al^{3+} sulfonate at 30°C are shown in Figure 4 which shows a mere doubling of the reaction rate, but with an unmistakable CMC of 2 g/liter.

In acid-micellar catalysis of a Diels-Alder condensation both reactants tend to be solubilized because they are weakly basic by virtue of their unsaturation, and the product is less unsaturated. However it is possible to have one reactant so preferentially solubilized that the other is not available in the micelle for the reaction. This was the case with cyclopentadiene and anthracene, where the reaction rate decreased by a factor of two upon addition of the aluminum sulfonate. It is believed that the anthracene was preferentially solubilized so that the two reactants were separated from each other.

ZIEGLER-CATALYZED DIENE POLYMERIZATION

Typical Ziegler catalysis of diene polymerization can be done in micelles as well as at the surface of colloidal precipitates.

The best known Ziegler catalysts are colloidal precipitates from the reaction of $TiCl_4$ and aluminum alkyls in a hydrocarbon solvent. A micellar version, described in the patent literature,[16] was made from titanium dinonylnaphthalene sulfonate (TiDNNS) or from the cobalt sulfonate (CoDNNS), obtained by cation exchange in an alcohol-benzene solvent, or by reaction of the metal chloride with the free sulfonic acid. These sulfonates combined with an aluminum alkyl (or aluminum alkyl chloride) co-catalyst are very active stereospecific catalysts for butadiene or isoprene polymerization, giving unusually high contents of cis 1,4 structures.

The aluminum alkyl chloride co-catalyst is solubilized into the polar core of the TiDNNS or CoDNNS micelles where the Ti^{4+} ions are reduced to Ti^{3+}. In the micelles the Al^{3+} and Ti^{3+} ions work with solubilized dienes to initiate and propagate the growth of the polymer chain, with the growing ionic end probably trapped in the micelle by virtue of the local high dielectric constant of the micellar core microenvironment.

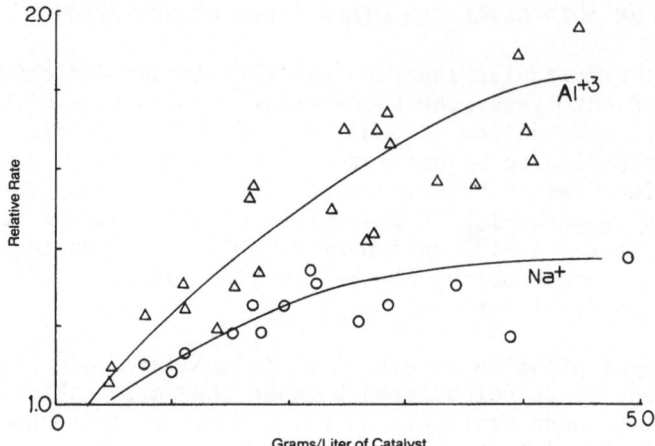

Figure 4. Ratio of catalyzed to uncatalyzed rates of Diels-Alder condensation of cyclopentadiene and benzoquinone at 30°C in benzene solutions of sodium Aerosol OT (O) and of aluminum Aerosol OT (Δ).

CATALYTIC MICELLES WITH MIXED CATIONS

The Ziegler micellar catalyst with two kinds of cations in each micelle may point the way towards effective catalysts for a variety of reactions. Two especial benefits may be expected: One of the cations may be introduced in a form so as to reduce or oxidize the other to an unstable but catalytic valence state (as in the Ziegler catalyst), or each cation might easily form a reaction intermediate with one of the two reactants of a bimolecular reaction.

We chose a model system for investigation of two kinds of cations in each micelle: a mixture of Na^+ Aerosol OT and of Cu^{2+} Aerosol OT in dry benzene solutions. ESR spectroscopy was used to measure exchange interactions between adjacent Cu^{2+} ions, for when Cu^{2+} ions are less than 6Å apart (center to center) the resonance line is very appreciably narrowed. Use of this principle allows measurement of the CMC or of the separation of cations in the micelle due to solubilization of water or methanol.[17,18] It was used in this study to measure the mixing of cations in the micellar core. As Na^+ Aerosol OT was added to Cu^{2+} Aerosol OT the ESR line width increased until at a $Cu^{2+}:Na^+$ ratio of unity there was no evidence of exchange interactions. This information is taken to mean that in micelles with mixed cations we can be sure that unlike cations are indeed in very close contact, an ideal situation for certain catalytic reactions.

ACKNOWLEDGMENT

Acknowledgment is made to the National Science Foundation and to the Donors of The Petroleum Research Fund, administered by the American Chemical Society, for partial support of this research.

REFERENCES

1. S. Kaufman and C. R. Singleterry, J. Colloid Sci. 10, 139 (1955); 12, 465 (1957).
2. A. Kitahara, T. Kabayashi, and T. Tachibana, J. Phys. Chem. 66, 363 (1962).
3. F. M. Fowkes, J. Phys. Chem. 66, 1843 (1962).
4. H. F. Eicke and V. Arnold, J. Colloid Interface Sci. 46, 101 (1974).
5. S. Kaufman, J. Colloid Sci. 17, 231 (1962).
6. H. F. Eicke, J. Colloid Interface Sci. 52, 65 (1975).
7. F. M. Fowkes, in "Solvent Properties of Surfactant Solutions," K. Shinoda, Editor, pp. 65-115, Marcel Dekker, New York, 1967.
8. L. P. Hammett and A. J. Deyrup, J. Amer. Chem. Soc., 54, 2721 (1932) see also F. M. Fowkes et al., Agric. Food Chem. 8, 203 (1960).
9. K. B. Wiberg and T. M. Shryne, J. Amer. Chem. Soc. 77, 2774 (1955).
10. K. L. Oliver and W. G. Young, J. Amer. Chem. Soc. 81, 5811 (1959).
11. A. Kitahara, K. Watanabe, K. Kon-No, and T. Ishikawa, J. Colloid Interface Sci. 29, 48 (1969).
12. F. M. Fowkes and D. Z. Becher, submitted to J. Colloid Interface Sci. (1976).
13. R. S. Drago, G. C. Vogel, and T. E. Needham, J. Amer. Chem. Soc. 93, 6014 (1971).
14. A. Wasserman, J. Chem. Soc. 1936, p. 1028; 1942, p. 618.
15. W. Rubin, H. Steiner, and A. Wasserman, J. Chem. Soc. 1949, p. 3046.
16. J. Boor, Jr. and F. M. Fowkes, U.S.P. 3,234,198 (1966).
17. C. C. Chao and F. M. Fowkes, to be published.
18. A. Kitahara, O. Ohashi, and K. Konno, J. Colloid Interface Sci. 49, 108 (1974).

SOLUBILIZATION AND CATALYSIS OF POLAR SUBSTANCES IN NONAQUEOUS SURFACTANT SOLUTIONS

Ayao Kitahara and Kijiro Kon-no

Science University of Tokyo, Kagurazaka

Shinjiku-ku, Tokyo, Japan

Solubilization of polar substances (n-propylamine, methanol and acetic acid) was investigated in toluene or carbon tetrachloride solutions of surfactants (didodecyldimethylammonium halides, tridodecylammonium halides, dodecylammonium propionate, polyoxyethylene nonylphenyl ethers, polyoxyethylene dodecyl ether and Aerosol OT) with gas chromatography. Solubilization isotherms relating the amount of polar substances solubilized to the remaining amount in the medium, obtained by gas chromatograms, could be classified into 4 types. Types A and B show strong interaction between solubilizates and surfactant micelle, the former suggesting complex formation. Types C and D show weak interaction. Most of cationic surfactant systems belonged to Type A or B irrespective of solubilizates. On the other hand, nonionic surfactants showed Type C or D.

Chemical shift or line width of NMR spectra was also investigated to study the interaction between solubilizates and surfactants. The results were consistent with that of gas chromatography within the same series.

The effect of propylamine, methanol or acetic acid on the catalysis of dodecylammonium propionate for hydrolysis of p-nitrophenyl acetate in toluene or carbon tetrachloride was investigated. The effect of three solubilizates was markedly different.

Propylamine promoted the rate of hydrolysis drastically, methanol decreased the rate to a slight extent and acetic acid decreased it markedly. The difference was discussed referring to solubilization behavior of the solubilizates.

The solubilization of acetone, ethylether or methylacetate was also studied. These showed considerably less solubization (belonging to type D) and decreased the rate of ester hydrolysis to a slight extent.

INTRODUCTION

It has been recently found out that oil-soluble surfactants such as alkylammonium carboxylate enhance the rate of chemical reaction in nonpolar media.[1,2] Since substances added to the surfactant solutions are solubilized into the reversed micelle, they may affect the rate of chemical reaction. The effect of water as substances added has been studied.[3,4,5] The effect of other substances added is worthy to be investigated with reference to the interaction of the substances with surfactants.

In this paper, polar substances such as acid, amine and alcohol were selected as the substances added because of the presence of functional groups. Solubilization behavior of the substances is studied with the gas chromatography and the NMR method to clarify the interaction between the substances and surfactants. The effect of the substances added on the rate of hydrolysis is also investigated. Solubilization behavior and the effect on the rate are compared to elucidate the mechanism of the effect of the substances.

EXPERIMENTAL

Materials

Didodecyldimethylammonium halides, tridodecylammonium halides and dodecylammonium propionate as cationic surfactants, polyoxyethylene dodecyl ether and polyoxyethylene nonylphenyl ethers as nonionic surfactants and sodium di-2-ethylhexyl sulfosuccinate

(Aerosol OT) as anionic surfactant were used. Tridodecylammonium halides were prepared by mixing equivalent amounts of tridodecyl- amine and aqueous hydrogen halide solution in methanol and recrystallizing a few times from acetone, cooled by dry ice and acetone mixture. Nonionic surfactants were homogeneous samples (single ethyleneoxide length) supplied from the Nippon Surfactant Co. Preparation and purification of the other surfactants were described elsewhere.[6] Melting point of cationic surfactants and their abbreviations are shown in Table I.

Acetic acid, methanol, and 1-propylamine were main solubili- zates and acetone, ethyl ether and methyl acetate were also used for comparison. These substances were used after drying the re- agents of Guarantee Grade through Molecular Sieves 4A.

Toluene and carbon tetrachloride used as solvents were puri- fied and dried as usual. The pure chemical grade of p-nitrophenyl acetate was used as a reactant of hydrolysis.

Table I. Melting Point and Abbrevation of Cationic Surfactants

	Abbre- viation	Melting Point (°C)	
		Measured	Literature*
Didodecyldimethyl- ammonium Chloride	DDAC	153–155	–
Didodecyldimethyl- ammonium Bromide	DDAB	169–170	–
Didodecyldimethyl- ammonium Iodide	DDAI	131.5–133	–
Tridodecylammonium Chloride	TDAC	80	84–85
Tridodecylammonium Bromide	TDAB	80	86–87
Tridodecylammonium Iodide	TDAI	55	52
Dodecylammonium Propionate	DAP	53–54	–

*G. Markovits and A. S. Kertes, "Solvent Extraction Chemistry" ed. by D. Dyrssen, J. O. Lilijenjin, J. Rydberg, North Holland Pub., Amsterdam (1967), p. 390.

Methods

The amount of substances solubilized into the non-aqueous surfactant solutions was obtained with gas chromatography. The solubilization of soluble substances was determined from the lowering of
activity by interaction with surfactant molecule or micelle.

The following systems were used as column fillers: Carbowax
(20%) on Chromosorb P for methanol, acetone, ethyl ether and methyl
acetate; Porapak Q plus Tween 80 (9%) and H_3PO_4 (1%) on Chromosorb
P for acetic acid; Porapak Q plus Diglycerol (15%), Tetraethylene
Pentamine (5%) and NaOH (2%) on Chromosorb P for 1-propylamine.

Vapor equilibrated with the solubilizate in the solvent or the
surfactant solution was analyzed with gas chromatography. The
amount of solubilization was obtained as the difference of vapor
pressure curves shown by the substance in the solvent and the
solution shown in Figure 1. The ordinate and abscissa of the figure
show the height of the chromatogram peak and the concentration of
the substance, respectively. The curves a and b are values of the
solvent and the solution, respectively. Hence $\overline{AB}$ and $\overline{OA}$ show the
amount of solubilized and free substances, respectively. Solubi-
lization isotherm was composed with use of both amounts, the ordinate being the former and the abscissa the latter (see Figures 2-4).

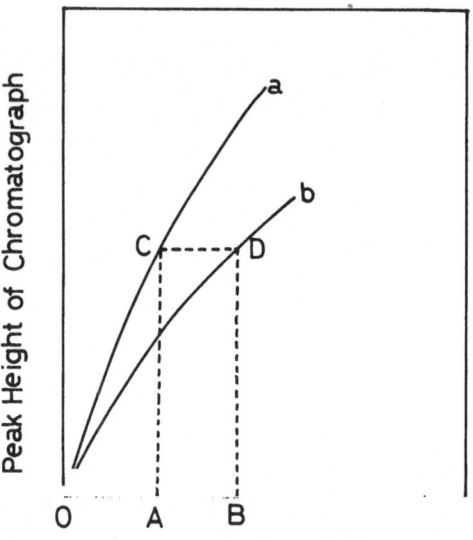

Figure 1. Principle obtaining the amount of solubilizate by gas
chromatography. a) in the solvent, b) in the surfactant solution.

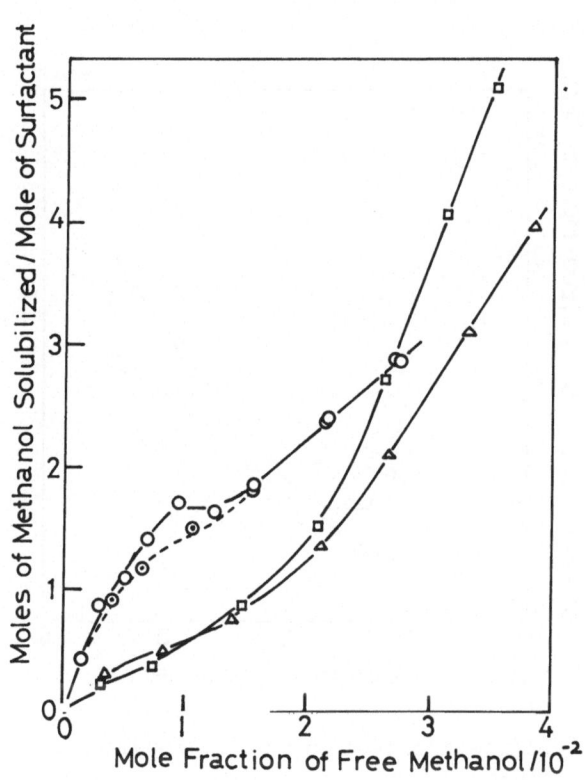

Figure 2. Solubilization isotherm of methanol. O: DDAC,
⊙: DDAB, □: DAP, Δ: NP-8.

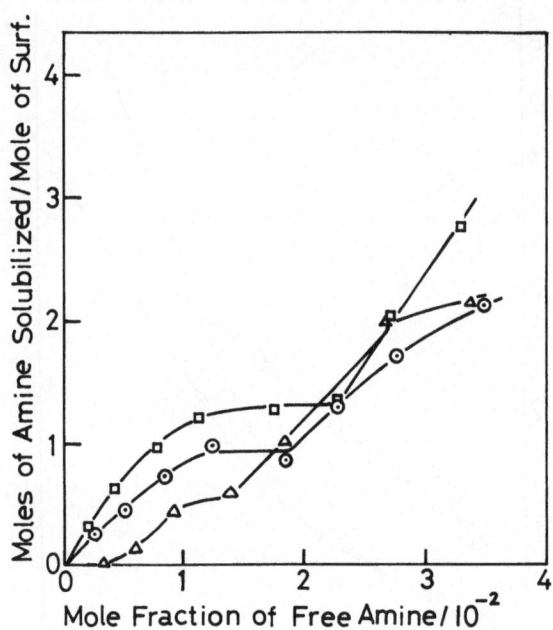

Figure 3.   Solubilization isotherm of 1-propylamine.   ⊙:   DDAB,
□: DAP,   Δ: NP-8.

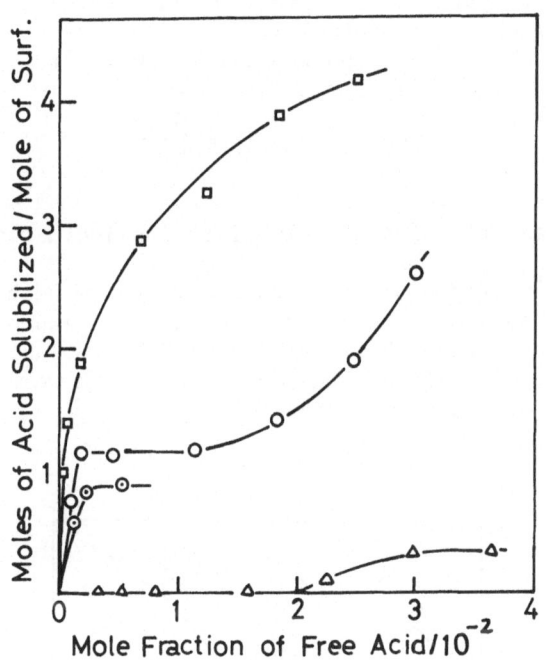

Figure 4.   Solubilization isotherm of acetic acid.   O: DDAC,
⊙: DDAB,   ☐: DAP,   Δ: NP-8.

The former was expressed as moles of substances solubilized per
mole of surfactants. Concentration of surfactants was fixed at
0.035M throughout gas chromatography measurement.

    The interaction of substances added with the surfactants was
also estimated by chemical shift and line width of NMR spectra.
NMR spectra were taken with the Hidachi R-24 NMR spectrometer
(60 MHz) relative to the external TMS standard. Concentration of
surfactants was fixed at 0.1M. Measurement temperature was 35°C
both for gas chromatography and NMR.

    The rate of hydrolysis of p-nitrophenyl acetate was measured
by tracing λ_{max} of p-nitrophenol liberated by the reaction with the
Shimadzu Double Beam Spectrophotometer UV 200.

RESULTS AND DISCUSSION

Classification of Solubilization Isotherms

    Various features of solubilization isotherm were observed with
varying surfactants or substances added. A few examples of solubilization isotherms were shown in Figures 2-4. Solubilization isotherms could be classified into four types which were schematically
depicted as A, B, C and D in Figure 5.

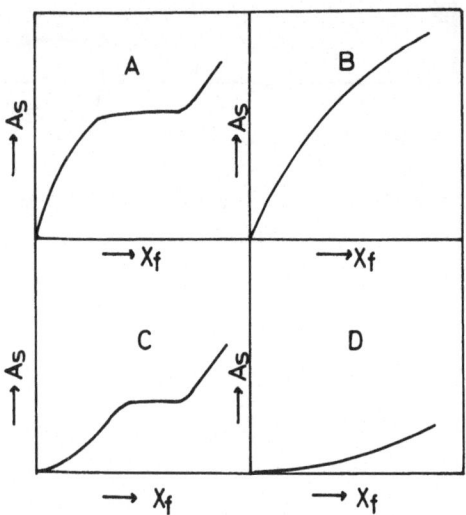

Figure 5. Four types of solubilization isotherms. A_s: the amount
of solubilizates; X_f: the concentration of free substances added.

Types A and B are characterized by the presence of strong interaction between substances and surfactants at lower concentration which is suggested by the steep initial slope of the curves. Type A shows further a flat region at medium concentration, suggesting formation of a kind of complex. Type C is characterized by the weak interaction at lower concentration and formation of complex at medium concentration and D the weak interaction over the whole range of concentration studied. The type of each system was described in Table II.

The following result for acetic acid, methanol and 1-propylamine emerges from columns in Table II. All of dodecyldimethylammonium halides belongs to Type A except a system. Most of tridodecylammonium halides showed Type B except chloride for acetic acid and iodide for methanol. Solubilization curves of acetone, ethyl ether and methyl acetate belonged to Type D for all surfactants.

Interaction at Initial Stage of Solubilization

The initial slope of solubilization isotherms at lower concentration was approximately evaluated to estimate the interaction between surfactants and substances added (solubilizates) and listed in Table II.

The magnitude of the interaction can be estimated from values of initial slope in Table II as follows (abbreviation: See Table I):

i)   DDAC>DDAB>DDAI for methanol, acetic acid or 1-propylamine.

ii)   TDAC>TDAB>TDAI for methanol.

iii)   DD series>TD series for the same halides.

iv)   DAP interact strongly with 1-propylamine and acetic acid.

v)   Acetone, ethyl ether and methyl acetate interact very weakly with all of the surfactants.

The result of i) and ii) suggests that the interaction of solubilizates with surfactants is ion-dipole interaction instead of acid-base or other interaction.

Table II　Analysis of Solubilization Isotherms

Solubilizates	Methanol			Propylamine			Acetic Acid			Acetone, Ethyl Ether, Methyl Acetate		
Surfactants	Type	Initial* Slope	Nc**	Type	Initial* Slope	Nc**	Type	Initial* Slope	Nc**	Type	Initial* Slope	Nc**
DDAC	A	3.0	1.6	-	-	-	A	6.7	1.1	-	-	-
DDAB	A	2.5	1.5	A	1.0	1	A	5.0	0.9	D	0	-
DDAI	A	0.9	0.8	A	0.8	1	C	0.6	0.9	-	-	-
TDAC	B	2.5	-	-	-	-	A	3.0	0.5	-	-	-
TDAB	B	1.0	-	-	-	-	-	-	-	-	-	-
TDAI	C	0.5	0.8	-	-	-	-	-	-	-	-	-
DAP	B∿C	0.7	(0.2)	A	1.6	1.3	B	20	-	D	0	-
NP-5†	A∿C	0.8	0.5	C	0	0.5	D	0	-	-	-	-
NP-8†	A∿C	0.8	0.5	C	0	0.5	D	0	-	D	0	-
DP-5†	A∿C	0.8	0.5	C	0	0.5	D	0	-	-	-	-
AOT	A	0.8	0.5	A	0.5	0.5	A	1.4	1	D	-	-

* Relative values.

** Nc : Co-ordination number

† NP-5: penta-oxyethylene nonylphenyl ether, NP-8: octa-oxyethylene nonylphenyl ether, DP-5: penta-oxyethylene dodecyl ether.

Interaction at Medium and Higher Concentration

A kind of complex is formed at the flat region of solubilization isotherm. The co-ordination number (Nc) of the complex (molar ratio of solubilizate to surfactant at the region) was also arranged in Table II. The number was approximately the order of unity for DD series and 0.5 for nonionic surfactants.

All of the solubilization isotherms were approximately linear at higher concentration. The slope of the linear region was shown in Table III. Differences were hardly recognized among the slope of the curves of same series (DD, TD or nonionic). This fact shows that the interaction between solubilizate and polar groups of each surfactant disappears in the region, because the interaction between each other of solubilizates may only survive.

Study on Interaction by NMR Spectra

Interaction of solubilizates with surfactants in carbon tetrachloride was also investigated by NMR spectra. Change of chemical shift of OH proton of methanol and acetic acid at their fixed concentration by the presence of surfactants of varying concentration was depicted in Figures 6 and 7.

Table III. The Slope of Solubilization Isotherms at Higher Concentration Range

	Methanol	Propylamine	Acetic Acid
DDAC	0.9	–	1.3
DDAB	0.9	≈0.7	?
DDAI	1.2	≈0.7	1.3
TDAC	0.5	–	?
TDAB	0.5	–	–
TDAI	≈0.5	–	–
DAP	3	1.4	0.5
NP-5	1.4	0.7	?
NP-8	1.4	0.7	?
DP-5	1.4	0.7	?

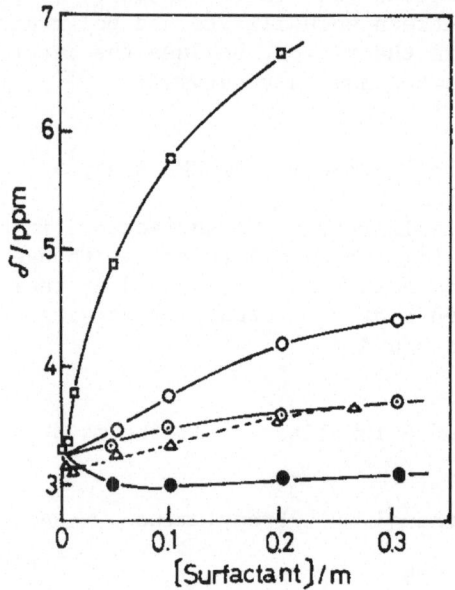

Figure 6.   Change of chemical shift of methanol O$\underline{H}$ proton with
surfactant concentration in CCl$_4$ (methanol: 0.25$\overline{0}$ m).   O: DDAC,
⊖: DDAB,   ●: DDAI,   Δ: TDAC,   □: DAP.

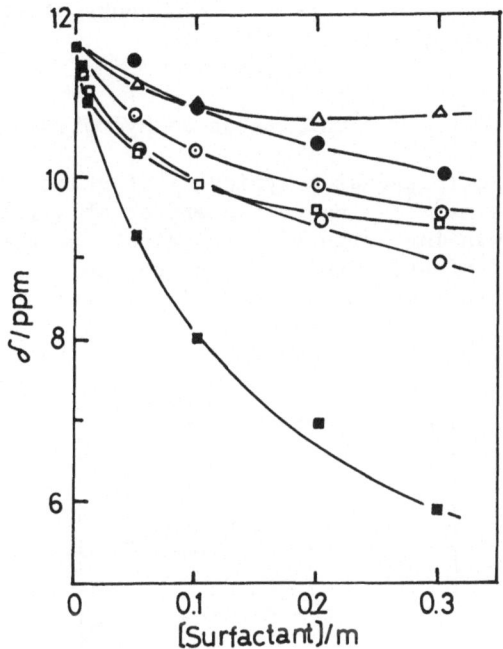

Figure 7. Change of chemical shift of acetic acid O$\underline{\text{H}}$ proton with
surfactant concentration in CCL$_4$ (Acetic acid: 0.12$\overline{5}$ m).
O: DDAC, $\odot$: DDAB, $\bullet$: DDAI, $\triangle$: TDAC, $\square$: DAP,
$\blacksquare$: DP-8.

Shift to higher and lower field was observed for acetic acid
and methanol, respectively. The apparent marked shift by the presence of DP-8 in Figure 7 seems to be due to exchange of proton
between OH of acid and OH of DP-8. Hence the case was excluded from
the comparison with other surfactants. The order of shift that
DDAC>DDAB>DDAI and that DDAC>TDAC for acetic acid and methanol
corresponds with the result of interaction by the gas chromatography.

The spectrum of proton of 1-propylamine in the presence of DAP
was hidden by DAP spectra, but the characteristic broadening of NH_3^+
proton of DAP was observed in the presence of a small quantity of
1-propylamine as shown in Figure 8. This shows strong interaction
between DAP and 1-propylamine.

Effect of Solubilizates on Hydrolysis

DAP has been utilized as a typical surfactant showing reversed
micellar catalysis.[2,5,8,9] Hence the effect of solubilizates on
catalysis of DAP was investigated. The result of the catalytic
effect on the rate of hydrolysis of p-nitrophenyl acetate in carbon
tetrachloride and toluene was depicted in Figures 9 and 10. DAP

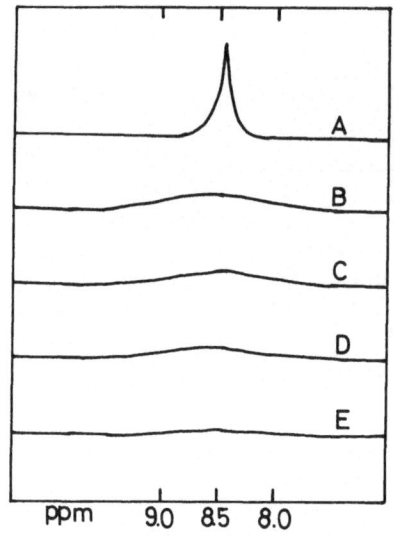

Figure 8. Line width of spectrum of NH_3^+ proton of DAP in the absence
or the presence of 1-propylamine in CCL_4 (DAP: 0.1 m). A: no amine.
B: amine: 0.009 m. C: amine: 0.026 m. D: amine: 0.045 m.
E: amine: 0.071 m.

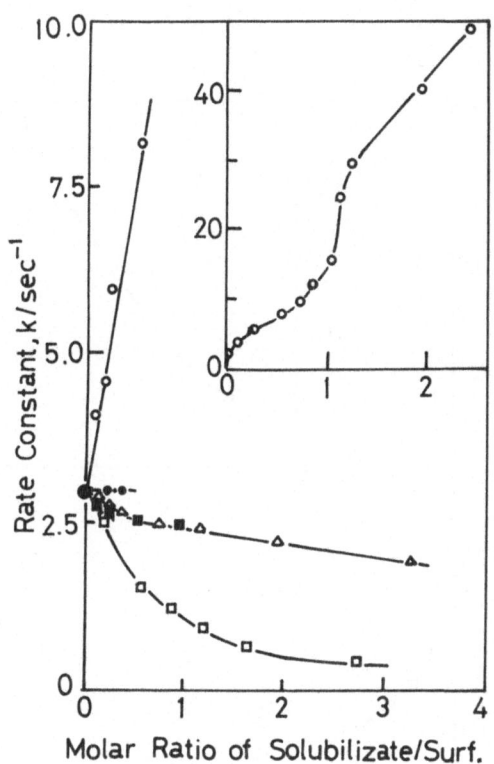

Figure 9. The effect of solubilizates on the rate of hydrolysis of p-nitrophenyl acetate in toluene solution of DAP at 30°C (DAP: 0.035 M). O: 1-propylamine, ▢: acetic acid, Δ: methanol, ■: acetone, ●: ethyl ether. Upper inserted figure has the same units.

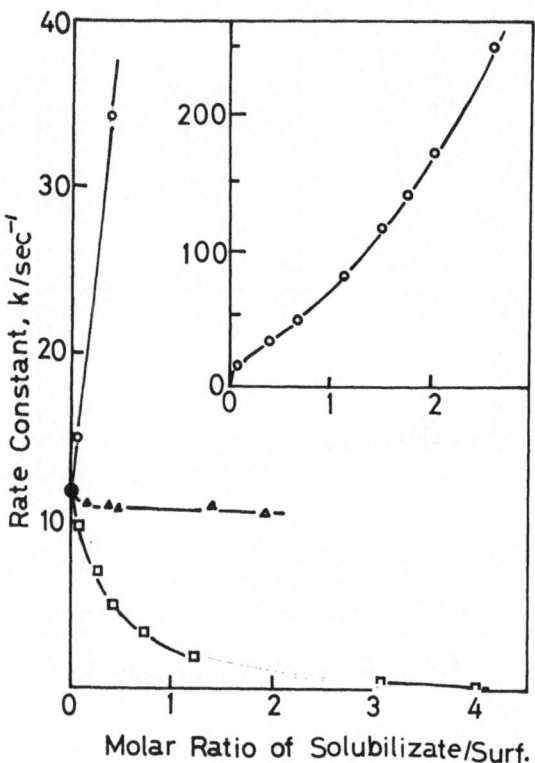

Figure 10. The effect of solubilizates on the rate of hydrolysis
of p-nitrophenyl acetate in CCl$_4$ solution of DAP at 35°C (DAP:
0.100 M).   O: 1-propylamine,   □ : acetic acid,   Δ: methanol.

concentration was fixed at 0.035 M in toluene and 0.100 M in CCl_4.
The inserted small figures show the range of higher rate constant
by 1-propylamine. The abscissa shows the amount of solubilized
substances which was calculated by subtracting the amount of free
substances from the total amount. The free amount was obtained from
the solubilization isotherm. The rate constant by propylamine in
toluene solutions was corrected for the contribution of free amine,
because free amine enhances the rate as shown in Figure 11 in which
the rate by amine in the presence or the absence of DAP was depicted.
Propylamine scarcely dissolved in CCl_4. Acetic acid and methanol
showed no effect on the rate over the concentration range studied
in the absence of DAP.

Figures 9 and 10 show marked differences existing among the
effect of amine, acid and alcohol. Methanol retarded the rate to a
minor extent* and acetic acid did to a great extent in the presence
of DAP. On the other hand, 1-propylamine enhanced the rate markedly.
1-octylamine showed behavior similar to that of 1-propylamine,
and propionic acid and lauric acid similar to that of acetic acid.
The effect of ether and acetone was also depicted in Figure 9.

Free amine may play aminolysis as mentioned by Menger et al.[10]
However, the rate enhancement shown by Figures 9 and 10 seems not
to be simple aminolysis accelerated by the co-existence of DAP. It
is suggested that the enhancement is the concerted effect of DAP and
1-propylamine, considering the strong interaction shown in Figure 8.

Solubilization and Catalysis

It was found out from the vapor pressure osometry that DAP in
CCl_4 initiates to form aggregate (small micelles) at 0.023 M and the
average aggregation number is 7. The concentration of DAP (0.100 M)
at which the rate study was carried out belongs to the micellar
region. It is the same in toluene solutions. Hence the interaction
between the solubilizate and the surfactant must take place in the
micelle. Since polar groups or interacting sites of the solubilizate and the surfactant are concentrated in the micelle, the
interaction behavior must vary widely depending on the pair of
solubilizates and surfactants. It reflects solubilization isotherm,
chemical shift or line width and the effect of the rate.

Amines interact strongly with the ammonium groups of DAP in
the micelle, resulting in line broadening of the NH_3^+ proton as shown
in Figure 8. This strong interaction can give rise to concerted

---

*Menger et al. have reported that methanol quenches catalysis of
trihexylammonium bromide.[10]

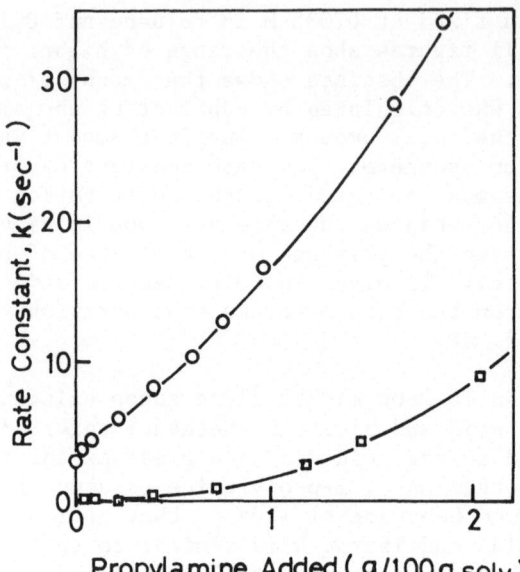

Figure 11.   The effect of 1-propylamine added on the rate of p-nitro-phenyl acetate in toluene at 30°C.   0: in the DAP solution (0.035 M);   □ : in the no DAP solution.

phenomena for rate enhancement, providing easier protonation and nucleophilic reaction.

Acetic acid solubilized in DAP micelles may block the ammonium group and retard the rate by prohibiting protonation of DAP. Metha-nol is also solubilized in DAP micelle, but it cannot have as much effect as shown by usual acid or base because of its very weak acidity.

ACKNOWLEDGMENTS

The excellent technical assistance of Mr. Fujiwara, Mr. S. Fujimoto, and Mr. K. Fujino in this laboratory is acknowledged.

REFERENCES

1.   S. Friberg and S. I. Ahmad, J. Phys. Chem., 75, 2001 (1971).
2.   J. H. Fendler, E. J. Fendler, R. T. Medary and V. A. Woods, J. Amer. Chem. Soc., 94, 7288 (1972).

3. C. J. O'Connor, E. J. Fendler and J. H. Fendler, J. Amer. Chem. Soc., 95, 600 (1973); 96, 370 (1974).
4. F. M. Menger, J. A. Donohue and R. F. Williams, J. Amer. Chem. Soc., 95, 288 (1973).
5. M. Seno, S. Shiraishi, K. Araki and H. Kise, Bull. Chem. Soc. Japan, 48, 3678 (1975); 49, 899 (1976).
6. A. Kitahara and K. Kon-no, J. Colloid Interface Sci., 29, 1 (1969); 35, 409 (1971).
7. S. Kaufman, J. Colloid Interface Sci., 25, 401 (1967).
8. K. Kon-no, K. Miyazawa and A. Kitahara, Bull. Chem. Soc. Japan, 48, 2955 (1975).
9. K. Kon-no, T. Matsuyama, H. Mizuno and A. Kitahara, Nippon Kagaku Kaishi, 1857 (1975).
10. F. M. Menger and A. C. Vitale, J. Amer. Chem. Soc., 95, 4931 (1973)

LIGAND EXCHANGE REACTIONS OF HEMIN AND VITAMIN $B_{12}a$ IN THE PRESENCE OF SURFACTANTS IN WATER AND IN NONPOLAR SOLVENTS

Janos H. Fendler

Department of Chemistry

Texas A&M University, College Station, TX 77843

Hemin is present as monomers in alcohol, aqueous micellar hexadecyltrimethylammonium bromide, CTAB, and sodium dodecyl sulfate, NaLS, as well as in the polar solvent pools in cyclohexane and in benzene, solubilized by polyoxyethylene(6) nonylphenol, Igepal CO-530. Conversely in water hemin exists as dimers. There are significant and specific micellar effects on the ligand exchange reactions of monomeric hemin with imidazole and cyanide ion. Vitamin $B_{12}a$ is completely insoluble in nonpolar solvents. In the presence of surfactants, capable of forming reversed micelles vitamin $B_{12}a$ is entrapped in the surfactant solubilized water pool and is shielded from the bulk apolar solvents by some 300 surfactant molecules. Depending on the size of the surfactant solubilized water pool, the effective polarity of the environment of vitamin $B_{12}a$ can vary considerably. Rate constants for the formation and decomposition of glycine, imidazole, azide and cysteine adducts of vitamin $B_{12}a$ in micellar environments differ dramatically and specifically from those in bulk water. Ligand exchange reactions with functionalized thiols are also affected. These data are discussed.

INTRODUCTION

Surfactant aggregates in water, micelles, and those in organic solvents, referred to as reversed or inverted micelles, have been extensively utilized as media for reacting substrates.[1-10]

Rates of numerous reactions are enhanced or retarded in micellar
environments with respect to those in water. These rate effects
are the consequence of the solubilization of the substrate by the
micelle and by differential reactivities in the micellar pseudo-
phase and in the bulk solvent phase. The increasing interest in
reaction kinetics in micellar systems is justified by the proposed
analogies between micelle-substrate and protein-substrate interactions as well as those between enzymatic and micellar catalysis.

The major thrust of our own research has been directed toward
the construction of increasingly more sophisticated model systems
which mimic the increasingly more complex functions of biochemical
ensembles. Our *modus operandi* has been that of judicious progress
from simple models and simple substrates to the more complex ones.
Thus, initially, we have studied simple hydrolyses and nucleophilic substitutions in simple micelles in aqueous solutions. Surfactant aggregates in nonpolar solvents reversed or inverted
micelles, have been our second generation model systems. The most
important properties of reversed micelles are that they are capable
of solubilizing water and polar substrates in their interior and
that substrate reactivities in these surfactant entrapped water
pools are often dramatically and specifically affected.[1] In
reversed micelles the polar substrate is isolated from the bulk
organic solvents by a layer of hydrocarbon; the surfactant aggregate. In this respect reversed micelles can be considered to be
simple membrane models. Recently we have initiated studies in
spherical single and multicompartment liposomes, which represent
our most sophisticated model system. Liposomes are smectic
mesophases of phospholipids in aqueous solutions into which substrates can be entrapped.[11-13] Here the substrate is localized
in an aqueous interior, surrounded by a hydrocarbon bilayer which
isolates it from bulk water. Liposomes are, of course, increasingly being utilized as membrane models.[11-13]

Our second generation substrates are represented by hemin and
vitamin B_{12}. Thus, our investigation of ligand exchange reactions
at these substrates in the presence of surfactants in water and in
organic solvents will serve as an overview of our current approach
to biochemical modelling. The present paper will discuss critical-
ly the effects of micelles on ligand exchange reactions at hemin
and at vitamin B_{12a} in addition to presenting new data and offering
some speculative generalizations.

Hemin (I), vitamin B_{12}(II) and their derivatives have important
biochemical roles. Hemoglobin, myoglobins, catalases, peroxidases
all contain the iron porphyrin moiety.[14-19] Kinetic investigations
of ligand exchange reactions involving hemin have been carried out
under a variety of experimental conditions[20-22] in order to obtain
insight into the biochemical roles of the more complex hemoproteins.

Hemin

Structure I

Structure II

Although these studies have provided important information, the
microscopic environments of the reaction center could not be assessed. Electrostatic, hydrophobic effects, stereochemical factors
and polarities at the active sites of hemoproteins can reasonably
be expected to be different than those observed in simple model
systems. Additionally hemin is present as dimers in water and
experiments with monomeric hemins had to be, therefore, carried out
in alcohol-water mixtures,[21-22] and in aqueous dimethylsulfoxide.[24-25]
Alternatively surfactants both in water[26-29] and in nonpolar solvents
provide media[30] in which hemin exists as monomers and which may
approximate the microenvironments of hemoprotein.

Rate constants for anation, governed by k_1, and those for
aquation, governed by k_{-1}, of vitamin $B_{12}a$, aquocobalamin
(bzm-Co-OH$_2$) have been determined for the ligands (L) N_3^-, OCN^-,
SCN^-, SO_3^{2-}, NCO^-, I^-, B_2, imidazole and glycine:[31-40]

$$bzm\text{-}Co\text{-}OH + L \xrightarrow[k_{-1}]{k} bzm\text{-}Co\text{-}L + H_2O \qquad (1)$$

Interest in vitamin B_{12} mediated enzymatic processes has prompted
these investigations. It is suggested that micelles approximate
the microenvironment of the seat of substitution *in vivo* better
than dilute aqueous solutions.

INTERACTIONS AND REACTIONS OF HEMIN

The common feature of micelle forming surfactants both in
water and in organic solvents is that they interrupt the association
of hemin. Apparently, hydrophobic interactions between hemin and
micelle overcome those between hemins. In aqueous anionic and
uncharged micelles, hemin is predominantly solubilized at the
micelle-water interface.[27] Penetration of hemin into cationic
micelles is deeper, however. Conversely, in reversed micelles
hemin is localized in the surfactant entrapped polar solvent pools,
both in benzene and in cyclohexane. This is hardly surprising
since hemin is completely insoluble in benzene or in cyclohexane
but it readily dissolves in water. Shifts in the absorption
maxima of the dicyanohemin complex, a chemically stable species,
in different solvents have been found to correlate well with the
solvent polarity parameter Z. Using this correlation as a ruler,
apparent environments of the surfactant solubilized hemin has been
established (Figure 1).[30]

It is seen that the effective polarity of hemin in sodium
dodecyl sulfate (NaLS) and Triton-X resembles that in water, while
that in hexadecyltrimethylammonium bromide (CTAB) is similar to

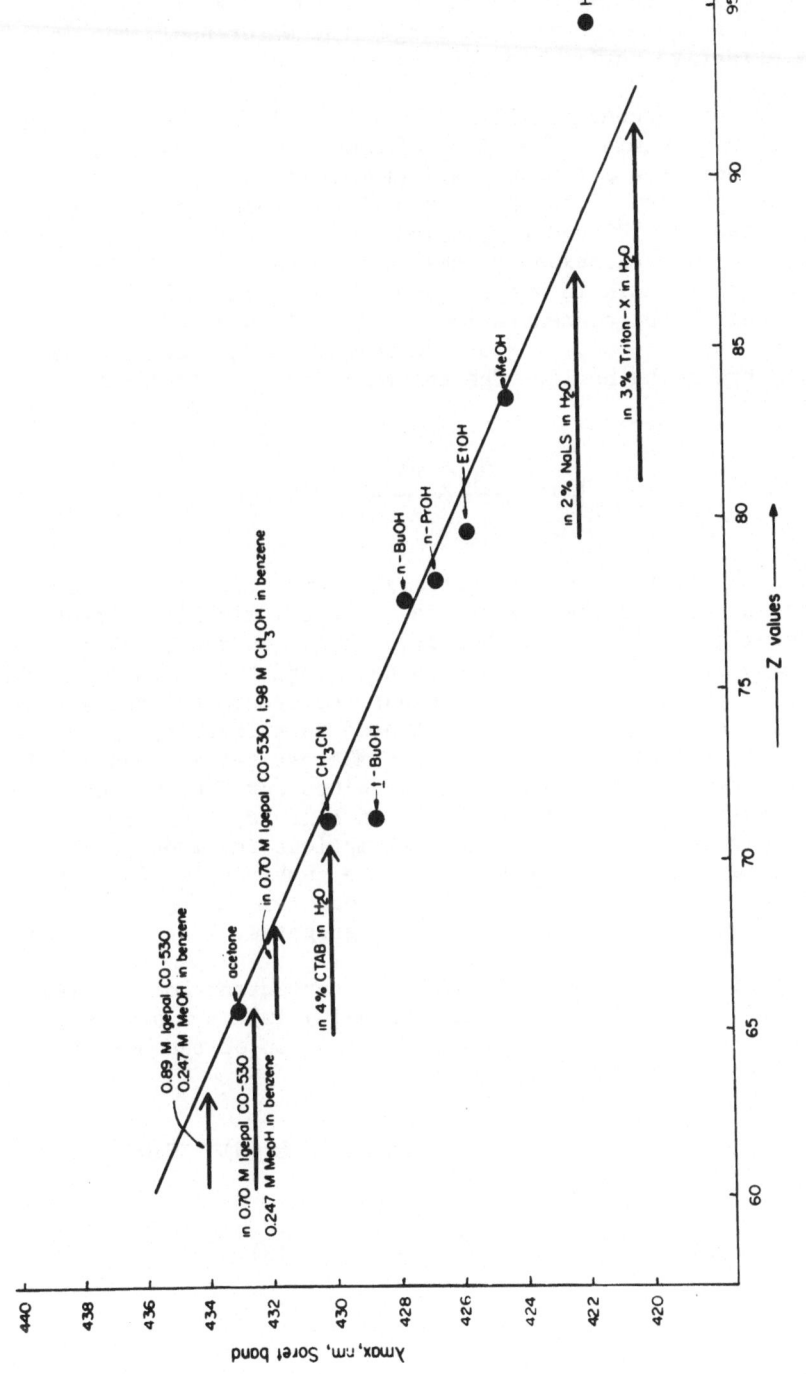

Figure 1. Correlation between the absorption maxima of the Soret band of the isolated hemin(CN)₂ complex and solvent polarity parameter Z.

the polarity of methanol. The microscopic environment of hemin in reversed micellar Igepal CO-530 depends on the size of the surfactant solubilized water pool. The larger the water pool, the more polar the effective environment of hemin.

The size of aqueous micelles increases only slightly upon the solubilization of hemin. The sizes of surfactant aggregates in nonpolar solvents are profoundly affected by substrate solubilization. In the absence of solubilizates the average aggregation number of Igepal CO-530 and alkylammonium carboxylate surfactants in benzene and in cyclohexane ranges between 4-10.[1] Solubility determinations of hemin in benzene and in cyclohexane in the presence of different concentrations of Igepal CO-530 have been utilized for the determination of the number of surfactant molecules, N, surrounding hemin. The concentration of micelles, M, is given[2] by

$$[M] = \frac{C_D - CMC}{N} \qquad (2)$$

where C_D is the stoichiometric surfactant concentration of the surfactant, and CMC is the operational critical micelle concentration. Assuming complete insolubility of hemin in the organic solvent and a 1:1 interaction between hemin and the surfactant, the solubility in given surfactant solutions represents the concentration of micelles. The hemin is being used to titrate the micelles. Plotting hemin solubilities against C_D-CMC resulted in good straight lines. From the slopes of these lines values for N have been calculated to be 2100±400 and 4000±800 in benzene and in cyclohexane, respectively. Although the assumptions involved in this calculation may not be entirely valid, the number of surfactant molecules needed to solubilize hemin in nonpolar solvents is considerably greater than the range of aggregates.

Interactions of ligands, L, with the surfactant solubilized hemin, hemin(OH$^-$,H$_2$O)(S) or hemin(MeO$^-$,MeoH)(S) (where S stands for surfactant), result in the equilibrium formation of the hemin(L)$_2$(S) complex in two consecutive steps:

$$\text{hemin(OH}^-\text{,H}_2\text{O)(S)} + \text{L} \underset{k_{-1}}{\overset{k_1}{\rightleftharpoons}} \text{hemin(OH}^-\text{,L)(S)} + \text{H}_2\text{O} \qquad (3)$$

$$\text{hemin(OH}^-\text{,L)(S)} + \text{L} \underset{k_{-2}}{\overset{k_2}{\rightleftharpoons}} \text{hemin(L) (S)} + \text{OH}^- \qquad (4)$$

If the ligand is present in protonated and unprotonated forms, equation 5 needs also to be considered:

$$L + H^+ \rightleftharpoons LH^2 \tag{5}$$

The kinetic rates are related to the apparent (i.e., pH dependent) equilibrium constant by equation 6:

$$K_{app} = k_1 k_2 / k_{-1} k_{-2} \tag{6}$$

and the observed rate constant for product formation, k_ψ, assuming the validity of steady state approximation, (i.e., $d[\text{hemin}(\overline{OH},L)(S)]/dt = 0$), is related to the ligand concentration by equation 7:

$$k_\psi = \frac{k_{-2}(1 + K_{app}[L]^2}{1 + (k_2/k_{-1})[L]} \tag{7}$$

At relatively high ligand concentration equation 7 simplifies to equation 8:

$$k_\psi = K_{app} k_1 [L] \tag{8}$$

while at low ligand concentration it reduces to equation 9:

$$k_\psi = k_{-2} \tag{9}$$

K_{app} can be determined from spectrophotometric data using equation 10:

$$\frac{Aun - A}{A - Ac} = \frac{1}{K_{app}[L]^2} \tag{10}$$

where Aun, A, and Ac are absorbances due to hemin in the absence of the ligand, that in the presence of different amounts of ligand, and that for the complex, hemin$(L)_2(S)$ at the appropriate wavelength. Figure 2 illustrates treatment of the data for the

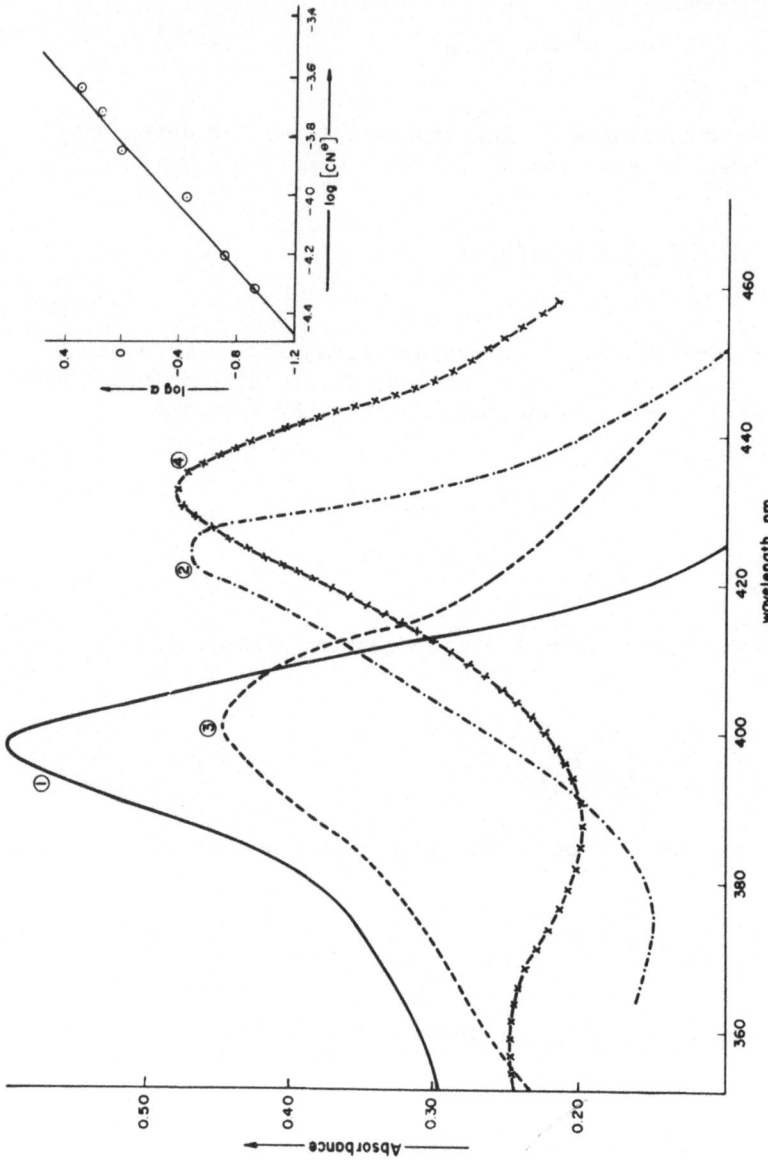

Figure 2. Absorption spectra of 6.0×10^{-6}M-hemin, containing 1.0×10^{-4}M-sodium methoxide: (1) in neat methanol; (2) in 2.8×10^{-3}M-NaCN in methanol; (3) in 0.70M-Igepal CO-530 and 7.2×10^{-4}M-NaCN in benzene. Data for the interaction of hemin with sodium cyanide in the 0.70M-Igepal CO-530 benzene system at 432.5 nm are plotted according to equation 10 in insert.

interaction of hemin with cyanide ion in reversed micellar Igepal CO-530 in benzene.

The slope in the insert of Figure 2 is 2.2 0.2; thus it substantiates the involvement of two molecules of ligand per hemin. Tables I and II summarize the available data for the interaction of hemin with cyanide ion and with imidazole. The salient points of these data are as follows:

(1) In micelles hemin is present as monomers.

(2) Dissociation constants of hemin and the ligand are altered in the micellar environment.

(3) The order of stability of the $hemin(CN)_2$ in the different media is NaLS < MeOH < Triton X-100 < Igepal CO-530 in cyclohexane < CTAB; while that of $hemin(imidazole)_2$ is Igepal CO-530 in cyclohexane and in benzene < CTAB < EtOH < NaLS. Electrostatic interactions, changes in pKa values of the reactants and that of their microenvironments are responsible for the reversal of the trend in the stabilities of these hemin complexes.

Table I. Kinetic and Thermodynamic Parameters for the Interaction of Hemin with Cyanide Ion at 24.0°

Conditions	$k_1, M^{-1}sec^{-1}$	k_{-2}, sec^{-1}	k_{app}, M^{-2}
MeOH, 1.0×10^{-4}M MeOH	380	5.0	21
4% CTAB, H_2O^a	11200		73000
2% NaLS, H_2O^a	3400	0.057	15.4
3% Triton X-100, H_2O^a	510		19.4
1.35M Igepal CO-530 in benzene, 5% MeOH (v/v), 1.0×10^{-4}M NaOMe[b]	90	0.0024	200

[a] Taken from reference 29.
[b] Taken from reference 30.

Table II.   Kinetic and Thermodynamic Parameters for the
            Interaction of Hemin with Imidazole at 25.0°

Conditions	$k_1, M^{-1}sec^{-1}$	k_{-2}, sec^{-1}	K_{app}, M^{-2}
EtOH-H$_2$O(8.9:11.1 w/w)[a]		8.6×10^6[b]	4.0×10^7[c]
0.05 M CTAB, H$_2$O		1.2×10^5[b]	1.0×10^6[c]
0.50 M NaLS, H$_2$O		1.0×10^7[b]	5.0×10^9[c]
0.25 M Igepal CO-530, cyclohexane[d]	3.4×10^3	5.0	8.0×10^2
0.80 M Igepal CO-530, cyclohexane	1.7×10^4	42	1.25×10^5
1.42 M Igepal CO-530, cyclohexane			8.0×10^5
0.69 M Igepal CO-530, cyclohexane[e]			1.1×10^7
0.80 M Igepal CO-530, cyclohexane[e]			1.4×10^6

[a] Estimated from data in reference 21 and 22.
[b] pH independent equilibrium constant, assuming
    both protonated and unprotonated imidazole as
    reactive species.
[c] pH independent equilibrium constants, assuming
    only the unprotonated imidazole as reactive
    species.
[d] Containing 0.082 M MeOH and 1.0×10^{-3}M MeONa.
[e] Containing 0.14 M H$_2$O and 3.0×10^{-4}M NaOH.

INTERACTIONS AND REACTIONS OF VITAMIN B$_{12}$

     Like hemin, vitamin B$_{12}$ is completely soluble in water and
completely insoluble in benzene and in cyclohexane. Unlike hemin,
vitamin B$_{12}$ is, however, not solubilized by simple surfactants in
aqueous solutions. Using a solvent polarity correlation similar
to that for hemin (Figure 1) microscopic polarities of vitamin B$_{12}$
in 0.50 M aqueous NaLS and in 5.0×10^{-2} M aqueous CTAB have been
established to be aqueous-like.[40] Not unexpectedly, therefore,

aqueous micellar CTAB and NaLS have only modest effects on the rate constants for the formation, k_1, and decomposition, k_{-1}, (in equation 1) of the glycine, imidazole and sodium azide adducts of vitamin $B_{12}a$ (Table III).[40] These micellar effects can be satisfactorily rationalized in terms of electrostatic effects and changes of dissociation constants of the different species.

Table III. Effects of Aqueous Micelles on the Apparent Kinetic and Thermodynamic Parameters for Ligand Exchange Reactions at Vitamin $B_{12}a$

Reaction	$\dfrac{k_1^{micelle}}{k_1^{H_2O}}$	$\dfrac{k_{-1}^{micelle}}{k_{-1}^{H_2O}}$	$\dfrac{K^{micelle}}{K^{H_2O}}$
Vitamin $B_{12}a$ + NaN_3			
5.0x10^{-2}M CTAB	0.55	1.8	0.46
5.0x10^{-1}M NaLS	0.078	0.52	0.25
Vitamin $B_{12}a$ + imidazole			
5.0x10^{-2}M CTAB	0.94	1.06	0.89
5.0x10^{-1}M NaLS	0.12	0.32	0.43
Vitamin $B_{12}a$ + glycine			
5.0x10^{-2}M CTAB	0.28	0.46	0.61
5.0x10^{-1}M NaLS	0.16	0.11	0.63
Vitamin $B_{12}a$ + L-cysteine			
0.10 M CTAB	0.83	25.5	0.033
0.10 M NaLS	0.66	3.41	0.19

The rate constant for anation of vitamin $B_{12}a$ by L-cysteine and that for the aquation of vitamin B_{12}-L-cysteine complex is only slightly affected by aqueous micellar CTAB and NaLS (Table III). There are more appreciable micellar effects, however, on the interaction of N-decanoyl-L-cysteine with vitamin $B_{12}a$. In 0.10 M CTAB k_1 and k_{-1} are 190 and 25 fold smaller than those in water respectively. Formation of mixed micelles in which the thiol group and the vitamin B_{12}-complex are less available for reaction is presumably the reason for the decreased reactivity.

Vitamin $B_{12}a$ is solubilized, however, by dodecylammonium propionate, DAP, aggregates in benzene. Solubility determinations have established that vitamin B_{12} is surrounded by some 300 molecules of DAP.[40] Solvent polarity correlations have indicated the environment of vitamin B_{12} in DAP solubilized water pools in benzene

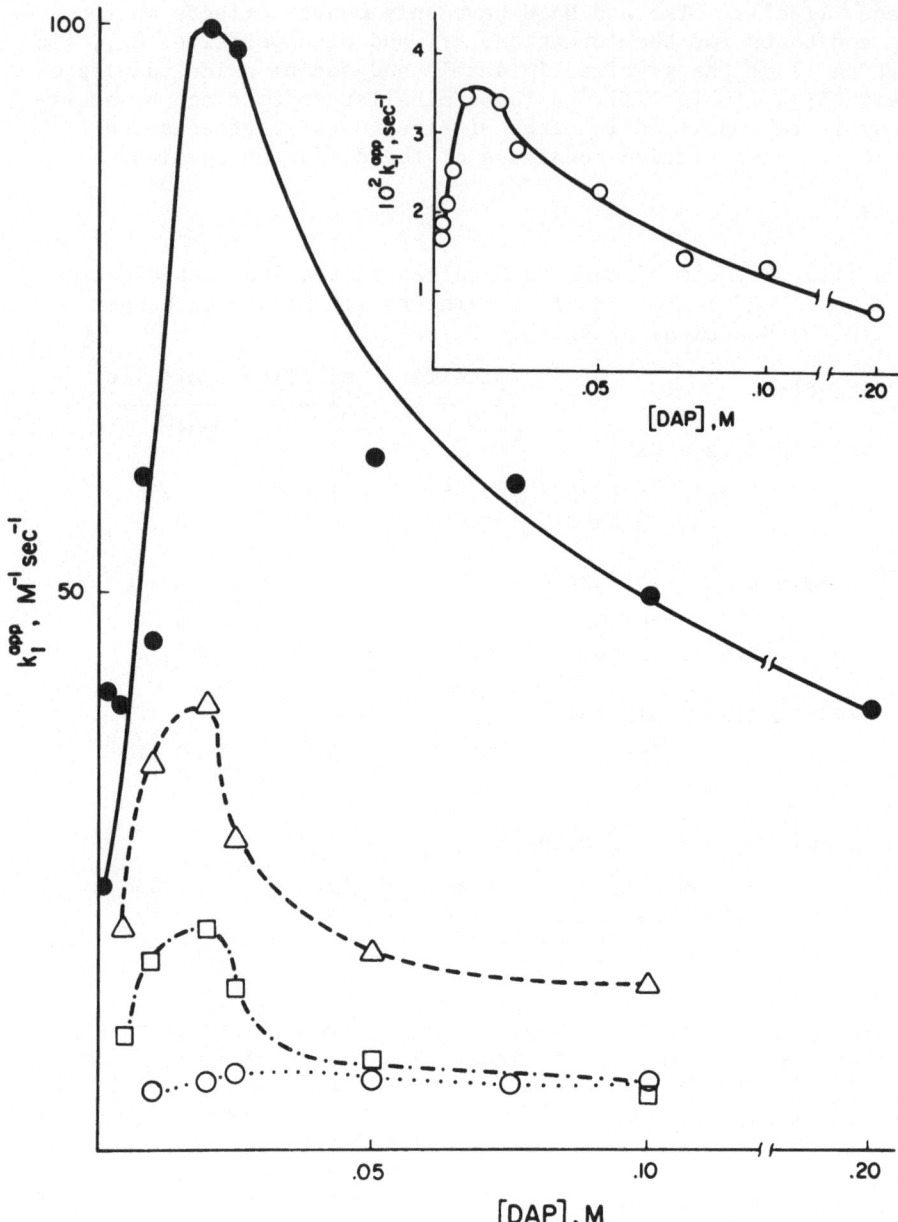

Figure 3.   Plot of k_1^{app} *vs.* DAP concentration for the interaction
of vitamin B_{12a} with glycine in benzene in the presence
of (●)0.010 M, (Δ)0.031 M, (□) 0.05 M, and (o) 0.10 M
solubilized water at 25.0°.   Insert shows a plot of k_{-1}^{app}
vs. stoichiometric DAP concentration for the same reaction
in benzene in the presence of 0.010 M solubilized water.

to vary, depending on the size of the water pool, from that being as polar as benzene to that being as polar as water. Rate constants for ligand exchange reactions of vitamin B_{12a} in this media differ dramatically from those in B_{12a} in bulk water.[40] Although a direct comparison is not straight forward,[40] the apparent rate constant for anation of vitamin B_{12a} by glycine in benzene in the presence of 0.02 M DAP and 0.01 M H_2O is 66,000-fold larger than that in pure water. The magnitude of rate enhancements depend on the ligand and on the concentration of cosolubilized water. Rate enhancements are most pronounced when the size of the DAP solubilized water pool is the smallest. Figure 3 illustrates the kinetic rate profiles for k_1 as functions of DAP concentrations in the presence of different amounts of cosolubilized water. The unique nature of the surfactant entrapped water pool, concentration and favorable orientation of the reactants, dipole-dipole interactions, ease of proton transfer are factors responsible for the observed rate effects in reversed micellar systems.[1]

CONCLUSIONS AND SPECULATIVE GENERALIZATIONS

The wheels of basic science, even these days, are turning relatively slowly. The relevance of probing some minor and often seemingly esoteric aspects of a complex picture is not always obvious. Likewise, the validity of chosen models mimicking biochemical functions is not always apparent. The chemist's approach is to extend the knowledge he gained in investigating simple molecules to increasingly larger and more complex ones. The biochemist and biophysicist, on the other hand, apply increasingly more sophisticated experimental techniques to biomacromolecules. The conceptual gap between these two approaches is getting narrower. Significant breakthroughs in understanding molecular biochemistry has mostly come from interdisciplinary approaches. The rejuvenation of colloid chemistry as an "interfacial playground" for organic, biological, physical and pharmaceutical chemists has contributed significantly to our present understanding of life. Pandora's box thus opened, will oblige us scientists to continue searching the mysteries of nature. These are some of the philosophies, we like to believe, which have governed our overall research program. The present contribution on the effects of micelles on ligand exchange reactions of hemin and vitamin B_{12} constitutes, it is felt, a small but essential part of the larger vistas.

Although no single model is perfect or is able to mimic all aspects of the biological ensemble, aqueous micelles, reversed micelles and liposomes have provided useful informations. It is relatively easy to design and execute experiments in these systems and they have the desired flexibility and built-in expandability.

The number of parameters involved in catalyses in reversed micelles are numerous and are as yet incompleted, understood.[1] Ligand exchange reactions at hemin and vitamin B_{12} are indeed different in the micellar environments than in the bulk solvents. The effects are specific and substantial. As our understanding of the models themselves and of the substrate interactions therein improves, the design of more pertinent models mimicking specific biochemical functions becomes possible. Additionally, it will become feasible to extend these investigations to larger porphyrins as well as to proteins.

We have initiated investigations of methyl transfer within the surfactant entrapped water pool and across structurally different surfactant bilayers. Our emphasis is on the elucidation factors which influence biological methylation. These studies are being extended to liposomes.

ACKNOWLEDGMENTS

Thanks are due to my coworkers whose names are given in the references and without whom this work would not have been possible. Support of this work by the Robert A. Welch Foundation, the National Science Foundation and the Energy Research and Development Administration is gratefully acknowledged.

REFERENCES

1. J. H. Fendler, Accounts Chem. Res., 9, 153(1976).
2. J. H. Fendler and E. J. Fendler, "Catalysis in Micellar and Macromolecular Systems:, Academic Press, New York, 1975.
3. W. P. Jencks, Adv. Enzymol. 43, 219 (1975).
4. I. V. Berezin, K. Martinek, and A. K. Yatsimirskii, Russ. Chem. Rev., 42, 787 (1973).
5. C. A. Bunton, Prog. Solid Stat Chem., 8, 239 (1973).
6. E. H. Cordes and C. Gitler, Prog. Bioorg. Chem., 2, 1 (1973).
7. E. H. Cordes, "Reaction Kinetics in Micelles", Plenum Press, New York, 1973.
8. T. C. Bruice, "Enzymes" 3rd ed., 2, 217 (1970).
9. E. J. Fendler and J. H. Fendler, Adv. Phys. Org. Chem., 8, 271 (1970).
10. E. H. Cordes and R. B. Dunlap, Accounts Chem. Res., 2, 329 (1969).

11. D. Papahadjopoulos and H. K. Kimelberg, Prog. Surf. Sci., 4, 141 (1973).
12. A. D. Bangham, Ann. Rev. Biochem., 41, 753 (1972).
13. A. D. Bangham, Prog. Biophys. and Mol. Biol., 18, 29 (1968).
14. R. J. Sundberg and R. B. Martin, Chem. Rev., 74, 513 (1974).
15. E. Antonini and M. Brunori, "Hemoglobin and Myoglobin and their Reactions with Ligands", North Holland Co., Amsterdam, 1971.
16. G. S. Marks, "Heme and Chlorophyll; Chemical, Biochemical and Medical Aspects", Van Nostrand Reinhold Company. London, 1969.
17. J. E. Falk, "Porphyrins and Metalloporphyrins", Elsevier, New York, 1964.
18. A. P. Adler, Ann. NY Acad. Sci., 206 (1973).
19. J. N. Phillips, Rev. Pure Appl. Chem., 10, 35 (1960).
20. E. B. Fleischer, S. Jacobs and L. Mestichelli, J. Amer. Chem. Soc., 90, 2527 (1968).
21. N. B. Auperman, B. B. Hasinoff, H. B. Dunford, and R. B. Jordan, Canad. J. Chem., 47, 3217 (1969).
22. B. B. Hasinoff, N. B. Dunford, and D. G. Howe, Canad. J. Chem., 47 3225 (1969).
23. T. H. Davis, Biochim. Biophys. Acta, 329, 108 (1973).
24. N. Ellfolk and K. Mattsson, Suomen Kem., B42, 319 (1969).
25. G. Cauquis and M. Georges, Bioelectrochem. Bioenerg., 1, 23 (1974).
26. J. Simplicio, K. Schwenzer, and F. Maenpa, J. Amer. Chem. Soc., 97, 7319 (1975).
27. J. Simplicio, Biochemistry, 11, 2524 (1972).
28. J. Simplicio, Biochemistry, 11, 2529 (1972).
29. J. Simplicio and K. Schwenzer, Biochemistry, 12, 1923 (1973).
30. W. Hinze and J. H. Fendler, J. C. S. Dalton Trans., 238 (1975).
31. J. M. Pratt, "Inorganic Chemistry of Vitamin B$_{12}$", Academic Press, New York, 1972.
32. D. G. Brown, Prog. Inorg. Chem., 18, 177 (1973).
33. G. N. Schrauzer, Pure Appl. Chem., 33, 545 (1973).
34. W. C. Randall and R. A. Alberty, Biochemistry, 5, 3189 (1966).
35. W. C. Randall and R. A. Alberty, Biochemistry, 6, 1520 (1967).
36. D. Thusius, Chem. Commun., 1183 (1969).
37. J. G. Heathcote and M. A. Slifkin, Biochim. Biophys. Acta, 158, 167 (1968).
38. J. G. Heathcote, G. H. Moxon, and M. A. Slifkin, Spectrochim Acta, A27, 1391 (1971).
39. D. Thusius, J. Amer. Chem. Soc., 93 2629 (1971).
40. J. H. Fendler, F. Nome and H. C. VanWoert, J. Amer. Chem. Soc., 96, 6745 (1974).
41. Y. Yanagi, I. Sekuzu, Y. Orii, and K. Okunulsi, J. Biochem., 71, 47 (1972).
42. N. Sutin, and J. K. Yandell, J. Biol. Chem., 247, 6932 (1972); A. Schejter and I. Aviram, Biochemistry, 8, 149 (1969).
43. N. Nanzyo and S. Sano, J. Biol. Chem., 243, 3431 (1968).

DISCUSSION

On the paper by F. M. Fowkes, D. Z. Becher,
M. Marmo, C. Silebi, and C. C. Chao

J. H. Fendler, *A & M University*: (1) Could you describe the experimental detail for the determination of Hamett's acidity function?
(2) How could you carry out experiments in Aerosol-OT Surfactants
in benzene at 160°?

F. M. Fowkes: (1) See L. P. Hammett and A. J. Deyrup, J. Amer.
Chem. Soc., 54, 2721 (1932).
(2) Benzene was used up to 45°C, n-decane to 126°C, but other
hydrocarbons can be used up to much higher temperatures.

A. Kitahara, *Science University of Tokyo*: (1) What is the content
of water in this study? Did you study the effect of water content?
(2) Is the solvent dodecane? What is the aggregation number of
the Na-salt in the solvent? Do you have any results of the solvent
effect?

F. M. Fowkes: (1) Much effort was spent in drying catalysts and
pumping off water vapor from freeze-dried catalysts at 110-115°C
under vacuum markedly increased the acid catalytic properties.
(2) The solvent is n-decane. The aggregation number of the Na-salt
in the solvent is 12 (at 75°C). I have no results of the solvent
effect.

Part VI

Microemulsions

Part VI

Microemulsions

THEORY FOR THE PHASE BEHAVIOR OF MICROEMULSIONS

Max L. Robbins

Exxon Research and Engineering Company

Linden, New Jersey 07036

General trends in water and oil uptake in saturated microemulsions are correlated by idealized ternary diagrams. Winsor's types of saturated microemulsions are assigned phase regions on the idealized diagram for the pseudo-3-component system: surfactant, oil and aqueous solution. Systematic shifts in these phase regions reflect changes in temperature, salinity, oil composition, surfactant head and chain size and other HLB parameters. Shinoda's phase diagrams for nonionic surfactants are explained by the idealized ternary diagrams stacked along a temperature axis. The systematic transition in microemulsion type with temperature is shown with stacked ternaries. Saturation water and oil uptake in microemulsions made with ethoxylated alkyl phenols are related to the ternary diagram. Water uptake increases and oil uptake decreases with increasing head/chain volume ratio and decreasing temperature or aromatics/paraffinics ratio in the oil. The interchangeability of the parameters surfactant head/chain volume ratio, surfactant/co-surfactant ratio (HLB), temperature, oil composition and aqueous phase salinity for controlling saturation water and oil uptake is demonstrated. A model is developed which quantitatively predicts saturation water and oil uptake as a function of surfactant head/chain volume ratio. The effect of oil composition on theoretical parameters is discussed to provide a basis for predicting shifts in water and oil uptake with changes in temperature and composition. The theory also correlates microemulsion/bulk phase interfacial tensions with saturation water and oil uptake and explains Shinoda's observed minimum in interfacial tension with temperature.

INTRODUCTION

It is well-known that microemulsions*[1-11] or solubilized systems*[12-17] can exist in equilibrium with excess oil, water or both[14]. Winsor[14,15] referred to these respective equilibria as Types I, II and III. He showed qualitatively that the transitions I $\rightleftharpoons$ III $\rightleftharpoons$ II are dependent on the hydrophilic vs. lipophilic character of the surfactants (HLB)[25], salinity, oil composition and temperature. Shinoda and co-workers experimentally defined these transitions as functions of temperature[26-28] and ethylene oxide content[20] for co-solubilized oil and water systems prepared with non-ionic surfact-ants. They described the single/2-phase boundaries for Type I [29] and Type II[30] systems on temperature-composition diagrams. They studied the shift in these boundaries with variation in HLB (ethy-lene oxide content, hydrocarbon chain length and added anionic surfactant), oil type and added salts[29-31]. Similar effects were noted by Kon-no and Kitahara for the solubilization of water and aqueous salt solutions in non-aqueous media (Type II systems) by cationic[32], non-ionic[33] and anionic[34,35] surfactants.

Robbins[36] quantized Type II systems by the saturation water uptake defined as the volume ratio of water to surfactant (V_w/V_s) in a microemulsion saturated with water. This quantity was shown both experimentally and theoretically to be a function of the volume fraction of heads in the surfactant (HLB). Saturation water uptake by ethoxylated octyl phenols was shown to depend on oil type. This author[37-39] extended the treatment to Type I systems defining saturation oil uptake (V_o/V_s) as a function of the volume fraction of chains. His model predicts an inverse correlation between interfacial tension and V_w/V_s in Type II systems and V_o/V_s in Type I systems.

Healy, Reed and Stenmark[40] demonstrated that Types I, II and III systems made with anionic surfactants could be characterized in terms of V_w/V_s and V_o/V_s which they called "solubility numbers". They showed that the transition I $\rightleftharpoons$ III $\rightleftharpoons$ II with increasing salinity could be described by an increase in the relative magni-tude of V_o/V_s to V_w/V_s. Their experiments verified that inter-facial tensions in saturated microemulsions correlated inversely with V_w/V_s and V_o/V_s. Their results were consistent with Winsor's qualitative predictions on the I $\rightleftharpoons$ III $\rightleftharpoons$ II transition. These authors assigned phase regions to Winsor's types of systems on a simple ternary diagram for the pseudo-3-component system: surfact-ant, oil and aqueous solution.

---

*Solubilized systems are often referred to as (swollen) micellar solutions.[19-21,24,62] References 19-21 consider Schulman's micro-emulsions to be (swollen) micellar solutions. References 22-24 differentiate between microemulsions and micellar or co-solubilized systems. The present discussion is conceptually consistent with both points of view. Reference 18 gives a review of the area.

The purpose of the current paper is to generalize the simple or idealized ternary diagram to include a description of the phase behavior of microemulsions made with non-ionic surfactants. For this purpose, the temperature-composition phase diagrams of Shinoda and co-workers[18,20,28] are correlated with the idealized ternary diagram[40]. The functional dependence on temperature of phase volumes in equilibrated systems are explored with ternary diagrams. Phase volume data are used to generate curves of water and oil uptake (V_W/V_S and V_O/V_S) vs. temperature, HLB of the surfactant and aqueous salinity. Shifts in water and oil uptake with the parameters HLB, salinity, oil composition and temperature are correlated with ternary diagrams and are shown consistent with Winsor's qualitative predictions. Applying Winsor's assumption[15] that Type III systems can be considered as equilibrium mixtures of Types I and II systems, curves of saturation water and oil uptake vs. HLB and interfacial tension vs. water and oil uptake are generated by the Robbins' model.

EXPERIMENTAL METHODS AND MATERIALS

Materials, experimental techniques and data for temperature studies are given by Shinoda and co-workers[20,28]. Equilibrium phase studies to determine HLB and salinity dependencies were run and data analyzed as described by Healy et al.[40]. Six ml each of aqueous and oil phase were shaken with a given volume of surfactant(s) and allowed to phase separate (usually 2 to 4 weeks) at room temperature. Elongated test tubes were used to permit accurate determination of the respective phase volumes from measured heights.

All materials were used as received from the supplier. The effect of varying average ethylene oxide content in the surfactant mixture was studied using a pair of ethoxylated octyl phenols, Triton X15 (1 EO) and Triton X114 (7.5 EO) supplied by Rohm and Haas. The effect of varying salinity was determined on an ethoxylated dodecyl phenol nominally containing 6 ethylene oxide groups (Igepal RC 520 supplied by GAF). The oil phase was reagent grade hexadecane and the aqueous phase prepared with house-distilled water redistilled through a Kontes Model WS 2 unit.

The dependence on oil type was studied using a titration method. Fifty ml each of the oil and surfactant(s) were mixed and titrated at 25°C with water to turbidity. Surfactants were varying ratios of Triton X15 and Triton X114 described above. Reagent grade hexadecane, decane and mixed xylenes were used as received. In some instances (at higher average ethylene oxide content) the surfactants were not completely miscible with the oil phase and required the addition of a small volume of water to clear the system before it could be titrated to turbidity. With this method,

water uptake was measured directly rather than by difference as in
the equilibration technique described above. Uptakes were lower than
measured on equilibrated systems because of a tendency to under-
estimate the turbidity end point. Data from both techniques agree
within 10% when correlated using model equations.

Surface pressure-area isotherms were measured using the
Wilhelmi plate technique. Surfactant-hexadecane mixtures were
spread on water from solutions in pentane. The weight ratio of
hexadecane to surfactant (mixtures of Triton X15 and X114) was
fixed and the surface pressure (π)-area (σ) isotherm measured on
the mixed monolayer. The ratio of hexadecane to surfactant was
increased until no further expansion was observed in the π vs. σ
isotherm. A freshly spread monolayer was used for each determination. No change was observed in the π–σ isotherm for hexadecane/
surfactant ratios exceeding 20/1. Monolayer compressibility
evaluated at 20/1 hexadecane/surfactant was assumed representative
of the surfactant-hexadecane interface.

RESULTS, DISCUSSION AND CONCLUSIONS

Microemulsions or micellar solutions are usually prepared with
a combination of surfactants of low and high Hydrophilic-Lipophilic
Balance (HLB)[25]. Even when a "single" surfactant is used, it is
rarely pure enough to be considered a single component. Nevertheless, it is often useful to treat the system, surfactant(s), oil
and water (or aqueous salt solution) as a pseudo-3-component system
on a triangular diagram[15,17,40-44]. Figure 1 shows a typical
ternary diagram for 2 immiscible components (O, W) both mutually
miscible in the third (S)[45]. (An example of a system exhibiting
such behavior[46] is the solution of 2 salts in water at constant temperature.) The 2-phase regions represent microemulsion (M) in
equilibrium with excess water (left lobe) and oil (right lobe).
All compositions lying on a tie line will split into 2 phases whose
compositions lie at either end of the tie line. Since the tie
lines connect 2 liquids of constant composition, they represent
lines of constant interfacial tension. Tie lines are drawn to the
water (W) and oil (O) apices because surfactant concentrations in
in the excess water and oil phases, though finite, are often low
enough to ignore[40]. The 3-phase region contains all compositions
which split into a micellar phase of constant composition in
equilibrium with excess oil and water as required by the phase
rule. This particular micellar composition is called the isothermal
invariant point (IIP)[46]. The above representation is highly
idealized and accurately depicts real micellar solution behavior in
only a limited number of cases. Because of the multicomponent
nature of most micellar systems, the phase boundaries depend to
some extent on the surfactant concentration and oil/water ratio in
the initial system[40]. The idealized ternary also ignores the

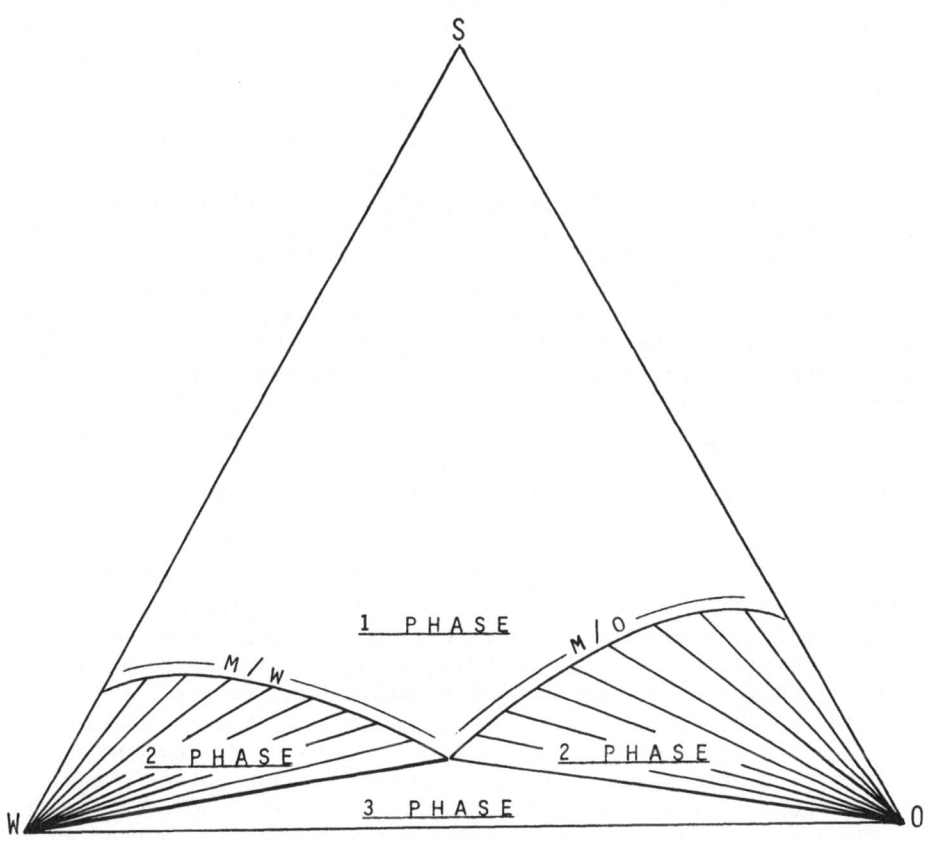

TIE LINES AT CONSTANT I.F.T

Figure 1. Idealized ternary diagram for microemulsions - Pseudo-3-component system, surfactant/oil/water; 2 immiscible components (O,W) mutually miscible in the third (S); Reference 45.

potential existence of more than 1 micellar phase in equilibrium
16,47. Nevertheless, these idealized diagrams serve to illustrate
and qualitatively correlate a wide variety of observations on the
phase behavior of microemulsions.

Figure 2 qualitatively relates the idealized ternary diagram
to experimental observation. At the top is depicted the test tube
appearance of Winsor's 3 types of equilibria[14]. The dashed lines
represent the I $\rightleftharpoons$ III $\rightleftharpoons$ II transitions. In this experiment, equal
volumes of water and oil are mixed with a given quantity of sur-
factant(s) and allowed to phase separate at constant temperature.
The overall composition is represented by the intersection of the
dashed lines on the ideal ternary diagrams. This composition falls
in the right lobe for Type I, in the 3-phase region for Type III
and in the left lobe for Type II systems. The transition I $\rightarrow$ III $\rightarrow$
II involves a shift in the IIP from the left to the right of the
ternary diagram. At the bottom is depicted Winsor's concept of
interfacial curvature corresponding to the 3 types of equilibria[15].
Winsor's S_1 micelles (Type I systems) are water external and S_2
micelles (Type II systems) are oil external. Winsor assumes Type
III systems to consist of an equilibrium mixture of the 2 micellar
types. This model has been quantitatively developed[36-39] and is
discussed later in this paper. The systematic shift in IIP des-
cribing the I $\rightleftharpoons$ III $\rightleftharpoons$ II transition will now be examined as a
function of temperature and correlated with variations in water and
oil uptake.

Figure 3 (upper left) by Shinoda and Kunieda[20] describes the
phase behavior of ethoxylated micellar systems. These investigators held constant the concentration of an ethoxylated nonyl
phenol containing 8.6 ethylene oxide groups and measured lower and
upper cloud points at various water/cyclohexane ratios. For
example, a composition, 25 wt.% cyclohexane in water containing
5 wt.% surfactant was heated following a path B'-B. At low temperatures the system split into 2 phases, a micellar phase in equilibrium with excess oil (II_{w-o}). In the vicinity of 50°C the system
became single phase (I_w). At higher temperatures the system split
into a micellar phase in equilibrium with excess water (II_{o-w}). If
a composition 50% cyclohexane in water was heated the system passed
through a 3-phase region in the vicinity of 55°C. Nomenclature is
that of Shinoda and Kunieda.

The shift in the IIP corresponding to Shinoda's et al. phase
transitions is given in the accompanying triangular diagrams.
These diagrams represent a semi-quantitative construct based on
their data; the position of the IIP is approximately determined but
the size and shape of the lobes are only qualitatively described.
The overall composition is shown on the ternary diagram by a constant surfactant concentration cut represented by the dashed line.
Both Shinoda's plot and the triangular diagrams show the transitions with increasing % oil:

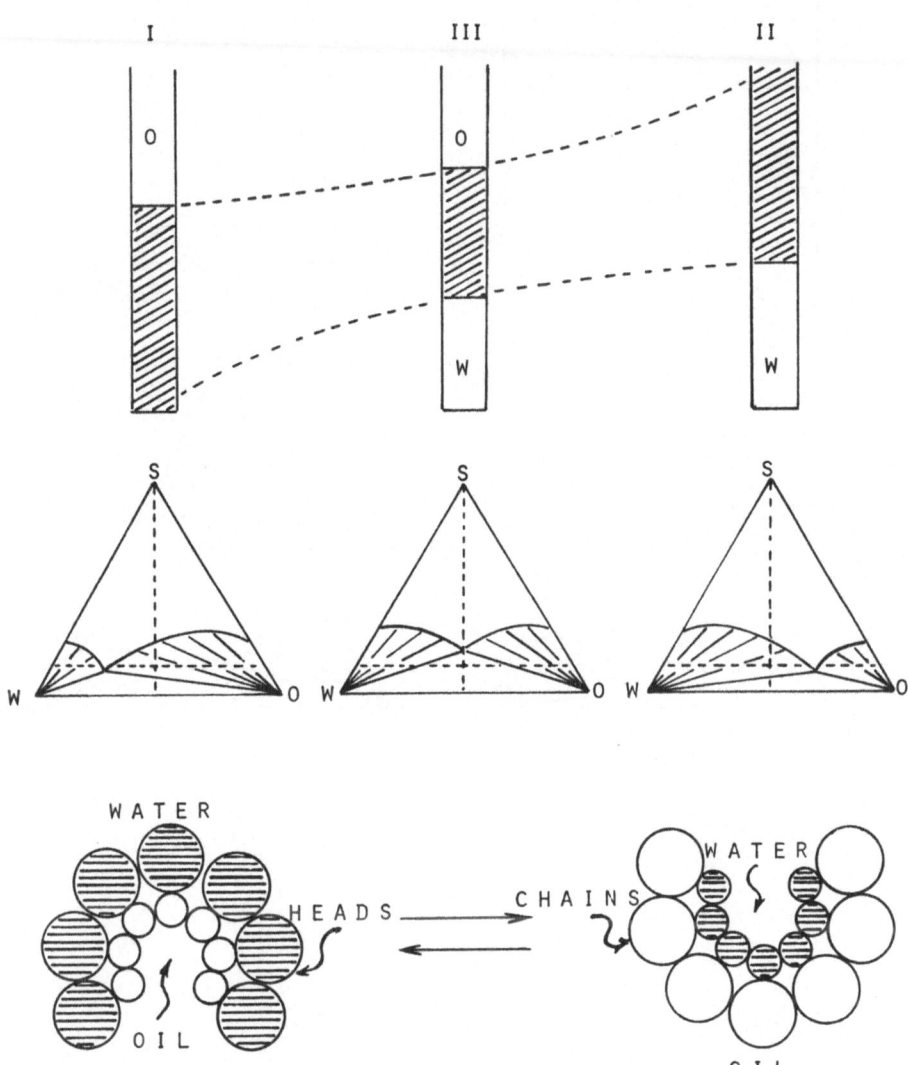

Figure 2. Phase behavior of microemulsions - Winsor's Type I⇄III⇄II transition; appearance of equilibrated system (top), representation on a ternary diagram (center) and interfacial curvature according to Winsor (S_1⇄S_2 type micelles); Reference 15.

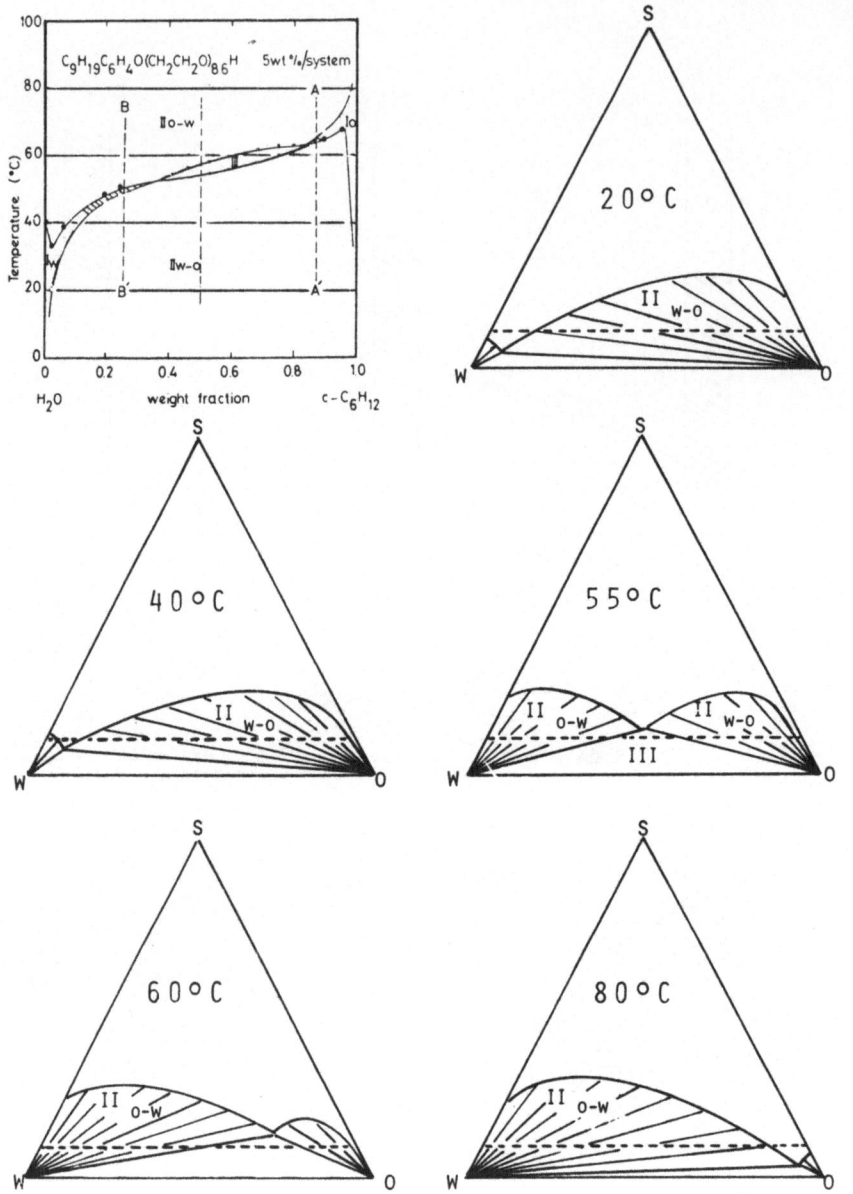

Figure 3. Temperature dependence in nonionic microemulsions - Temperature vs. composition according to Shinoda and Kunieda;[20] lower and upper cloud points at varied water/cyclohexane ratios with 5 wt.% ethoxylated nonyl phenol (8.6 EO); I_w = single phase (O/W), I_o = singlephase (W/O), II_{w-o} = 2-phase (micellar phase in equilibrium with excess oil, II_{o-w} = 2-phase (micellar phase in equilibrium with excess water, III = 3-phase (micellar phase in equilibrium with oil and water). Ternary diagrams show corresponding shift in phase regions and invariant point with temperature.

1. At 20°C: 1-phase → 2-phase (Microemulsion/Oil)
2. At 40°C: 2-phase (M/W) → 1-phase → 2-phase (M/O) → 1-phase
3. At 60°C: 2-phase (M/W) → 3-phase → 2-phase (M/O) → 1-phase
4. At 80°C: 2-phase (M/W) → 1-phase

For systems where the size, shape and position of the 2-phase lobes are determined, the triangular diagram obviously provides a more complete picture of the phase behavior of micellar systems.

Figure 4 qualitatively summarizes the interrelation of Shinoda's and co-workers' phase diagrams with the idealized ternary diagram. Shinoda's plot in the plane of the paper is a constant surfactant concentration slice through a set of stacked ternaries. The 1, 2 and 3-phase regions pass through corresponding regions on the stacked ternaries. Note that the IIP shifts to the right with rising temperature.

Figure 5 (upper left) represents phase equilibration experiments by Saito and Shinoda[28]. In these experiments, equal weights of water and cyclohexane are mixed with 5 wt.% ethoxylated nonyl phenol (8.6 EO) and allowed to phase separate at constant temperature. The relative phase volumes are then plotted vs. temperature. This plot represents the actual appearance of the system at a given temperature. For example, at 40°C the system splits into 2 phases, a bottom micellar phase occupying ∿55% and a top oil phase occupying ∿45% of the total volume. The nomenclature is that of Saito and Shinoda. The overall composition of this system is represented by the intersection of the dashed lines on the ternary diagrams. These diagrams qualitatively show the shift in phase regions and invariant point relative to this fixed composition. These diagrams are the same as those shown in Figure 3 except that they focus on a single composition rather than a range of oil/water ratios. The ternary diagrams thus correlate the 2 types of phase behavior experiments done by Shinoda and co-workers.

Figure 6 (top) repeats the data of Saito and Shinoda on an expanded scale. The bottom figure gives corresponding water (V_w/V_s) and oil (V_o/V_s) uptake curves calculated by material balance from the phase separation data assuming all the surfactant is in the micellar phase. The overall composition based on weight was converted to volume fraction using literature values for densities. Based on volume, the overall composition was 42.0% water, 53.8% cyclohexane and 4.2% ethoxylated nonyl phenol (8.6 EO). This composition fixes the maximum $V_w/V_s = 10.0$ and $V_o/V_s = 12.8$

Water and oil uptake curves give the composition of the micellar phase at a given temperature. Equal water and oil uptake occurs at 56°C (crossover temperature); $V_w/V_s = V_o/V_s = 8.9$. At equal water and oil uptake the invariant point (IIP) is in the middle of the ternary diagram (Figure 5 at 55°C). At temperatures

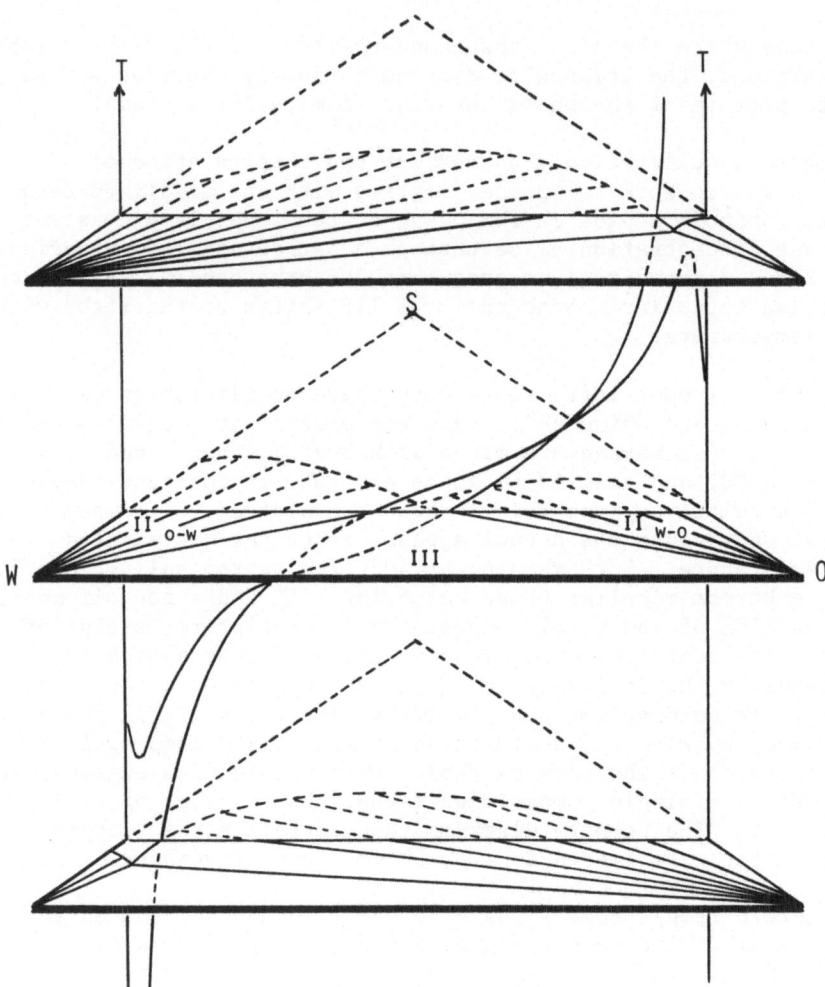

Figure 4.   Temperature dependence in nonionic microemulsions –
Qualitative superposition of Shinoda's phase diagram on ternary
diagrams stacked along the temperature axis; a constant surfactant
composition slice through the stacked ternaries.

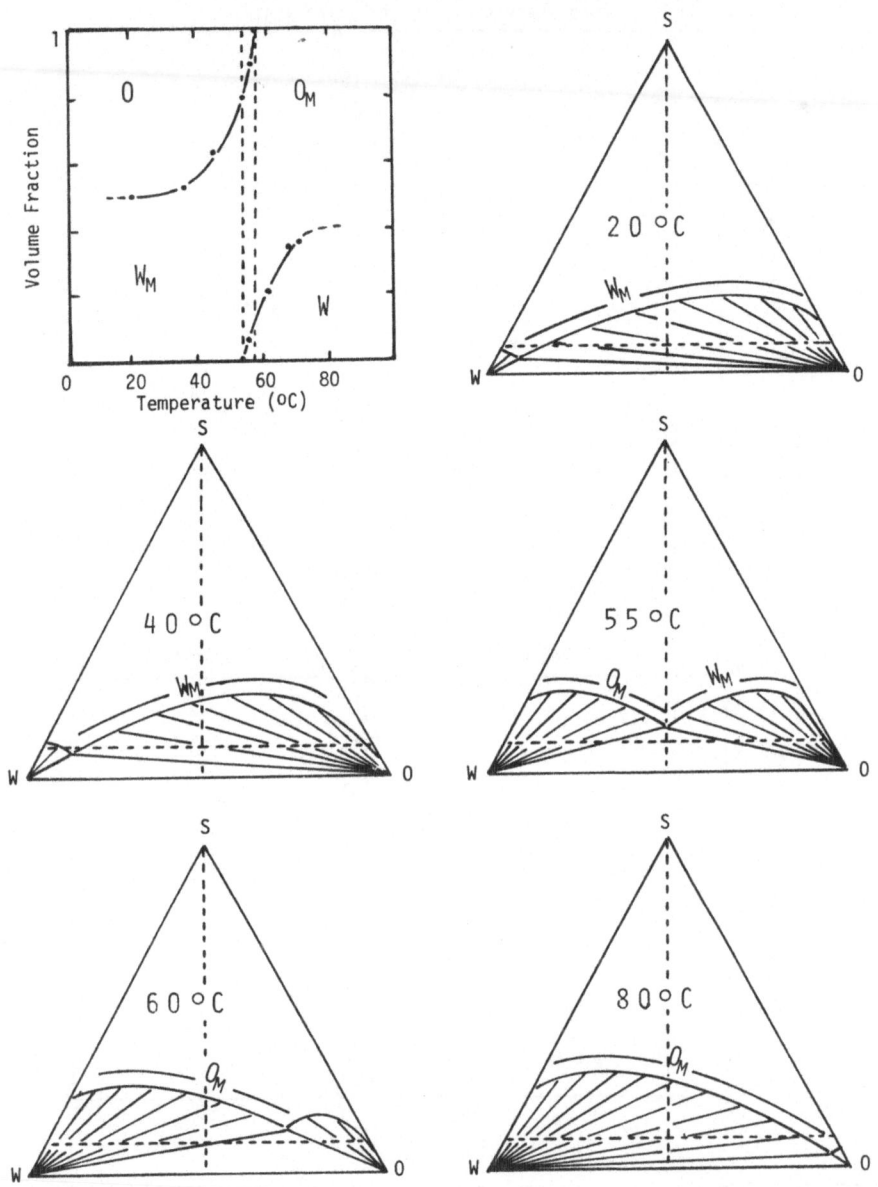

Figure 5. Temperature dependence in nonionic microemulsions – Phase volume fraction vs. temperature in equilibrated systems according to Saito and Shinoda;[28] phase split at 1/1 wt. ratio water/cyclohexane and 5 wt.% ethoxylated nonyl phenol (8.6 EO); W_M = water-continuous micellar phase in equilibrium with excess oil (O), O_M = oil-continuous micellar phase in equilibrium with excess water (W). Ternary diagrams show corresponding shift in phase regions and invariant point with temperature.

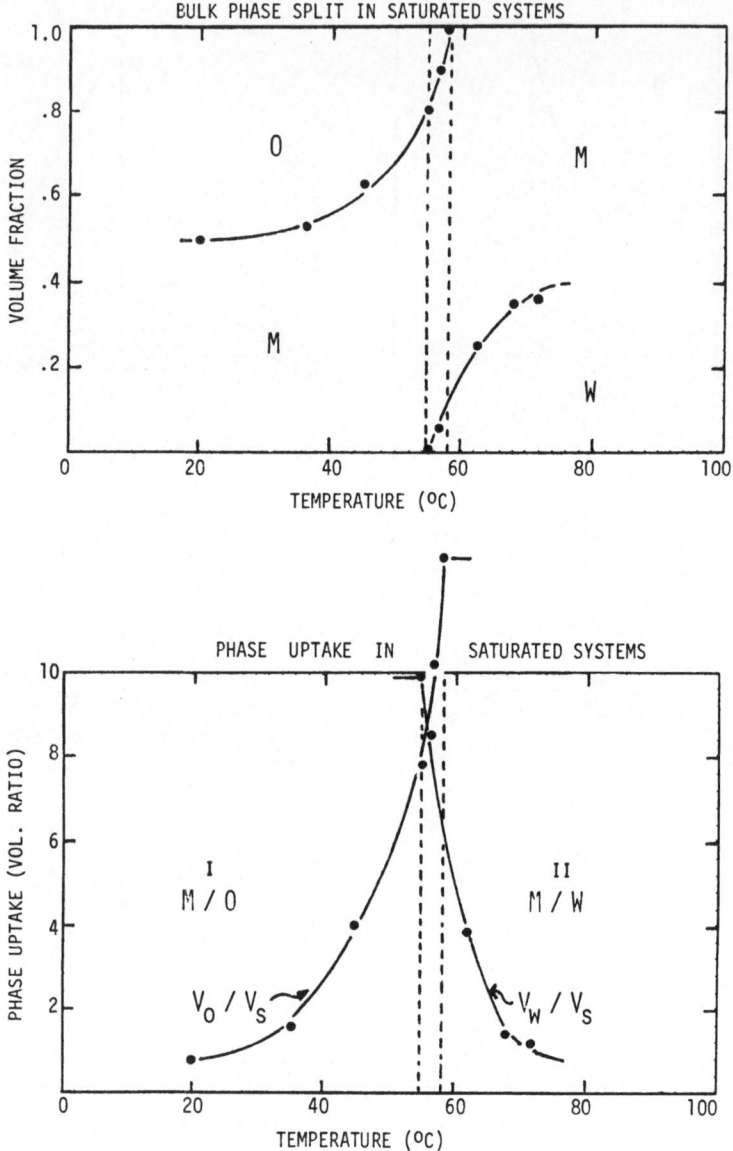

Figure 6.   Temperature dependence in nonionic microemulsions – Top: phase split at 1/1 wt. ratio water/cyclohexane and 5 wt.% ethoxylated nonyl phenol (8.6 EO) according to Saito and Shinoda.[28]   Bottom: Corresponding water (V_W/V_S) and oil (V_O/V_S) uptake curves assuming all the surfactant is in the micellar phase; M = micellar phase, O = excess oil, W = excess water; Roman numerals refer to Winsor's equilibrium types; densities used to convert weights to volumes: $\rho_O = 0.779$, $\rho_W = 1.00$, $\rho_S = 1.040$.

appreciably below the crossover temperature, $V_W/V_S > V_O/V_S$, the IIP
lies to the left on the ternary diagram and the system splits 2-
phase M/O (Figure 5 at 40°C). At temperatures much above the crossover temperature, $V_W/V_S < V_O/V_S$, the IIP lies to the right on the
ternary diagram and the system splits 2-phase M/W (Figure 5 at
80°C).

In the vicinity of the crossover temperature the system is 3-
phase (at the given surfactant concentration) and V_W/V_S passes from
greater than to less than V_O/V_S. Shinoda and co-workers have shown
that in this narrow temperature region solubilization of both water
and oil is a maximum[20] and that the system forms coarse emulsions
which invert from O/W to W/O as the temperature is raised[48]. They
called this region the phase inversion temperature (PIT). The PIT
can be more precisely defined as the crossover temperature, i.e.,
the temperature at which $V_W/V_S = V_O/V_S$. Shinoda and co-workers
have linked the PIT to HLB[26] and aqueous phase salinity[31] in systems
containing nonionic surfactants.

The variables, surfactant HLB and aqueous phase salinity are
completely analogous to temperature in their effect on phase uptake
in microemulsions. Figure 7 gives the phase volume and water and
oil uptake variation with HLB defined in terms of the volume ratio
of Hydrophilic heads to Lipophilic chains (V_H/V_L). Micellar systems
were prepared with equal volumes of hexadecane and water and 9 vol.%
of a mixture of two ethoxylated octyl phenols, Triton X15 and
Triton X114. Triton X15 contains 1 ethylene oxide (EO) group per
octyl phenyl chain and Triton X114 contains 7.5 EO groups per chain.
Average number of EO groups per chain in the surfactant mixture
varied according to the weight ratio of the two component surfactants. The molar volume of an EO group (39.0 cc) was determined
from the slope of a plot of specific volume vs. molecular weight for
an homologous series of ethoxylated octyl phenols of varying EO
content (Triton X series). The contribution of the OH group to the
head volume (8.5 cc) was estimated from extrapolated values of MW/
density = fn (# CH_2 groups) for homologous series of alcohols and
glycols by the methods of Fedors[49]. The volume of the octyl phenyl
chain (205.5 cc), was estimated from the MW/density of octyl phenol.

The same trends in water and oil uptake as observed with
decreasing temperature in nonionic microemulsions are noted with
increasing head/chain volume ratio. Water uptake increases and oil
uptake decreases as the surfactant mixture becomes more hydrophilic.
Thus a decrease in temperature corresponds to an increase in HLB
and the 2 variables can be traded against each other. The transitions, II → III → I type systems are noted with increasing V_H/V_L.
Crossover ($V_W/V_S = V_O/V_S$) occurs at $V_H/V_L = 0.78$. For the overall
composition containing 9% surfactant, maximum values of V_W/V_S and
V_O/V_S are 5.1.

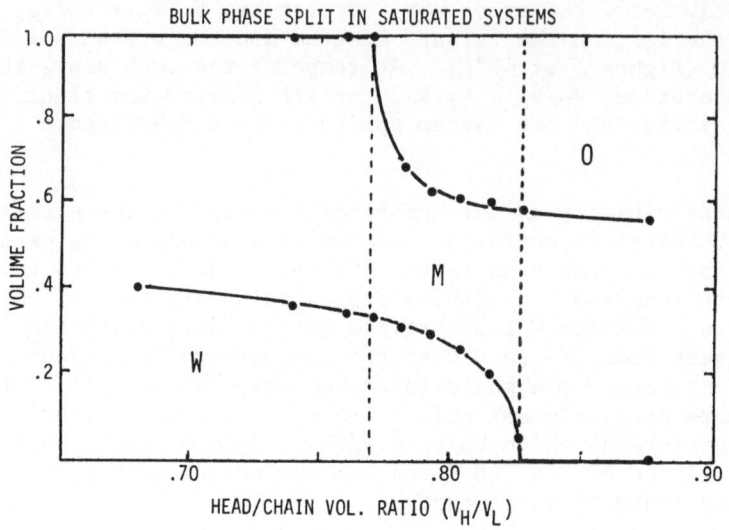

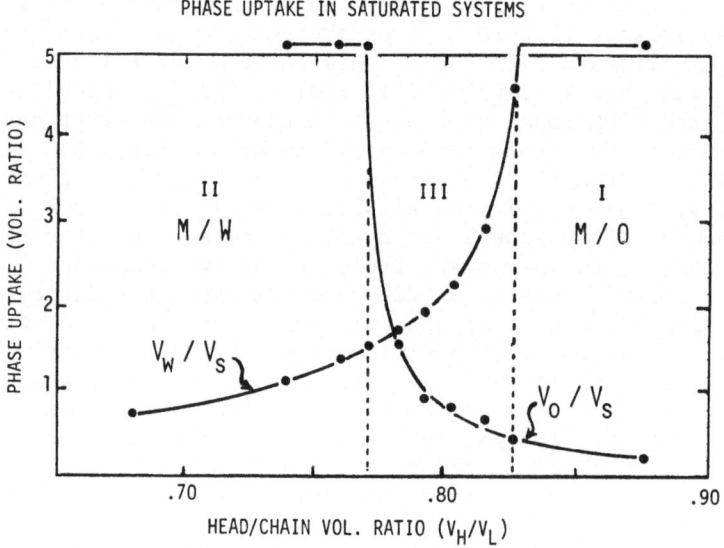

Figure 7.   HLB dependence in nonionic microemulsions – Top: Phase
split at 1/1 volume ratio water/hexadecane and 9 vol.% ethoxylated
octyl phenols: Triton X15 (1EO) and Triton X114 (7.5 EO); V_H/V_L
calculated from corresponding mole ratio of X15/X114 assuming molar
volumes of 39.0 ml/EO group, 8.5 ml/OH group and 205.5 ml/octyl
phenyl chain.   Bottom: Corresponding water (V_W/V_S) and oil (V_O/V_S)
uptake curves.   Temperature = 25°C.

A similar variation in phase uptake with HLB is noted for anionic
surfactants. The data in Figure 8 are supplied by Winsor (Reference 14, Table II, p. 385). In this case the HLB of the mixture is
decreased by adding the lipophilic cyclohexanol to the alkyl
sulfates. The transitions are I → III → II with increasing alcohol,
V_O/V_S increases and V_W/V_S decreases. (The surfactant volume, V_S,
used to evaluate phase uptakes was taken as the sum of the alcohol
and alkyl sulfate volumes.) Crossover occurs at approximately 1.7
vol. ratio alcohol/alkyl sulfate.

The analogous picture with varying aqueous salinity is shown
in Figure 9. Phase separation data were obtained with equal volumes
of hexadecane and aqueous phase and 2 vol.% of an ethoxylated dodecyl phenol (6 EO). Crossover ($V_W/V_S = V_O/V_S$) occurs at 5% aqueous
NaCl. The system is 2-phase (M/O) in pure water and 3-phase for
salinities ranging up to 8% salt. (Above 6% salt the systems did
not phase split cleanly and a coarse, stiff emulsion remained in
the aqueous phase even after 6 months storage.) Healy and co-
workers[40] report similar results for a system containing an anionic
surfactant.

Phase split and corresponding phase uptake curves describe the
I ⇌ III ⇌ II transitions with variation in temperature, surfactant
HLB (head/chain size, surfactant/co-surfactant ratio), aqueous
salinity and oil composition. They measure trends in the overall
hydrophilic vs. lipophilic behavior of the system. They demonstrate
that for the nonionic surfactants described herein, increasing temperature, salinity and decreasing head group size shift the system
more lipophilic and that these variables are interchangeable and
mutually compensating. These trends are summarized in Figure 10.
Movement to the right in this figure signifies increasingly lipophilic behavior in systems containing ethoxylated surfactants.
(Other surfactant types may show different response to temperature
and aromatic solvents[15,40].) This is shown by a decrease in relative
water/oil uptake, a shift in the invariant point to the right and a
I → III → II transition in type of micellar system. The micellar
model corresponding to these trends is given at the top. The shift
in interfacial curvature from water external to oil external is
represented by dehydration and shrinkage of the heads and solvation
and swelling of the chains. This is only a pictorial representation
of the energetic interactions in the interface among surfactant
molecules, water and oil but serves to rationalize the observed
trends. An increase in temperature tends to reduce hydration of the
ethylene oxide groups. (This is the cause of the appearance of a
cloud point at elevated temperatures in aqueous solutions of ethoxylated surfactants.) An increase in salinity tends to "salt out"
the head groups likewise resulting in dehydration. More aromatic
oils, presumably, can better solvate the aromatic chains on the
surfactants described herein. These trends all favor the I → III →
II transition and a decrease in relative water/oil uptake.

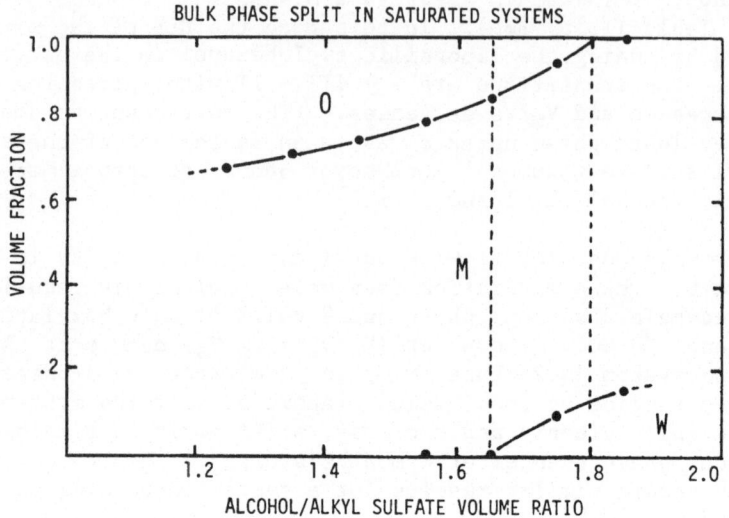

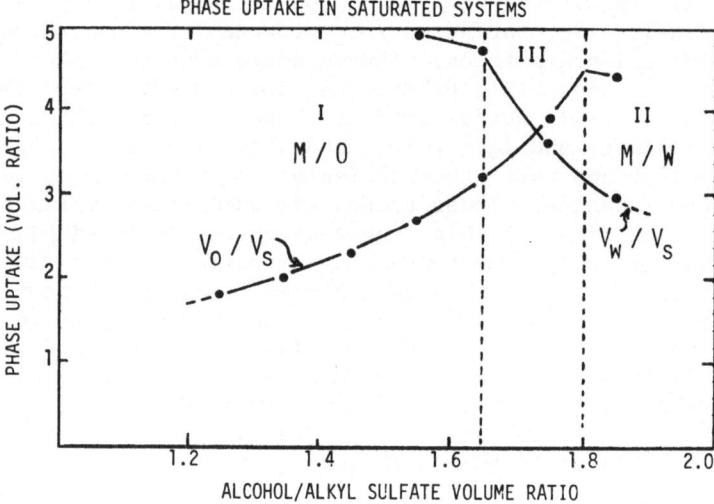

Figure 8. HLB dependence in anionic microemulsions – Top: Phase split at 1/1 volume ratio petroleum ether/2.4 wt.% aqueous Na_2SO_4 and 4 wt.% C_{10}-C_{12} alkyl sulfates according to Winsor[14] with increasing volume ratios of cyclohexanol/alkyl sulfate; density alkyl sulfate = 1.0. Bottom: Corresponding water and oil uptake curves normalized to total alcohol + alkyl sulfate.

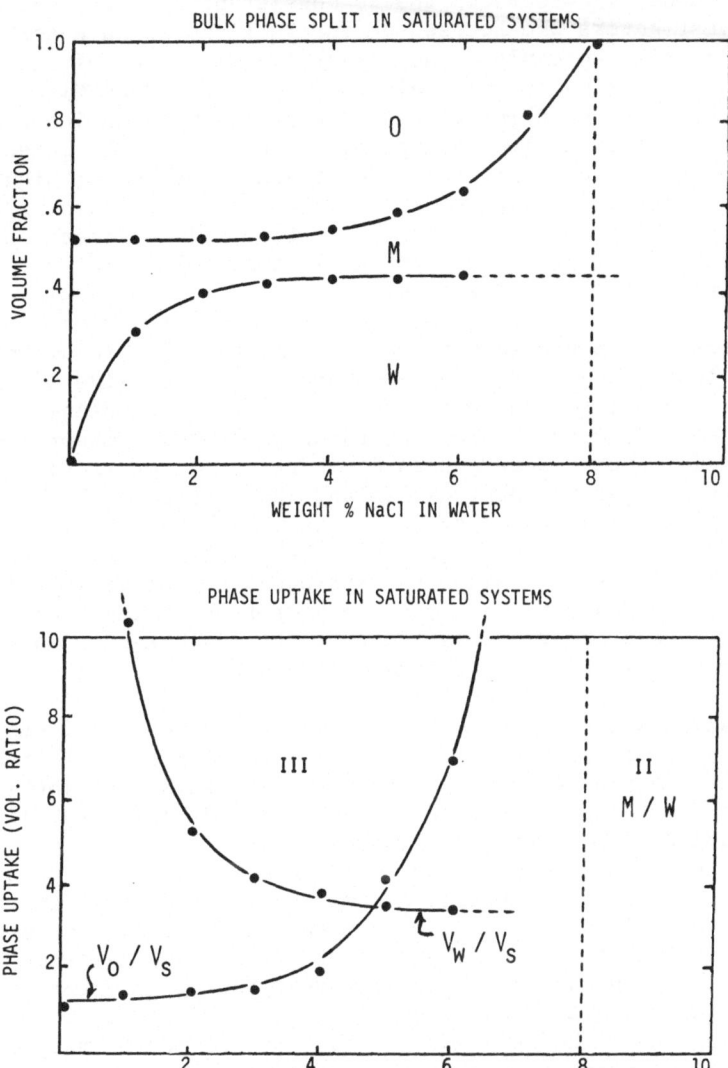

Figure 9.   Salinity dependence in nonionic microemulsions - Top:
Phase split at 1/1 volume ratio water/hexadecane and 2 vol.%
ethoxylated dodecyl phenol (~6 EO): Igepal RC520.   Bottom:
Corresponding water and oil uptake curves.   Temperature = 25°C.

MICROEMULSION THEORY

This model incorporates and is consistent with the concepts of those investigators who focused on interactions in an adsorbed interfacial film to explain the direction and extent of interfacial curvature. Treating coarse emulsions, Bancroft[50] and Clowes[51] considered the adsorbed film as duplex in nature with an inner and an outer interfacial tension acting independently[52]. The interface would then curve such that the inner surface was one of higher tension. Bancroft's rule is often stated as "that phase will be external in which the emulsifier is most soluble". Beerbower and Hill[53] stated Bancroft's concept in terms of solubility parameters for each half of the duplex film.

Winsor[14] paraphrased this concept for micellar solutions in terms of independent spreading pressures on either side of the interface. The interface would curve convex to the phase against which it exhibited the higher spreading pressure. Winsor also phrased this concept more generally in terms of interactions between each half of the duplex film and the contacting bulk phase. The interface would curve convex to the phase with which it had the greatest affinity. Winsor expressed the relative affinities in terms of his R ratio. Shinoda[26] attributed curvature in films of nonionic surfactants to variation in hydration of the hydrophilic moiety with temperature. Low temperatures would result in increased hydration with a convex curvature toward water.

Schulman and co-workers[7,11] applied the concepts of Bancroft and Clowes to microemulsions to explain the direction of interfacial curvature. They postulated a negative interfacial tension[9,10] to explain spontaneous emulsification and thermodynamic stability in microemulsions. The concept of negative interfacial tension was reviewed and modified by Rosano[54] and Gerbacia[55]. They proposed that diffusion of surface active components across the interface could temporarily drop the dynamic interfacial tension to zero while the equilibrium IFT remained positive. Prince[56,57] extended the treatment by Schulman and co-workers attributing curvature to a differential surface pressure on either side of the interface. The side having the higher pressure would curve convex to the external phase. Prince postulated that the interface would curve until the surface pressure on either side of the interface was equalized, establishing a criterion for degree of curvature.

Robbins[36-39] considered the stress gradient within the interfacial film as determining the direction of curvature and the "Laplace" pressure difference across the curved interface. Relating the pressure difference to the fugacity of water in a W/O microemulsion, he established thermodynamic criteria for spontaneous water uptake without postulating a negative interfacial tension. He showed

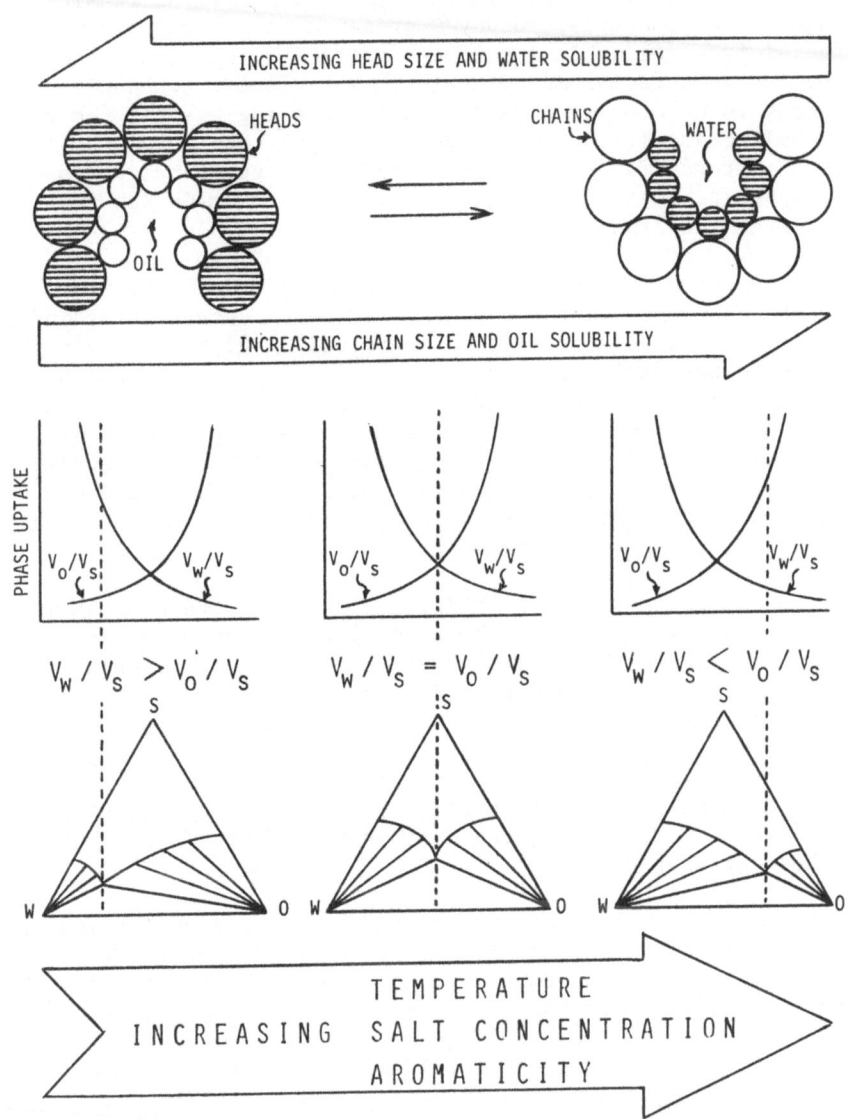

Figure 10. Microemulsion theory - The correlation of a micellar model with trends in water and oil uptake and phase boundaries on the idealized ternary diagram. A shift to the right signifies increasingly lipophilic behavior.

quantitatively that at saturation the stress gradient tending to
bend the interface was just balanced by the interfacial tension.
Applying similar concepts, Murphy[58] presented a general analysis of
interfacial free energy in interfaces of finite thickness. He
described an interfacial model having 2 interfacial zones with an
intervening interface, each zone also interfacing against the contacting bulk phase (tripartite model). Murphy considered the interactions of interfacial stretching, bending and torsional stresses
and developed conditions for mechanical equilibrium in low tension,
highly curved interfaces.

All the above approaches have one feature in common, they treat
the interface as an oriented duplex film[59,60]. Curvature is induced
by interfacial forces differing in magnitude on either side of the
film. A somewhat different approach to equilibrium in micellar
systems was taken by Adamson and co-workers[61,62]. They postulated
a balance between osmotic and Laplace (capillary) pressure for a
water droplet in a W/O "micellar emulsion". Their model required
differences in electrolyte concentration between the bulk and
micellar aqueous phases and is not applicable to the solubilization
of pure water by nonionic surfactants.

The model for a W/O microemulsion presented here assumes mono-
dispersity, each water droplet covered by an oriented monolayer of
surfactant molecules. Adjacent to the water core is a shell of
hydrophilic heads. Surrounding the heads is a concentric shell of
lipophilic chains. The heads and chains act as separate uniform
liquid phases with water dissolved in the heads and oil in the
chains. Curvature is imposed by the differential tendency of water
to swell the heads vs. oil to swell the chains. The magnitude of
the interfacial stresses induced by this differential swelling controls the degree of curvature.

The geometric relationships for this model are summarized in
Figure 11. In the figure, R_W is the radius to the water interface,
R the radius to the head-chain junction and R_O the radius to the oil
interface, τ_H and τ_L are the thicknesses of the Hydrophilic and
Lipophilic shells. The upper 4 equations relate the total volumes
to the droplet volumes and the total area to the droplet area. For
example, the two equations on the upper right state that 1) the
total volume of water and heads is equal to the number of droplets
times the droplet volume to the head-chain junction and 2) the total
area is equal to the number of droplets multiplied by the droplet
area. Ratios of the upper 4 equations together with the definitions
for τ_H and τ_L yield the bottom 3 equations. The thickness/radius
ratio for both shells is defined in terms of 2 bulk parameters, the
volume fraction of heads in the surfactant molecule and the water
uptake. The behavior of the thickness/radius ratios, τ_H/R and τ_L/R,
will now be examined in saturated microemulsions prepared with
nonionic surfactants.

$$V_W = N(4/3)\pi R_W^3$$

$$V_W + V_S = N(4/3)\pi R_0^3$$

$$V_W + V_H = N(4/3)\pi R^3$$

$$N_S\sigma = N(4\pi R^2)$$

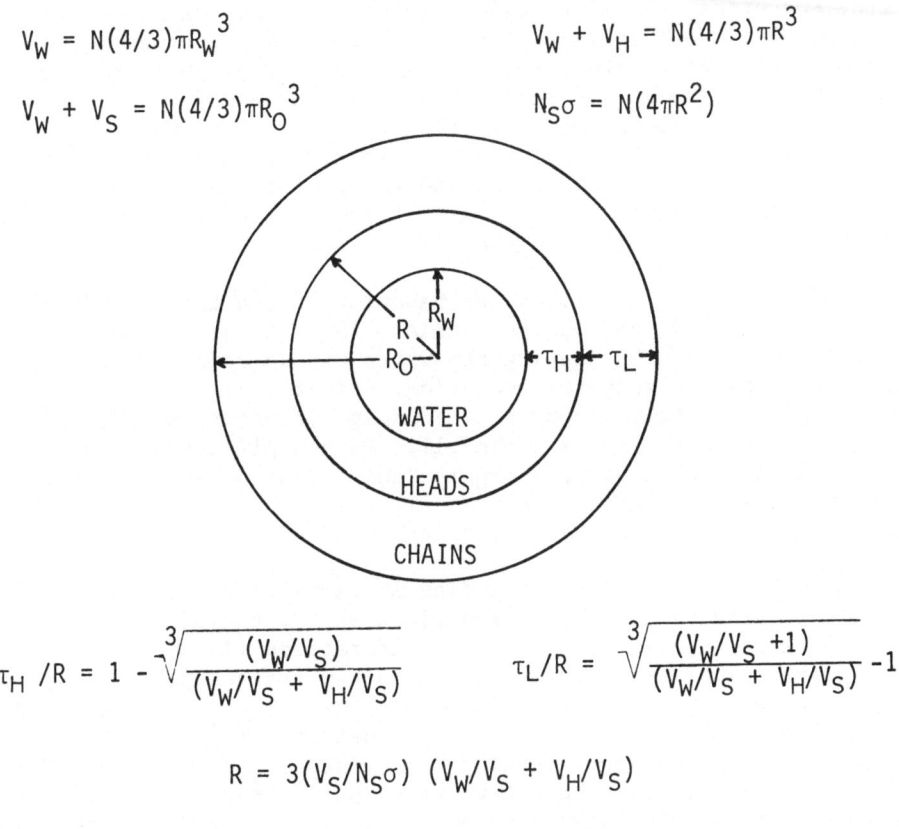

$$\tau_H/R = 1 - \sqrt[3]{\frac{(V_W/V_S)}{(V_W/V_S + V_H/V_S)}}$$

$$\tau_L/R = \sqrt[3]{\frac{(V_W/V_S + 1)}{(V_W/V_S + V_H/V_S)}} - 1$$

$$R = 3(V_S/N_S\sigma)\ (V_W/V_S + V_H/V_S)$$

V_W = WATER VOLUME		τ_H = HEADS THICKNESS	
V_S = SURFACTANT VOLUME		τ_L = CHAINS THICKNESS	
V_H = HEADS VOLUME		R = DROPLET RADIUS	
σ = MOLECULAR AREA		N_S = NO. MOLECULES	
		N = NO. DROPLETS	

Figure 11. Geometry of a water droplet – The surfactant orients in concentric shells of heads and chains around the water core. Thickness/radius ratios, τ_H/R and τ_L/R for each shell are functions of water uptake and the volume fraction of heads.

The upper plot in Figure 12 describes water uptake in 3 different oils by varying ratios of 2 ethoxylated octyl phenols, Triton X15 (1 EO) and Triton X114 (7.5 EO). The average volume fraction of ethylene oxide was calculated by methods already described for Figure 7. Water uptake rises assymptotically with increasing EO fraction depending on the specific oil. The curves shift toward higher EO fractions in the order cetane, decane and xylene. These data imply the following physical picture. Since xylene tends to solvate the alkyl phenyl chains more strongly, it tends to curve the interface more tightly around the water droplets. Going to higher EO fractions tends to counteract the tighter curvature by putting more volume in the EO shell.

The bottom plot describes the behavior of the thickness/radius ratios for the heads and chains shells. The ratios, τ_H/R and τ_L/R, were calculated by substituting the water uptake data for the 3 oils (top) into the geometric equations (Figure 11). Each oil gives an approximately linear relationship between τ_H/R and τ_L/R and τ_H/τ_L is constant for a given oil. This implies that for a fixed surfactant chemistry, τ_H/τ_L is independent of surfactant composition (HLB), and depends only on oil composition and, presumably, temperature and aqueous phase salinity.

The significance of this finding in terms of the theory is clarified in Figure 13. The 2 equations at the top of the figure are derived from purely geometric considerations (Appendix A). The left equation applies to water uptake in W/O systems and the right equation to oil uptake in O/W microemulsions. These equations are plotted below holding τ_H/τ_L and δ_H/δ_L constant. This plot has all the features of the experimentally determined phase uptake vs. HLB curves (Figures 7, 8). The water uptake curve assymptotes to $V_H/V_L = \tau_H/\tau_L$; the oil uptake curve assymptotes to $V_H/V_L = \delta_H/\delta_L$. Crossover occurs at $V_H/V_L = \sqrt{(\delta_H\tau_H)/(\delta_L\tau_L)}$. The closer the values of δ_H/δ_L and τ_H/τ_L, the higher the water and oil uptake at crossover. The phase uptake curves truncate depending on the overall surfactant concentration. The higher the surfactant concentration, the lower the maximum water and oil uptake. If the overall surfactant concentration is high enough to cause truncation below the crossover point, then a 3-phase system will not form. The transition with increasing V_H/V_L (HLB) would be II → IV → I. Winsor's Type IV systems are homogeneous and isotropic[14]. If the overall surfactant concentration changes with changing HLB (see Figure 8), the truncating lines will not be horizontal as indicated here. This figure was drawn for equal overall water and oil concentrations. If the overall water concentration is greater than the overall oil concentration, the water uptake curve truncates higher than the oil uptake curve and vice versa.

Implicit in this description is the assumption that τ_H/τ_L and δ_H/δ_L are independent of the overall surfactant, water and oil

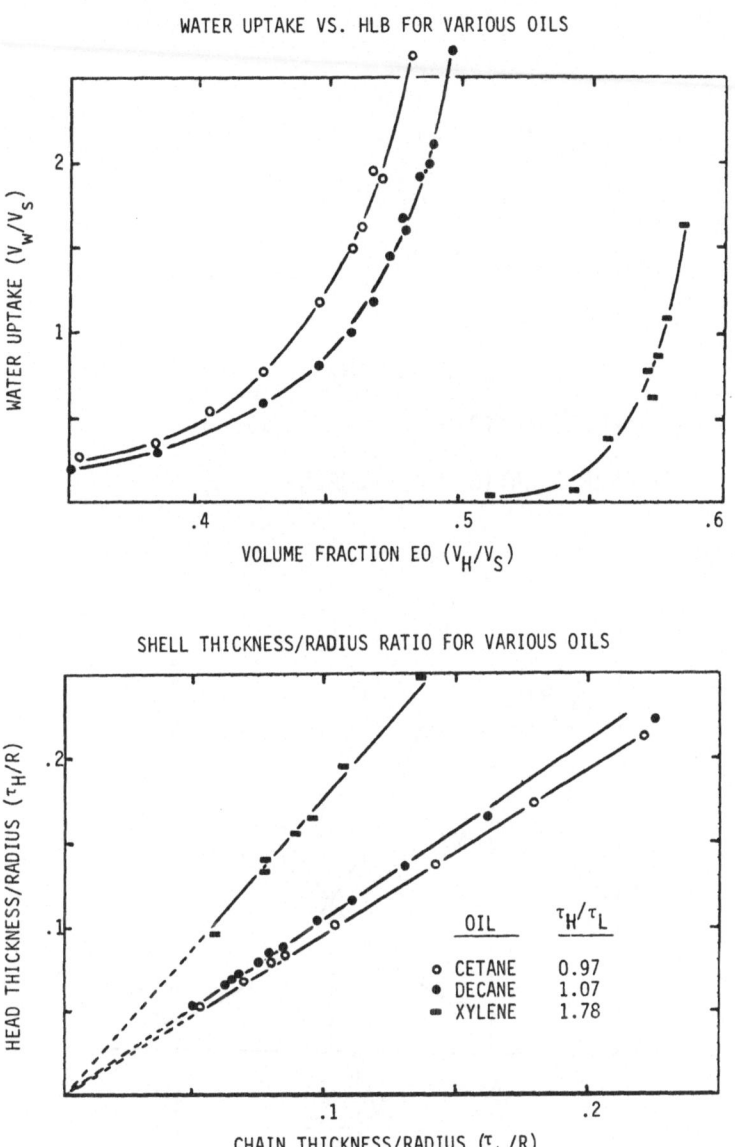

Figure 12.   Correlation of water uptake with thickness/radius ratio –
Top: Water uptake by titration of 50 ml of the given oil + 50 ml of
varying ratios of Triton X15 and Triton X114; V_H/V_S calculated as in
Figure 7. Bottom: Corresponding plots of head vs. chain thickness/
radius ratios from model equations.

$$V_W/V_S = 1 \Big/ \left[\left(\frac{\tau_H/\tau_L}{V_H/V_L} \right)^3 - 1 \right] \qquad V_O/V_S = 1 \Big/ \left[\left(\frac{V_H/V_L}{\sigma_H/\sigma_L} \right)^3 - 1 \right]$$

V_W/V_S = AQUEOUS PHASE UPTAKE

V_O/V_S = OIL PHASE UPTAKE

V_H/V_L = VOLUME RATIO HEADS/CHAINS

τ_H/τ_L = THICKNESS RATIO HEADS/CHAINS (W/O)

σ_H/σ_L = THICKNESS RATIO HEADS/CHAINS (O/W)

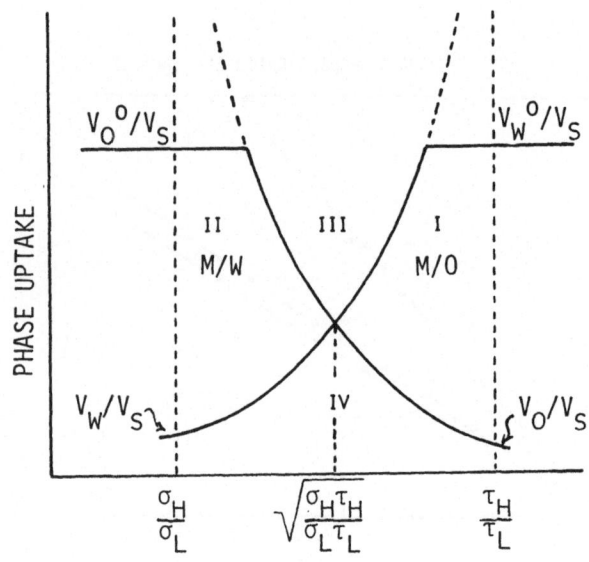

VOLUME HEADS/CHAINS (V_H/V_L)

Figure 13.   Theory for saturated microemulsions: Geometry – Theoretical equations for W/O (left) and O/W (right) systems with plot for corresponding phase uptake curves.

concentrations. With this assumption, the model yields symmetrical uptake curves and the crossover point does not change position with changing overall composition. Though systems exist which give symmetrical uptake curves, most real systems do not behave this well. In Figure 7, the oil uptake curve rises more steeply than the model predicts. For real systems, the crossover point may shift with changing overall surfactant concentration. The model could accomodate such systems if τ_H/τ_L and δ_H/δ_L were known as functions of overall composition. Nevertheless, even the simple model is a very useful semi-quantitative description of real systems in that if the position of the crossover point is known, an approximation to the entire phase uptake plot can be drawn.

The foregoing geometric considerations demonstrate that phase uptake is controlled by the thickness and volume ratios of heads/ chains. Experimental evidence (Figure 12) indicates that these parameters are mutually independent in saturated micellar systems. The geometric model also shows that the relative magnitude of the thickness and volume ratios determines the droplet radius and hence interfacial curvature. It is inherent in the duplex film assumption that interfacial curvature should result from differing stresses on each side of the interfacial film. This concept is explored in Appendix B. Figure 14 presents the results of this analysis for saturated micellar systems. The top equations describe the balance between interfacial tension and bending stresses in saturated microemulsions. Bending stresses are defined in terms of the relative interfacial compressibilities of the heads vs. chains. It is evident from these equations that the value of the head/chain thickness ratio in saturated microemulsions is con- trolled by the interfacial tension and the head and chain inter- facial compressibilities.

Interfacial compressibility is defined as the fractional change in molecular area with interfacial pressure (Reference 52, p. 131). As defined it allows for the change in shell volume on swelling with solvent. This definition treats the interfacial shell surrounding a water droplet as a double osmotic cell[63]. Water equilibrates between the core and heads shell; oil equili- brates between the external phase and chains shell. The stress in each shell provides the counterbalancing osmotic pressure to con- tinued solvent dilution of the interfacial monolayer. A similar picture holds for an oil droplet.

Interfacial compressibility for a monolayer can be evaluated from surface pressure (π) vs. molecular area (σ) isotherms. The slope of a plot of ln σ vs. π gives the compressibility of the mono- layer. It can be shown by considering the reciprocal of the com- pressibility as a kind of spring constant, that the reciprocal compressibility for the film is the sum of the individual reciprocal compressibilities for the heads and chains. It can also be shown

$$C_H \gamma_W = 4\,(T_r - V_r)\,(T_r - C_r)\ /\ (T_r + V_r)\,(T_r + 1)$$

$$C_H \gamma_0 = 4\,(V_r - \delta_r)\,(C_r - \delta_r)\ /\ (\delta_r + V_r)\,(\delta_r + 1)$$

γ_W = INTERFACIAL TENSION WATER DROPLET VS. OIL

γ_0 = INTERFACIAL TENSION OIL DROPLET VS. WATER

C_H = INTERFACIAL COMPRESSIBILITY OF HEADS

C_r = COMPRESSIBILITY RATIO HEADS/CHAINS = C_H/C_L

V_r = VOLUME RATIO HEADS/CHAINS = V_H/V_L

τ_r = THICKNESS RATIO HEADS/CHAINS = τ_H/τ_L (W/O)

δ_r = THICKNESS RATIO HEADS/CHAINS = δ_H/δ_L (O/W)

INTERFACIAL COMPRESSIBILITY: FRACTIONAL CHANGE IN MOLECULAR AREA WITH
                              INTERFACIAL PRESSURE

$$C_H = -(1/\sigma_H)\,d\sigma_H/d\pi \qquad\qquad C_L = -(1/\sigma_L)\,d\sigma_L/d\pi$$

HIGH (C_H/C_L): WATER SWELLS HEADS, CURVES O/W

LOW (C_H/C_L): OIL SWELLS CHAINS, CURVES W/O

EVALUATION:

$1/C_H + 1/C_L = 1/C_S$:   FILM BALANCE

$C_H/C_L = V_H/V_L$: THEORY

Figure 14. Theory for saturated microemulsions: Force Balance –
Top: Theoretical equations relating interfacial tensions γ_W (W/O)
and γ_0 (O/W) to volume, thickness and compressibility ratios of
heads/chains. Bottom: Significance and evaluation of head and
chain interfacial compressibilities.

that the compressibility ratio (C_r) equals the volume ratio (V_r). The assumption (Appendix C) that the tensional free energy is equal for water and oil droplets in equilibrium,

$$\gamma_w(4\pi R^2) = \gamma_o(4\pi A^2)$$

applied to the force balance equations, leads directly to the result: $C_r \approx V_r$. This is illustrated by numerical examples in the following table calculated using the equations in Appendix C.

Equal Phase Uptake

Fig.	Uptake	δ_r	V_r	τ_r	C_r	$C_H\gamma_w$	$C_H\gamma_o$
6	8.9	1.50	1.55	1.61	1.55	.0015	.0016
7	1.7	0.67	0.78	0.91	0.78	.022	.020
9	3.5	.82	.89	.97	.89	.0070	.0062

The values for model parameters are derived from the crossover points given in the figures listed in the first column. Numerical substitution shows that the approximation $C_r \approx V_r$ depends somewhat on both the level of phase uptake and the value of V_r at the crossover point. For $V_r > 1$, C_r is slightly larger than V_r and for $V_r < 1$, C_r is slightly smaller. The deviation of C_r from V_r is greatest at equal phase uptake being $\sim$1% for a phase uptake $\sim$1 and dropping to <0.1% for a phase uptake $\sim$10. Healy and co-workers[40] have shown that at equal water and oil uptake the interfacial tensions of the micellar phase vs. both bulk phases in Type III systems are approximately equal. This conclusion is supported by the data of Saito and Shinoda[28]. The theory predicts similar behavior for the interfacial tensions of water and oil droplets vs. their respective continuous phases.

The behavior of the model parameters is summarized in Figure 15. Head/chain thickness (τ_r, δ_r) and compressibility (C_r) ratios are plotted vs. the volume ratio (V_r). The theory requires the following conditions to hold:

$$\delta_r \leq V_r \leq \tau_r \ , \qquad \delta_r \leq C_r \leq \tau_r \ .$$

The functional relationships depicted in Figure 15 are consistent with these conditions. The qualitative behavior is described of phase uptake and interfacial tension over ranges of the volume ratio. Interfacial tension varies inversely with phase uptake.

The conclusion that $C_r = V_r$ can now be applied to measured monolayer compressibilities to estimate the individual head and chain compressibilities. The measurement of interfacial compressibility

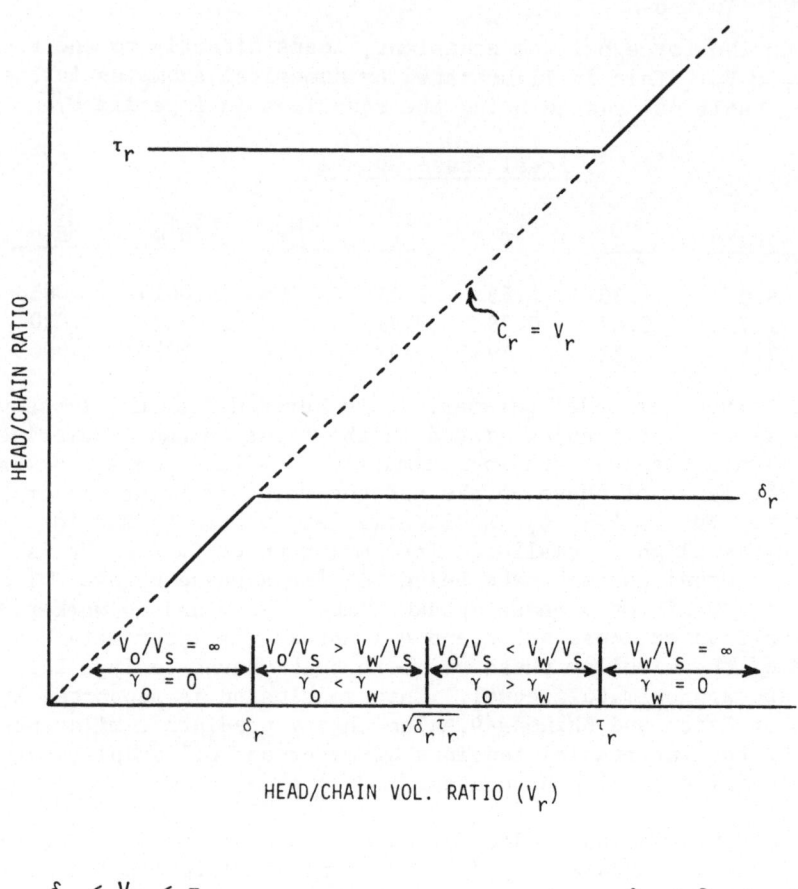

$$\delta_r \leq V_r \leq \tau_r \qquad\qquad\qquad \delta_r \leq C_r \leq \tau_r$$

Figure 15. Theory for saturated microemulsions: Model Parameters – Dependence of thickness ratios (δ_r, τ_r) and compressibility ratio (C_r) on head/chain volume ratio; correlation of interfacial tensions with phase uptake.

by the surface balance techniques described in the experimental
section is discussed in Figure 16. Data are presented for mixed
monolayers of hexadecane with the ethoxylated octyl phenols, Triton
X15 (1 EO), Triton X114 (7.5 EO) and an equal weight ratio of the
two. The corresponding average volume ratios, V_r, are given in the
top figure together with the slopes, C_s. These compressibility
values are very reasonable for liquid-expanded monolayers[64].
Assuming that the interfacial compressibilities, C_s, are representative of those at the hexadecane/water interface, the individual
head and chain compressibilities can be calculated using the
equations at the bottom of Figure 14. The results are plotted in
Figure 16 (bottom). The values for $C_H = f(V_r)$ and the values
$\delta_r = 0.67$ and $\tau_r = 0.91$ estimated from the crossover point in Figure 7 can be used to evaluate interfacial tensions from the
equations:

$$C_H \gamma_w = 4(\tau_r - V_r)^2 / (\tau_r + V_r)(\tau_r + 1)$$

$$C_H \gamma_o = 4(V_r - \delta_r)^2 / (\delta_r + V_r)(\delta_r + 1)$$

These equations are plotted in Figure 17 (top). Interfacial
tension at equal water and oil uptake is predicted to be 0.17 dyn/
cm. The heads compressibility was estimated at 0.136 cm/dyn.
Interfacial tensions for water (γ_w) and oil (γ_o) droplets against
contacting bulk phases are plotted vs. phase uptake in the bottom
figure. The scale is not expanded enough to distinguish between
$\gamma_w = f(V_w/V_s)$ and $\gamma_o = f(V_o/V_s)$ though small differences exist. Interfacial tensions less than 0.02 dyn/cm are predicted for phase
uptakes greater than 5. These predictions are in accord with the
findings of Healy and co-workers[40].

Figure 18 (top) gives model predictions of droplet sizes
assuming a mono-disperse system[9]. Droplet radii were calculated
assuming a constant hydrocarbon chain length of 10.2Å. This chain
length was estimated by the ratio of the chain volume (341Å^3) to
molecular area (33.4Å^2). Molecular area was estimated from a plot
of σ vs. V_H/V_L (data from Figure 16) extropolated to $V_H/V_L = 0$.
The predicted radii range from ∿50Å to 500Å for phase uptakes
ranging from ∿0.5 to 10. This size range is reasonable according
to the findings of Schulman and co-workers[1-5,7-9,11]. The geometric
model also predicts the number of surfactant molecules coating a
droplet. This is given in Figure 18 (bottom). Predicted values of
molecules of surfactant per droplet (N_s/N) range from 10^3 to 10^5
over the range of phase uptakes from 0.5 to 10.

Prediction of the response of theoretical parameters to the
variables temperature, aqueous salinity and oil composition is
largely empirical at present. The response to temperature will
serve as an example. Figure 19 gives empirical equations for esti-
mating the dependence of the thickness ratios, δ_r and τ_r, on

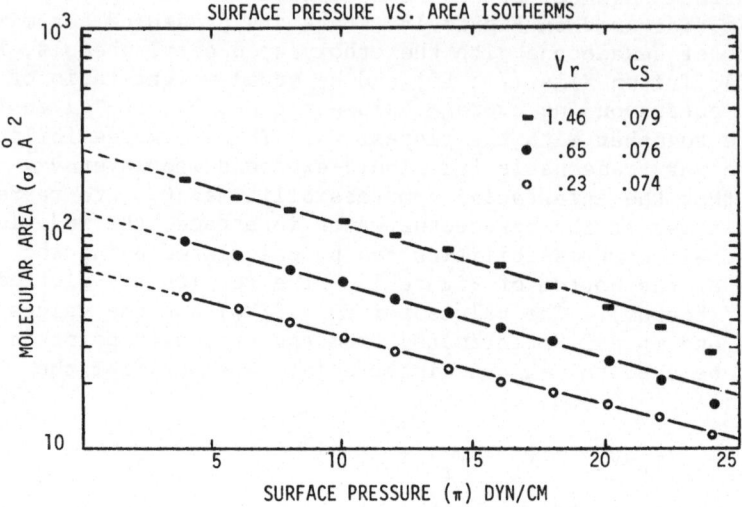

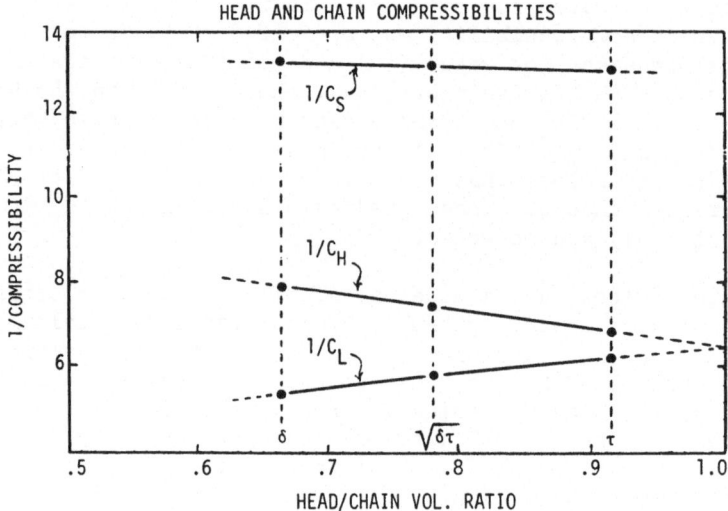

Figure 16. Measurement of interfacial compressibility - Top: Surface pressure vs. molecular area (semi-logarithmic plot) for 20/1
(by wt.) mixed monolayers of hexadecane/ethoxylated octyl phenols
(o Triton X15, ■ Triton X114, ● 1/1 X15/X114); table lists volume
ratio heads/chains (V_r) and interfacial compressibility (C_s).
Bottom: Interfacial compressibility (C_s) and calculated head (C_H)
and chain (C_L) compressibilities vs. head/chain vol. ratio (V_r).

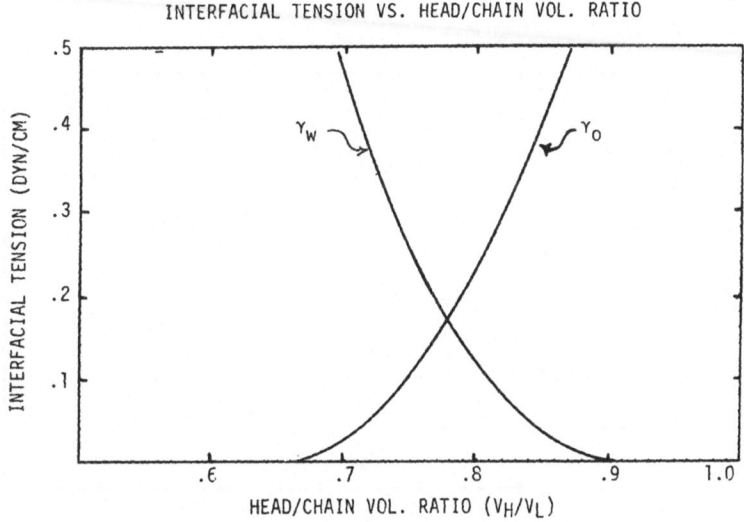

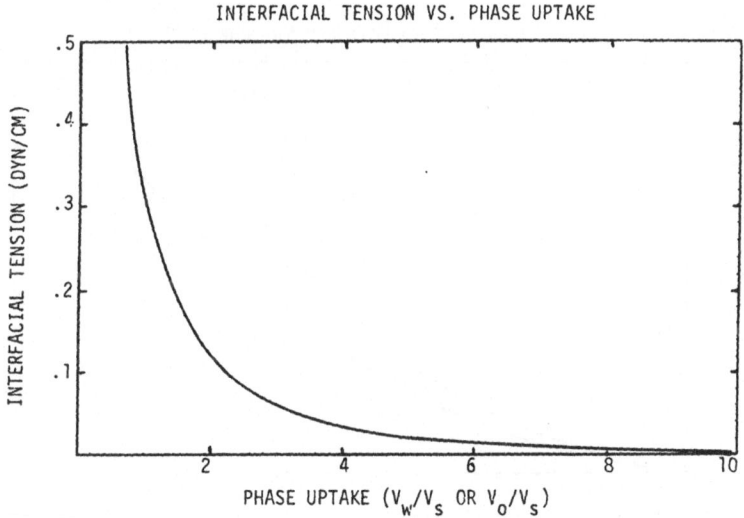

Figure 17. Theoretical predictions – Top: Interfacial tension vs. head/chain vol. ratio; γ_w = IFT water droplet vs. oil, γ_o = IFT oil droplet vs. water; calculations based on Figures 14 and 16 with τ_r = 0.91 and δ_r = 0.67. Bottom: Interfacial tension vs. water and oil uptake.

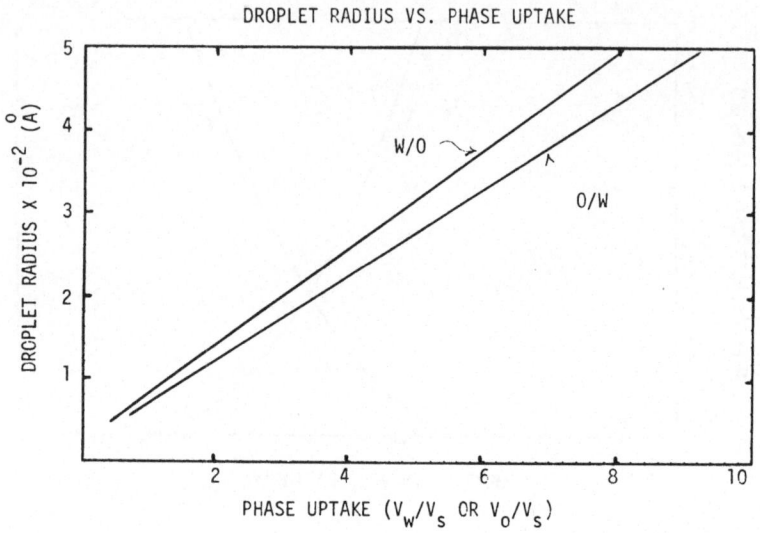

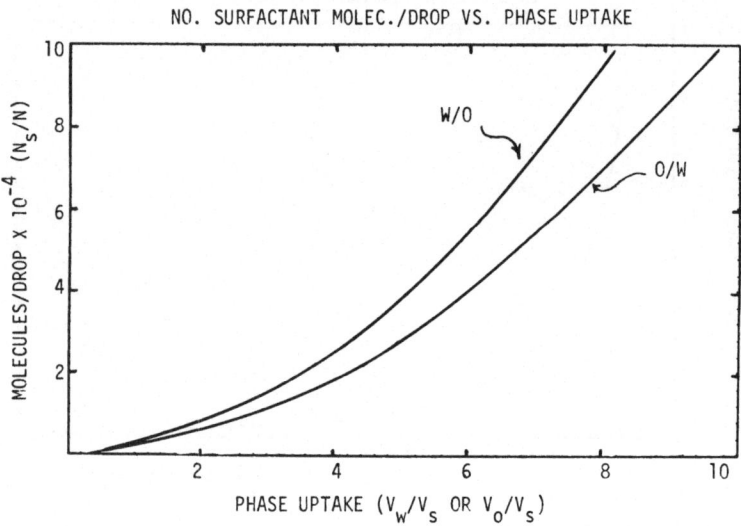

Figure 18. Theoretical predictions - Top: Droplet radius vs. phase uptake; calculations based on Figure 11 with $\tau_L = 10.2$Å. Bottom: Number of surfactant molecules/drop vs. phase uptake.

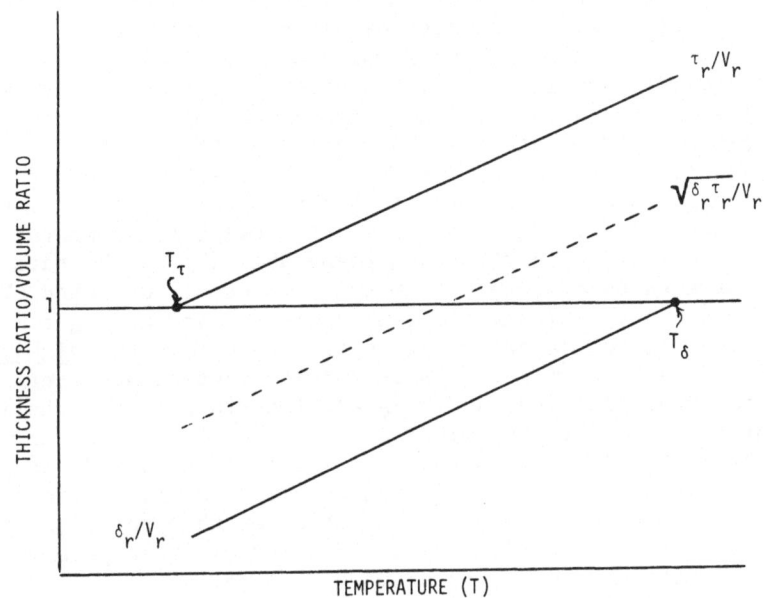

$$(\tau_r / V_r) - 1 = K_\tau \left[(T/T_\tau) - 1 \right] \qquad 1 - (\delta_r / V_r) = K_\delta \left[1 - (T / T_\delta) \right]$$

τ_r/V_r = HEAD TO CHAIN THICKNESS RATIO/VOL. RATIO (W/O)

δ_r/V_r = HEAD TO CHAIN THICKNESS RATIO/VOL. RATIO (O/W)

T_τ = TEMPERATURE FOR "INFINITE" WATER UPTAKE

T_δ = TEMPERATURE FOR "INFINITE" OIL UPTAKE

T_τ = 50.5°C T_δ = 62.0°C $K_\tau = K_\delta = 0.34$

Figure 19. Theory for saturated microemulsions: Temperature Dependence - Variation of thickness ratio/volume ratio with temperature for W/O (τ_r/V_r) and O/W (δ_r/V_r) microemulsions. Values for empirical parameters T_τ, T_δ and K chosen to fit the data of Saito and Shinoda (Figure 6).

temperature. This figure is analogous to Figure 15 with temperature
replacing volume ratio as the abscissa. Values of the empirical
parameters given at the bottom of the figure satisfy the phase uptake data of Saito and Shinoda (Figure 6). In Figure 20 the
relative interfacial tensions calculated from the model are compared
with interfacial tensions reported by Saito and Shinoda[28]. The
system containing equal weights of water and cyclohexane and 5 wt.%
ethoxylated nonyl phenol (8.6 EO) was allowed to phase separate
(Figure 6) and interfacial tensions between the micellar and bulk
phases measured. The data are represented by the points
whose values are given on the left axis. Calculated values of $C_H \gamma_w$
and $C_H \gamma_o$ are given on the right axis for the theoretical curves
(dashes). Interfacial tensions are plotted against temperature
(top) and phase uptake (bottom). A reasonable fit to the data is
obtained for a value of C_H = 0.1 cm/dyn. It should be noted that
γ_w and $\gamma_{m/w}$ are not strictly the same entities, nor are γ_o and $\gamma_{m/o}$.
The quantities γ_w and γ_o are interfacial tensions in the <u>highly
curved</u> interface between the droplet and its contacting external
phase. The quantities $\gamma_{m/w}$ and $\gamma_{m/o}$ are interfacial tensions in the
<u>planar</u> interface between the micellar phase and the contacting bulk
phases. It is fairly obvious that the γ's for the curved and planar
interfaces must be related since both result from interactions among
the surfactant and bulk phases. The nature of this relationship is
not evident. Yet, the similarity in behavior of these parameters is
so striking that the comparison seems warranted.

SUMMARY

   A model has been developed which quantitatively predicts phase
behavior in micellar systems or microemulsions. This model treats
the interface between a droplet and its contacting external phase
as a duplex monolayer of oriented surfactant molecules. The
respective hydrophilic heads and lipophilic chains sides of the
interface are treated as independent interphases, water interacting
with the heads and oil with the chains. Direction and degree of
curvature are imposed by a lateral stress gradient in the interface
resulting from differences in interaction on either side of the interface. This stress gradient is expressed in terms of physically
measurable quantities, surfactant molecular volume, interfacial
tension and compressibility.

   Surfactant volume and compressibility are split into separate
contributions by heads and chains. (This lays a quantitative
physical basis for describing Hydrophilic vs. Lipophilic interactions
in the system (HLB).) When the heads volume and compressibility are
larger than that of the chains, the interface curves convex to the
water phase, the microemulsion is O/W. The microemulsion is W/O when
the chains volume and compressibility is larger. Equations are
developed for both oil droplets (O/W) and water droplets (W/O) which

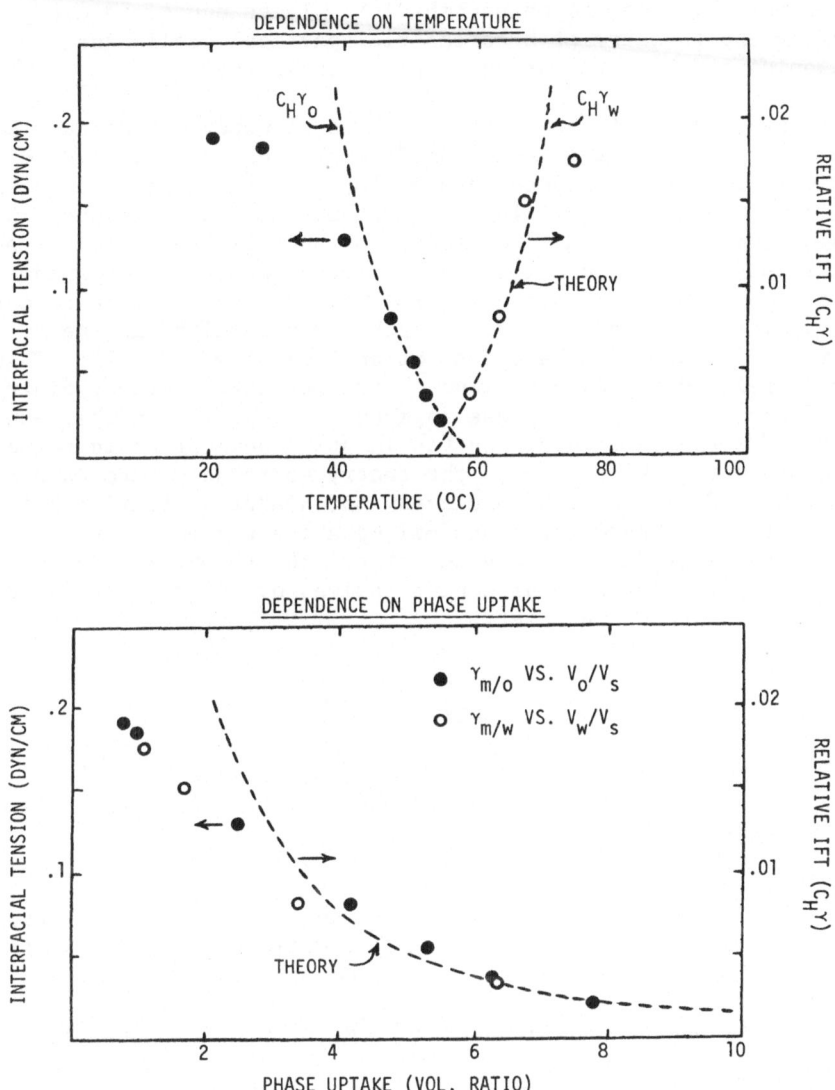

Figure 20. Experiment vs. theory: Interfacial Tension - Top: IFT vs. temperature according to Saito and Shinoda[28] corresponding to phase separation data of Figure 6; left axis refers to ● IFT Microemulsion/excess Oil, o IFT Microemulsion/excess Water; right axis refers to theoretical values of $C_H\gamma_O$ (O/W) and $C_H\gamma_W$ (W/O). Bottom: IFT vs. water and oil uptake.

relate interfacial tension and surfactant head and chain volumes
and compressibilities to phase behavior. These equations are generated by the geometry of 2 concentric spherical shells surrounding
the droplet and by force balance across these shells.

Phase behavior is expressed in terms of water and oil uptake
in saturated microemulsions and is described on an idealized ternary
phase diagram. The responses are explored of water and oil uptake
to temperature and composition. For ethoxylated surfactants, increasing temperature, salt concentration and oil aromaticity result
in increased oil uptake and decreased water uptake. Decreasing
head/chain volume and compressibility ratios have the same effect.
The system tends to become more lipophilic and shifts in the
direction of Winsor's micellar solution Types I → III → II. This
shift is related to phase regions on the idealized ternary diagram.
Types I and II fall in 2-phase regions and Type III in the 3-phase
region. The theory predicts shifts in phase boundaries in response
to the above variables. Thus, the theory correlates water and oil
uptake, the idealized ternary diagram and Winsor's transition in
micellar types. Where water and oil uptake are known, interfacial
tension can be predicted and vice versa. The theory also predicts
droplet size and interfacial concentrations of adsorbed surfactant
in terms of molecules/droplet.

REFERENCES

1. T. P. Hoar and J. H. Schulman, Nature, 152, 102 (1943).
2. J. H. Schulman and T. S. McRoberts, Trans. Faraday Soc., 42B,
    165 (1946).
3. J. H. Schulman, T. S. McRoberts and D. P. Riley, Proc. Physiol.
    Soc., 107, 49P (1948).
4. J. H. Schulman and D. P. Riley, J. Colloid Sci., 3, 383 (1948).
5. J. H. Schulman and J. A. Friend, J. Colloid Sci., 4, 497 (1949).
6. J. H. Schulman, R. Matalon and M. Cohen, Disc. Faraday Soc.,
    11, 117 (1951).
7. J. E. L. Bowcott and J. H. Schulman, Z. Electrochem., 59, 283
    (1955).
8. J. H. Schulman, W. Stoeckenius and L. M. Prince, J. Phys. Chem.,
    63, 1677 (1959).
9. W. Stoeckenius, J. H. Schulman and L. M. Prince, Kolloid Z.,
    169, 170 (1960).
10. J. H. Schulman and J. B. Montagne, Ann. N. Y. Acad. Sci., 92,
    366 (1961).
11. C. E. Cooke and J. H. Schulman, in "Surface Chemistry, p. 231,
    Munksgaard, Copenhagen, Academic Press, N.Y. (1965).
12. J. W. McBain, Advances Colloid Sci., 1, 99 (1942).
13. J. W. McBain and P. W. Richards, Ind. Eng. Chem., 38, 642
    (1946).
14. P. A. Winsor, Trans. Faraday Soc., 44, 376 (1948).

15. P. A. Winsor, "Solvent Properties of Amphiphilic Compounds", pps. 7, 57-60, 68-71, 190, Butterworths, London, 1954.

16. P. A. Winsor, Chem. Reviews, 68, 1 (1968).

17. S. R. Palit, V. A. Moghe and B. Biswas, Trans. Faraday Soc., 55, 463 (1959).

18. K. Shinoda and S. Friberg, Adv. Colloid Interface Sci., 4, 281 (1975).

19. G. Gillberg, H. Lehtinen and S. Friberg, J. Colloid Interface Sci., 33, 40 (1970).

20. K. Shinoda and H. Kunieda, J. Colloid Interface Sci., 42, 381 (1973).

21. S. I. Ahmad, K. Shinoda and S. Friberg, J. Colloid Interface Sci., 47, 32 (1974).

22. D. O. Shah, A. Tamjeedi, J. W. Falco and R. D. Walker, Jr., AIChE J., 18, 1116 (1972).

23. D. O. Shah, "On Distinguishing Microemulsions from Cosolubilized Systems", preprints for the 48th National Colloid Symposium, p. 173, Austin, Texas, June 1974.

24. L. M. Prince, J. Colloid Interface Sci., 52, 182 (1975).

25. W. C. Griffin, J. Soc. Cosmet. Chem., 1, 311 (1949); 5, 249 (1954).

26. K. Shinoda, J. Colloid Interface Sci., 24, 4 (1967).

27. K. Shinoda and H. Saito, J. Colloid Interface Sci., 26, 70 (1968).

28. H. Saito and K. Shinoda, J. Colloid Interface Sci., 32, 647 (1970).

29. H. Saito and K. Shinoda, J. Colloid Interface Sci., 24, 10 (1967).

30. K. Shinoda and T. Ogawa, J. Colloid Interface Sci., 24, 56 (1967).

31. K. Shinoda and H. Takeda, J. Colloid Interface Sci., 32, 642 (1970).

32. K. Kon-no and A. Kitahara, J. Colloid Interface Sci., 33, 124 (1970).

33. K. Kon-no and A. Kitahara, J. Colloid Interface Sci., 34, 221 (1970).

34. K. Kon-no and A. Kitahara, J. Colloid Interface Sci., 37, 469 (1971).

35. K. Kon-no and A. Kitahara, J. Colloid Interface Sci., 41, 47 (1972).

36. M. L. Robbins, "A Model for Oil Continuous Microemulsions", presented at the Symposium on Microemulsions, ACS National Meeting, Washington, D.C., September 1971.

37. M. L. Robbins, "Theory of Microemulsions", preprint for the Symposium on Interfacial Phenomena in Oil Recovery, AIChE National Meeting, Tulsa, Oklahoma, March 1974.

38. M. L. Robbins, "Theory of Microemulsions", preprints for the 48th National Colloid Symposium, p. 174, Austin, Texas, June 1974.

39. M. L. Robbins, "Theory of Microemulsions I and II", submitted
     to J. Colloid and Interface Sci.
40. R. N. Healy, R. L. Reed and D. G. Stenmark, "Multiphase Micro-
     emulsion Systems", preprint 5565 for the Society of Petroleum
     Engineers of AIME Meeting, Dallas, Texas, September 1975.
41. L. W. Holm, J. Pet. Tech., 23, 1475 (1971).
42. H. Al-Rikabi and J. S. Osaba, Oil and Gas Journal, 10/22/73,
     p. 87.
43. F. Harusawa, S. Nakamura and T. Mitsui, J. Colloid and Polymer
     Sci., 252, 613 (1974).
44. S. Friberg and I. Lapczynska, Prog. Colloid and Polymer Sci.,
     56, 16 (1975).
45. R. Haase and H. Schonert, "Solid-Liquid Equilibrium" in the
     "International Encyclopedia of Physical Chemistry and Chemical
     Physics", Topic 13, Volume 1, p. 152, Pergamon Press (1969).
46. S. Glasstone and D. Lewis, "Elements of Physical Chemistry",
     2nd ed., Van Nostrand, Princeton, N.J. (1960), p. 410.
47. R. N. Healy and R. L. Reed, "Physicochemical Aspects of Micro-
     emulsion Flooding", preprint SPE 4583 for the 48th Annual Fall
     Meeting of the SPE of AIME, Las Vegas, Nevada, September 1973.
48. K. Shinoda and H. Arai, J. Phys. Chem., 68, 3485 (1964).
49. R. F. Fedors, Polymer Engineering Sci., 14, 147 (1974).
50. W. D. Bancroft, J. Phys. Chem., 17, 501 (1913).
51. G. H. A. Clowes, J. Phys. Chem., 20, 407 (1916).
52. A. W. Adamson, "Physical Chemistry of Surfaces", p. 393,
     Interscience, 1960.
53. A. Beerbower and M. W. Hill, "McCutcheon's Detergents and
     Emulsifiers", p. 223, Allured Publishing, Ridgewood, N. J.,
     1971.
54. H. L. Rosano, J. Soc. Cosmet. Chem., 25, 609 (1974).
55. W. Gerbacia and H. L. Rosano, J. Colloid Interface Sci., 44,
     242 (1973).
56. L. M. Prince, J. Colloid Interface Sci., 23, 165 (1967).
57. L. M. Prince, J. Colloid Interface Sci., 29, 216 (1969).
58. C. L. Murphy, "Thermodynamics of Low Tension and Highly Curved
     Interfaces", Ph.D. Dissertation, Univ. of Minnesota (1966),
     University Microfilms, Ann Arbor, Michigan.
59. I. Langmuir, J. Amer. Chem. Soc., 39, 1848 (1917).
60. W. D. Harkins, E. C. H. Davies and G. L. Clark, J. Amer. Chem.
     Soc., 39, 541 (1917).
61. A. W. Adamson, J. Colloid Interface Sci., 29, 261 (1969).
62. W. C. Tosch, S. C. Jones, and A. W. Adamson, J. Colloid
     Interface Sci., 31, 297 (1969).
63. A. W. Adamson, "Physical Chemistry of Surfaces", 2nd ed.,
     pp. 100, 145, Interscience, 1967.
64. W. D. Harkins, "Physical Chemistry of Surface Films", p. 136,
     Reinhold, New York, 1952.

APPENDIX A

Derivation: $V_W/V_S = f[(\tau_H/\tau_L)/(V_H/V_L)]$

$V_H = N(4\Pi/3)(R^3 - R_w^3)$ $V_L = N(4\Pi/3)(R_o^3 - R^3)$

$V_H/V_L = [1-(R_w/R)^3]/[(R_o/R)^3 - 1]$

$\quad\quad = [1-(1-\tau_H/R)^3]/[(1+\tau_L/R)^3 - 1]$

$\quad\quad = \dfrac{(\tau_H/R)^3 - 3(\tau_H/R)^2 + 3(\tau_H/R)}{(\tau_L/R)^3 + 3(\tau_L/R)^2 + 3(\tau_L/R)}$

$\quad\quad \approx (\tau_H/\tau_L)(1-\tau_H/R)/(1+\tau_L/R); \quad (\tau_H/R)^2 \ll 1, \ (\tau_L/R)^2 \ll 1$

$\quad\quad = (\tau_H/\tau_L)\sqrt[3]{(V_w/V_s)/(V_w/V_s+1)}; \quad \text{see Figure 11}$

$V_w/V_s = 1/\{[(\tau_H/\tau_L)/(V_H/V_L)]^3 - 1\}$

Analogous treatment for an oil droplet in water leads to:

$V_o/V_s = 1/\{[(V_H/V_L)/(\delta_H/\delta_L)]^3 - 1\}$

APPENDIX B

Derivation of Force Balance Equations

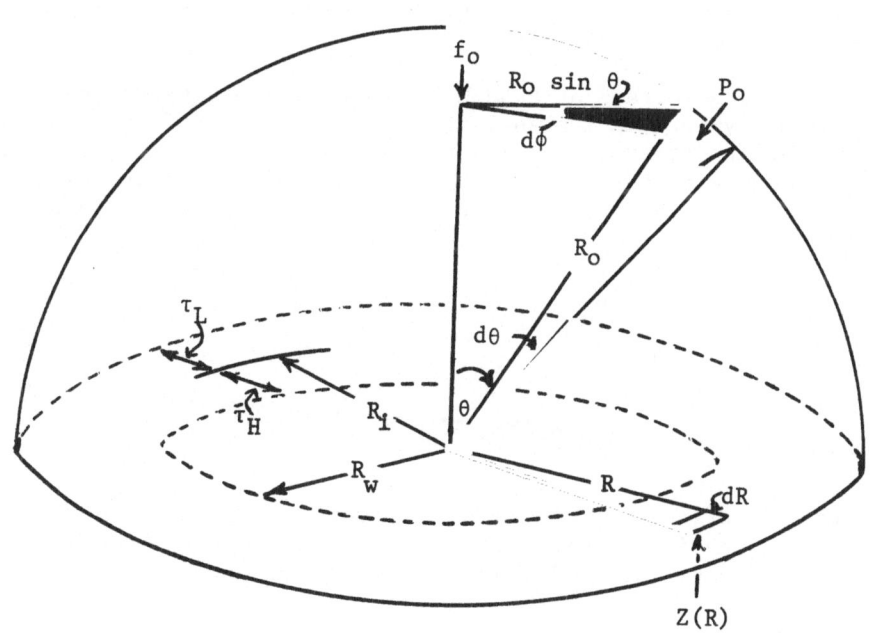

Consider a static isolated droplet with its shell of surfactant. Imagine a plane slicing the droplet through its center. The
force component perpendicular to a hemispherical cross section
caused by the external pressure P_o is given by:

$$f_o = \int_0^{2\pi} \int_0^{\pi/2} P_o R_o^2 \sin\theta \cos\theta \, d\theta \, d\phi = \pi R_o^2 P_o$$

Likewise, the force component due to the internal pressure P_w is
given by:

$$f_w = \int_0^{2\pi} \int_0^{\pi/2} P_w R_w^2 \sin\theta \cos\theta \, d\theta \, d\phi = \pi R_w^2 P_w$$

The force acting across the interfacial cross section is given by:

$$\Delta f = \pi (R_o^2 P_o - R_w^2 P_w)$$

This force must be balanced by the lateral stress, $Z(R)$, acting
over the cross section of the shell otherwise the droplet would
collapse. The lateral stress is a function of the radius, R, that
is, a stress gradient exists across the thickness of the interfacial
shell. The stress, $Z(R)$, arises from the resistance to lateral compression of the surfactant molecules. It is this same resistance
to compression which gives rise to the spreading or surface pressure,
π, in surfactant monolayers.

Choosing a surface at R_i such that $R_w < R_i < R_o$, we can relate
the stress anywhere in the shell to the reference surface at R_i. We
can express the stress variation relative to the reference surface
as a constant stress, Z, across the interface plus the incremental
stresses, ΔZ_H and ΔZ_L, across the heads and chains shells respectively. Thus,

$$R_o^2 P_o - R_w^2 P_w = 2\int_{R_w}^{R_o} Z(R) R \, dR$$

$$= 2[Z\int_{R_w}^{R_o} R \, dR + \int_{R_i}^{R_w} \Delta Z_H R \, dR + \int_{R_i}^{R_o} \Delta Z_L R \, dR]$$

We now define the interfacial pressure, π, by the expression
$\Delta Z = d\pi/dR$, the interfacial compressibility by $C = -(1/\sigma)d\sigma/d\pi$
(Reference 52, p. 131) and the interfacial tension by

$$\gamma_w = \int_{R_w}^{R_o} (P_o - Z) \, dR.$$

The definition for π states that interfacial pressure equals stress multiplied by interfacial thickness. The interfacial compressibility is the fractional change in molecular area with interfacial pressure. As defined, it allows for the change in shell volume on swelling with solvent. The interfacial tension relates the stress at the reference surface to atmospheric pressure, it is the excess stress integrated over the interfacial thickness. The definitions for interfacial pressure and tension are mutually consistent leading to the result $d\gamma = -d\pi$.

The assumption that the heads and chains act as separate, uniform liquid phases implies that the compressibilities are constant over the respective shells. For a spherical surface, $C = -(2/R)dR/d\pi$. We substitute into the force balance equation the definitions for interfacial pressure and compressibility and integrate holding compressibilities constant.

$$R_o{}^2 P_o - R_w{}^2 P_w = 2[Z\int_{R_w}^{R_o} R dR - (2/C_H)\int_{R_i}^{R_w} dR - (2/C_L)\int_{R_i}^{R_o} dR]$$

$$= Z(R_o{}^2 - R_w{}^2) + 4 \tau_H/C_H - 4 \tau_L/C_L$$

Algebraic manipulation and the definition for interfacial tension lead (with the subscript dropped from R_i) to:

$$P_w - P_o = \frac{\gamma_w(2 + \tau_L/R - \tau_H/R)}{R(1 - \tau_H/R)^2} - \frac{4[(\tau_H/RC_H) - (\tau_L/RC_L)]}{R(1 - \tau_H/R)^2}$$

It is obvious that this equation reduces to the Laplace equation $\Delta P = 2\gamma/R$ when the thickness/radius ratios approach zero, that is, for large drops. Applying the thermodynamic identity, $P_w - P_o = RT/v_w \ln p_w{}^m/p_w{}^o$ and setting the fugacity of water in the microemulsion equal to the fugacity of bulk water for a saturated system:

$$\gamma_w(2 + \tau_L/R - \tau_H/R) = 4[(\tau_H/RC_H) - (\tau_L/RC_L)]$$

Analogous treatment for an oil droplet dispersed in water gives:

$$\gamma_o(2 + \delta_H/A - \delta_L/A) = 4[(\delta_L/AC_L) - (\delta_H/AC_H)]$$

Algebraic manipulation yields:

$$\tau_L/R \approx (\tau_H/\tau_L - V_H/V_L)/[(\tau_H/\tau_L)^2 + V_H/V_L] \quad \text{(see Appendix A)}$$

$$= 2C_H\gamma_w/[4(\tau_H/\tau_L - C_H/C_L) + C_H\gamma_w(\tau_H/\tau_L - 1)]$$

$$C_H \gamma_w = \frac{4(\tau_H/\tau_L - V_H/V_L)(\tau_H/\tau_L - C_H/C_L)}{(\tau_H/\tau_L + V_H/V_L)(\tau_H/\tau_L + 1)}$$

$$C_H \gamma_o = \frac{4(V_H/V_L - \delta_H/\delta_L)(C_H/C_L - \delta_H/\delta_L)}{(\delta_H/\delta_L + V_H/V_L)(\delta_H/\delta_L + 1)}$$

APPENDIX C

Evaluation of the Compressibility Ratio

Since a saturated droplet is in equilibrium with its bulk phase, $P_w - P_o = 0$ (see Appendix B) and no PV work is done to form a saturated droplet from its bulk phase. At constant temperature, only interfacial work is done.

Work to form a water droplet: $W_w = \gamma_w(4\Pi R^2)$

Work to form an oil droplet: $W_o = \gamma_o(4\Pi A^2)$

In a micellar solution saturated with both water and oil (Type III) water and oil droplets are assumed in equilibrium. Therefore, $W_w = W_o$ and:

$$\frac{\gamma_w}{\gamma_o} = \frac{(\tau_r - V_r)(\tau_r - C_r)(\delta_r + V_r)(\delta_r + 1)}{(V_r - \delta_r)(C_r - \delta_r)(\tau_r + V_r)(\tau_r + 1)} = \frac{A^2}{R^2}$$

$$C_r = (\tau_r + X\delta_r)/(1 + X)$$

$$X = \frac{(V_r - \delta_r)(\tau_r + V_r)(\tau_r + 1)A^2}{(\tau_r - V_r)(\delta_r + V_r)(\delta_r + 1)R^2}$$

$$A/R = (V_o/V_s + V_L/V_s)/(V_w/V_s + V_H/V_s) \quad \text{(see Figure 11)}$$

NOMENCLATURE

P = Pressure

Z = Stress

R = Radius (water droplet)

π = Interfacial Pressure

C = Interfacial Compressibility

γ = Interfacial Tension

σ = Molecular Area

τ = Shell Thickness (water droplet)

δ = Shell Thickness (oil droplet)

A = Radius (oil droplet)

V = Volume

H = Hydrophilic Heads

L = Lypophilic Chains

W = Water

0 = Oil

S = Surfactant

r = Head/Chain Ratio

STABILITY, PHASE EQUILIBRIA, AND INTERFACIAL FREE ENERGY IN MICROEMULSIONS

E. Ruckenstein

State University of New York at Buffalo, Faculty of Engineering and Applied Sciences, Buffalo, NY 14214

The stability of emulsions is examined from the points of view of thermodynamics as well as dynamics. First a statistical thermodynamical treatment is developed, which demonstrates that, in contrast to usual emulsions, microemulsions can be stable from a thermodynamic point of view. The expression established for the free energy of formation of microemulsions predicts phase inversion from one type of microemulsion to the other one as well as phase separation. Several basic types of multiphase systems are found theoretically: (1) microemulsion in equilibrium with another (in particular dilute) microemulsion of oil-in-water; (2) microemulsion in equilibrium with another (in particular dilute) microemulsion of water-in-oil, and (3) a dilute microemulsion of water-in-oil in equilibrium with a dilute microemulsion of oil-in-water. Similar multiphase systems have been previously identified experimentally by Winsor and by Healy et al. The present expressions also explain the low values of the interfacial free energy between two microemulsion phases in equilibrium. Second, a transport equation is established for a concentrated colloidal system. Based on this equation conditions are identified when the system is stable or unstable to small perturbations and information about its time evolution obtained. This dynamic approach provides a time scale for significant changes to occur in an unstable system.

INTRODUCTION

The stability of ordinary emulsions can be treated in the framework of the DLVO theory[1,2]. The repulsive double layer forces and attractive van der Waals forces between two particles generate a potential barrier between them; the rate of flocculation decreases exponentially with the height of the barrier. There are other cases however for which the stability problem is less simple. For instance, water in oil (W/O) emulsions of moderate concentration always flocculate, at variance with the traditional DLVO theory, which predicts high potential barriers for these cases. This behavior can be understood by a simple extension of the DLVO theory. The double layer forces are short range in the case of oil in water (O/W) emulsions, where the thickness of the double layer is generally less than 100 Å, and therefore the interaction forces between two particles control stability. For W/O emulsions, the thickness of the double layer can be as large as several microns and therefore the double layer forces are long range. The cooperative interaction between such particles strongly diminishes the potential barrier between two neighboring particles, thus promoting flocculation[3,4]. Ordinary emulsions are always unstable from a thermodynamic point of view since a decrease in the area between the two phases results in a decrease of the Gibbs free energy. The apparent stability is only due to the slow kinetics of their return to the initial state.

There are, however, concentrated or dilute emulsions of W/O and O/W types which are completely stable and whose stability cannot be understood in terms of van der Waals and double layer forces alone. They are the so called microemulsions which are isotropic, optically transparent and thermodynamically stable dispersions of oil-in-water or water-in-oil. Such dispersions form spontaneously by mixing oil, water, a surfactant (such as an alkali metal soap) and a co-surfactant (such as an alkyl alcohol)[5-10].The occurrence of these thermodynamically stable emulsions was attributed[9-12] to a negative interfacial tension due to the high film pressure established by the mixed surfactant and co-surfactant at the hydrocarbon-water interface. One should note, however, that even if the concept of a negative interfacial tension is accepted, it is not sufficient to account for the finite size of globules. The existence of a finite size of particles of 100 to 1000 Å implies also at least a positive contribution to the free energy of the system. Together with a negative surface free energy, this will allow the formation of a negative minimum of the free energy of formation of the microemulsions for a given finite radius of particles. The free energy of formation of the double layer, the entropy of dispersion, and the van der Waals interaction forces have a negative contribution to the free energy of formation of a microemulsion. If these effects are associated with the positive contributions provided by a sufficiently

small but positive surface free energy and by the double layer repulsive forces, one can, at least in principle, predict stable microemulsions without assuming negative values for the surface free energy.
Such a point of view was developed by the author[19]. For the usual
emulsions the surface free energy is large and therefore thermodynamically stable states can never be achieved. A high free energy barrier, which opposes coalescence, can only occur in these cases.

Microemulsions have been studied extensively from an experimental point of view[5-18]. The results of interest here can be summarized as follows: (a) Microemulsions have practically an infinite
life. (b) An inversion can occur from one type to another. Shah
et al.[12], studying the hexadecane-water-potassium oleate-hexanol
system, suggest the occurrence of the following structural changes
as the ratio water/oil is increased: water-in-oil microemulsions →
water cylinders in oil → lamellar structure of surfactants, oil and
water → oil-in-water microemulsions. (c) Phase separation occurs
in some conditions. Winsor[16] and Healy et al.[17] identified three
basic types of multiphase systems: (1) microemulsions in equilibrium with excess water; (2) microemulsions in equilibrium with
excess oil containing molecularly dispersed surfactant; and (3)
excess of water and excess of oil in equilibrium with a phase
placed between the first two and containing oil, water, and surfactant. (d) The interfacial free energy between the demixed phases
is very low. This is of particular significance in oil recovery
because only a large decrease of the capillary forces enables to
the injected water to displace the oil droplets from the small pores
in which they are trapped.

Since microemulsions appear to be stable from a thermodynamic
point of view, it is natural to inquire if thermodynamics is able
to predict their occurrence. The answer is positive. Ruckenstein
and Chi[19] have developed such a treatment and the main goal of the
present paper is to explain on its basis the above listed experimental facts. To gain insight about some of the thermodynamic
results as well as information about the length of time in which
appreciable changes occur in an unstable system, the dynamic
response of colloidal systems to small perturbations is examined
in the final section of the paper.

THERMODYNAMIC STABILIY

The treatment is based on the simplifying assumption that the
surfactant and co-surfactant determine only the surface properties
of the droplets of one phase in the other one. In other words,
instead of considering a system formed of four components, we
treat the problem assuming only two components, oil and water, with
only the surface free energy and other surface properties influenced by the surfactant and co-surfactant. A more complete treatment

accounting for the separate existence of the surfactant and co-surfactant will be published elsewhere.

The free energy change due to the mixing of the liquids consists of three major contributions

$$\Delta G_M = \Delta G_1 + \Delta G_2 + \Delta G_3 \quad . \tag{1}$$

The first, ΔG_1, represents the free energy of formation of the
interface and contains the surface free energy (a positive quantity)
and the free energy of formation of the double layer (which is negative because double layers form spontaneously). An expression for
ΔG_1 was established previously, in the framework of the Debye–Hückel
approximation, assuming large distances between globules[19]:

$$\Delta G_1 = 4\pi R^2 m [f_s - \frac{\psi_o^2}{8\pi R} \{\epsilon_1 (1 + K_1 R) + \epsilon_2 (K_2 R \text{ ctnh } K_2 R - 1)\}] \quad , \tag{2}$$

where R is the radius of the globule and m is the number of globules per cm^3 ($m = \phi_2/v_2$). Note that if the specific surface free
energy is small enough and/or the surface potential ψ_o is large
enough, ΔG_1 becomes negative. It will be shown latter that the
stability of microemulsions is strongly affected by this term.

The second term in Equation (1) represents the contribution of
the interaction forces between the globules and contains a negative
term $\Delta G_2''$ due to the attractive van der Waals forces and a positive
term $\Delta G_2'$ due to the repulsive double layer interactions. Numerical computations have shown that, in general, $\Delta G_2'' \ll \Delta G_2'$. The
following expression was deduced[19] for $\Delta G_2'$:

$$\Delta G_2' = \frac{9m \, \epsilon_1 \, \psi_o^2 \, R^2}{K_1^3 \, (R_1 - 2R)^3 \, R_1} [1 + 1.48 \frac{1 + 1.3 \, K_1 R_1}{(K_1 R_1)^2} e^{-0.3 \, K_1 R_1}] \times$$

$$\{[K_1(R_1 - 2R) - 1] + [K_1(R_1 - 2R) + 1] e^{-2K_1(R_1 - 2R)}\}, \tag{3}$$

where $R_1 = 2R (0.74/\phi_2)^{1/3} \quad . \tag{4}$

Here R_1 is the average distance between globules. Equation (3) assumes: (1) a particular globule is surrounded by twelve nearest
neighbor globules which can occupy any arbitrary positions on a
spherical shell of radius R_1; (2) all the other globules are homogeneously distributed in the space outside a sphere of radius 1.3 R_1;

(3) the center of the globule considered can be located with equal probability within the sphere of radius $R_1 - 2R$.

The van der Waals interactions have been computed previously on the basis of the Lifshitz approach. For the sake of simplicity, an expression is used here which is based upon the more traditional Hamaker treatment:

$$\Delta G_2'' = - m A\left[\frac{2}{(R_1/R)^2 - 4} + \frac{2}{(R_1/R)^2} + \ln \frac{(R_1/R)^2 - 4}{(R_1/R)^2}\right]. \qquad (5)$$

Equation (5) involves the simplifying assumption that each globule interacts only with 12 nearest neighbors placed a distance R_1 from the globule.

The last term in Equation (1) represents the entropic effect and has obviously a negative contribution to ΔG_M. Its importance increases with the decrease of the radius R. Various expressions have been established by us in the previous paper[19]. One of these is

$$\Delta G_3 = k\, T m \ln (v\, m) \quad , \qquad (6)$$

where v is the volume of one water molecule. Other expressions were also used in the present computations, but no significant change occurred in ΔG_M.

The radius of the microemulsion corresponds to the minimum of ΔG_M. Obviously, the corresponding value of ΔG_M, denoted by ΔG^*_M, has to be negative. Using the above expressions, ΔG^*_M and the corresponding equilibrium radius R* of the globules can be computed as a function of the volume fraction ϕ_{oil}, ionic strength, Hamaker constant, surface potential, surface free energy and temperature.

Traditional thermodynamic considerations can be used to look for the occurrence of stable microemulsions, phase inversion or phase separation. For instance if $\frac{d^2 \Delta G^*_M}{d\phi^2_{oil}} < 0$ over a range of ϕ_{oil}, then the corresponding microemulsion will be unstable with respect to phase separation and will separate into two phases in equilibrium. The equilibrium curve is obtained from the equality between the chemical potentials of the components in the two phases in equilibrium. Here the chemical potential is defined as the derivative of the free energy with respect to the volume of the component in the respective phase. The usual formalism, in terms of mols or molar fractions is replaced here by one in terms of volumes and volume fractions. Obviously this implies that the volumes are additive. The values ϕ'_{oil} and ϕ''_{oil}, of the volume fraction in the two phases in equilibrium are such that $\Delta G^{*'}_M$ and $\Delta G^{*''}_M$ have a

common tangent. In Figure 1 a phase diagram for a binary mixture
is shown in terms of temperature vs. volume fraction. The interval
between ϕ' and ϕ'' contains two regions, the unstable region in which
$\dfrac{d^2 \Delta G^*_M}{d \phi^2_{oil}} < 0$ and the metastable region in which this second derivative is positive. The spinodal curve separates the two regions.
A hypothetical mixture corresponding to a point between the spinodal and equilibrium curves also separates into the two equilibrium
phases. In the stable region $\dfrac{d^2 \Delta G^*_M}{d \phi^2_{oil}} > 0$. Concerning phase inversion, that type of microemulsion for which $- \Delta G^*_M$ is larger will
exist; therefore, the value of ϕ_{oil} for which phase inversion occurs
can be computed from the equality of the two ΔG^*_M's for the two
types of microemulsions. The transition from one type of microemulsion to the other involves various liquid crystal structures
which may be stable in a given range of ϕ_{oil}. They cannot be predicted by the present model which assumes spherical globules.

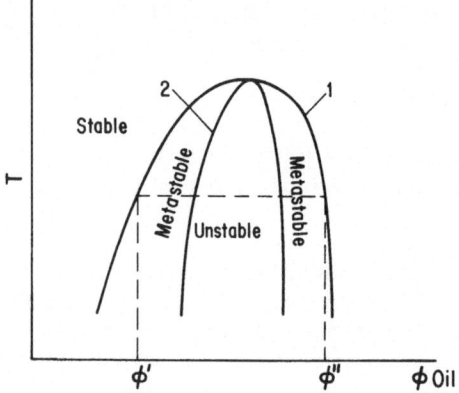

Figure 1. Temperature - Volume Fraction Phase Diagram. 1 - the
equilibrium curve; 2 - the spinodal curve.

     The equilibrium free energy of formation of a microemulsion,
ΔG^*_M, and the corresponding radius, R^*, have been computed numerically as a function of the relevant parameters. Some of the
results are plotted in Figure 2 in terms of $- \Delta G^*_M$ vs. ϕ_{oil} and in
Figure 3 in terms of R^* vs. ϕ_{oil}. Before discussing Figure 2 fully,
one should note that all the curves representing O/W microemulsions
in this figure start in reality from the origin, grow abruptly to
large values at a small (or even very small) volume fraction ϕ_{oil},
attain a maximum and then decrease. The curves representing W/O
microemulsions start from $\Delta G^*_M = 0$ and $\phi_{oil} = 1$. The values for ψ_o
are in the range identified experimentally for emulsions by Albers
and Overbeek[3].

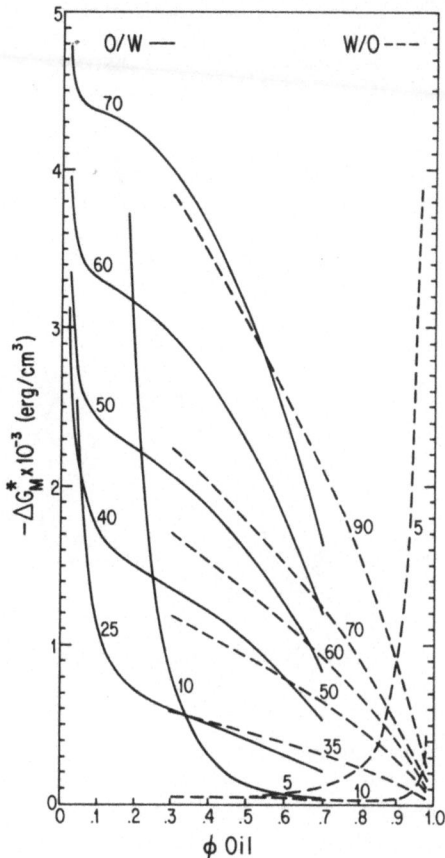

Figure 2. Negative Free Energy of Formation of Microemulsions vs.
Volume Fraction of Oil, ϕ_{oil}. The numbers on the curves give the
surface potential in mV. The values of the other parameters are:
$A = 10^{-14}$ erg; $f_s = 10^{-4}$ dyne/cm; $K_w^{-1} = 4 \times 10^{-6}$ cm; $K_{oil}^{-1} = 10^{-4}$
cm; $T = 300°$ K.

     The curve for O/W microemulsion with $\psi_o = 70$ mV has three re-
gions: A region very near the origin where dilute microemulsions
form with R* below 100 Å; a region which extends up to about
$\phi_{oil} \approx 0.3$ where separation into two O/W phases occurs because a
common tangent is possible between $\phi'_{oil} \approx 0$ and $\phi''_{oil} \approx 0.3$; and
a region for $\phi_{oil} > 0.3$ in which microemulsions stable to phase
separation exist (the radius in this region is of the order of
10^3 Å). The present model, based upon the assumption that the
globules are spherical, is not expected to hold for O/W microemulsions with large values of ϕ_{oil}. For values of ϕ_{oil} larger than
about 0.6 if no phase inversion occurs, various kinds of structures

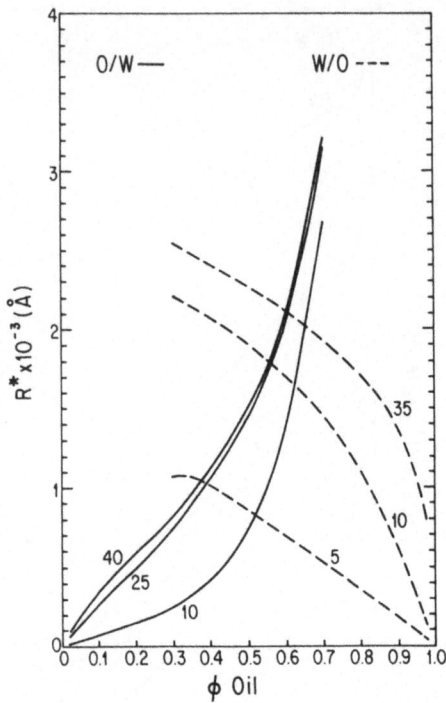

Figure 3.   Equilibrium Radius, R*, vs. Volume Fraction of Oil, ϕ_{oil}.
The numbers on the curves give the surface potential in mV.   The
values of the parameters are as for Figure 2.   R* is almost independent of ψ_o for $\psi_o > 40$ mV.

probably originate before phase inversion.

The picture can change if one associates the corresponding W/O
curve with the O/W curve.   Since there is no reason to assume that
the surface potential is the same for the corresponding O/W and W/O
microemulsions, let us associate the W/O curve for $\psi_o = 90$ mV with
the O/W curve for $\psi_o = 70$ mV.   Then any mixture with $0.45 \lesssim \phi_{oil} \lesssim$
0.7 will separate into two phases: one is an O/W phase with $\phi_{oil} \approx$
0.45 and the other a W/O phase with $\phi_{oil} \approx 0.7$.   If the associated
W/O curve has $\psi_o = 70$ mV, the two curves have no common tangent and
a phase inversion is possible for $\phi_{oil} \approx 0.73$.

Let us now assume that the O/W curve for $\psi_o = 40$ mV can be
associated with the W/O curve for $\psi_o = 60$ mV.   In this case phase
separation will occur for $0 < \phi_{oil} \lesssim 0.9$.   One of the phases has
an O/W structure but is very dilute in oil ($\phi_{oil} \approx 0$), and the other
has a W/O structure ($\phi_{oil} \approx 0.9$).   For $\phi_{oil} \gtrsim 0.9$, W/O structures
occur which are stable against phase separation.

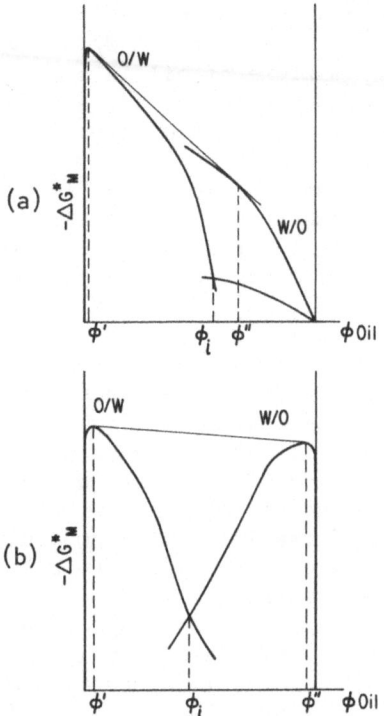

Figure 4a, 4b. Negative Free Energy of Formation of Microemulsions vs. ϕ_{oil}. ϕ' and ϕ'' are the volume fractions of oil for the two phases in equilibrium. ϕ_i is the volume fraction of oil at the inversion point. In (a) the inversion corresponds to stable microemulsion; in (b) to an unstable one.

As a last example, let us combine the O/W structure for ψ_o = 25 mV with the W/O structure for ψ_o = 10 mV. In this case the mixtures will separate into two dilute phases, one of the O/W type and the other of the W/O type. One should note that the interpretation of Figures 3 and 4 in our previous paper[19] has to be revised. Any mixture of oil and water in those figures will separate into two very dilute phases of the O/W and W/O type.

Figures 4a and 4b summarize some of the important features of the computations from a qualitative point of view. In Figure 4a, a microemulsion phase of the W/O type with $\phi_{oil} = \phi''$ is in equilibrium with excess water containing a very small amount of oil droplets. It represents the first basic type of multiphase system found experimentally by Winsor[16] and Healy et al.[17]. If the roles of water and oil are reversed, the same figure represents the second basic type of microemulsions. (Note that the dilute phases are not pure. For the pure phases $\Delta G*_M$ = 0, while for very dilute ones

$-\Delta G*_M$ grows abruptly passing through maxima). Figure 4a also shows
a phase inversion at $\phi_{oil} = \phi_i$ for the W/O curve below the previous
one. In Figure 4b any mixture of oil in water will separate into
two phases: one of the O/W type and the other of the W/O type,
each of them having a small amount of dispersed phase.

The interface between the two phases in equilibrium is expected
to be diffuse and the transition between the two phases continuous.
The transition region between the equilibrium volume fractions ϕ'
and ϕ'' passes through the metastable and unstable states. Therefore, near the excess water phase there is a microemulsion of oil-
in-water, and near the excess oil interface a microemulsion of
water-in-oil. Of course the inversion from globules of oil in
water to globules of water in oil requires some intermediary changes
(through cylinders of oil-in-water and lamellar structures of oil,
water and surfactant) which have not been accounted for in the
model. This representation of the diffuse layer between the two
phases is, however, not complete. Indeed, the curve $\Delta G*_M$ vs. ϕ
obtained on the basis of traditional thermodynamics ignores the
fact that the diffuse layer is non-uniform as concerns the concentration of various components, the surface properties of the interface between the phases and the topology of the distribution of one
phase in the other one. Compared with uniform systems, these non-
uniformities change the interaction potential acting in a given
point, generating an additional field. This field in a given "point"
is the difference between the interaction potential with the non-
uniform system and that with a uniform system having the properties
of the considered point. At least in principle (the difficulties
are enormous) it is possible to obtain information about the distribution of the volume fraction and topology of dispersion using
improved version of the thermodynamics of non-uniform molecular
systems[20,21]. Hence, the states in the diffuse layer do not correspond to the metastable and unstable states of traditional thermodynamics, but are new states probably stabilized by the field
generated by non-uniformity. From a qualitative point of view we
expect the globules to be spherical in the bulk phase of a microemulsion and somewhat deformed near the interface of two phases of
the same kind. For the system in Figure 4b the globules are probably deformed near the excess phases. In the middle of the diffuse
layer the structure is complex and it is probably impossible to
identify which of the phases is continuous and which dispersed. If
instead of treating microemulsions as a binary system the presence
of surfactants is also taken into account, one can suspect that the
diffuse layer of the system in Figure 4b will transform into a new
phase. This phase has diffuse layers at the two interfaces and a
very complex structure of oil, water and excess of surfactant in
its middle. It is not possible to identify which of the components
is dispersed and which is continuous in the intermediate structure.
The phase is bi-dispersed and bi-continuous.

It is of interest to evaluate the relative contribution of various effects to the free energy of formation, ΔG_M, of microemulsions. First we mention that numerous computations have shown that microemulsions can be obtained if the specific free energy $f_s < 10^{-2}$ dyne/cm. For unusually large values of ψ_0, however, the value of f_s can be larger. Values of $f_s < 10^{-4}$ dyne/cm have no influence on ΔG^*_M and R^*. The van der Waals interactions have a very small effect (generally negligible) upon ΔG^*_M and R^*. The entropic effect is important only when the equilibrium radius is smaller than about 200 Å. The most important effects for the present computations have been the free energy of formation of the double layer and the free energy contribution of the double layer interactions. The former has a negative contribution and the latter a positive one to the free energy of formation of microemulsions. It should be noted

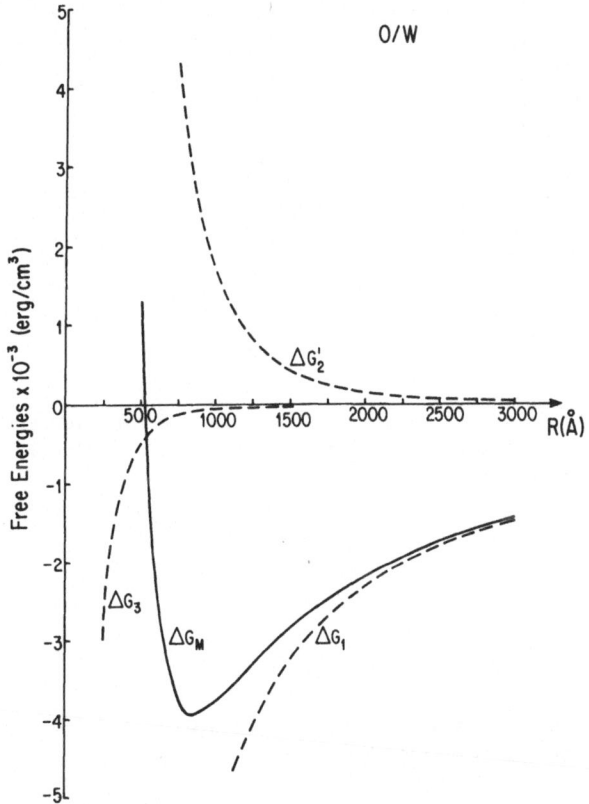

Figure 5. Comparison Between Various Contributions to the Free Energy of Formation of the O/W Microemulsions. The values of the parameters are: $\psi_0 = 70$ mV; $\phi_{oil} = 0.3$; $f_s = 10^{-3}$ dyne/cm; A = 10^{-13} erg; $K_w^{-1} = 4 \times 10^{-6}$ cm; $K_{oil}^{-1} = 10^{-4}$ cm; T = 300° K. The microemulsion is stable in this case.

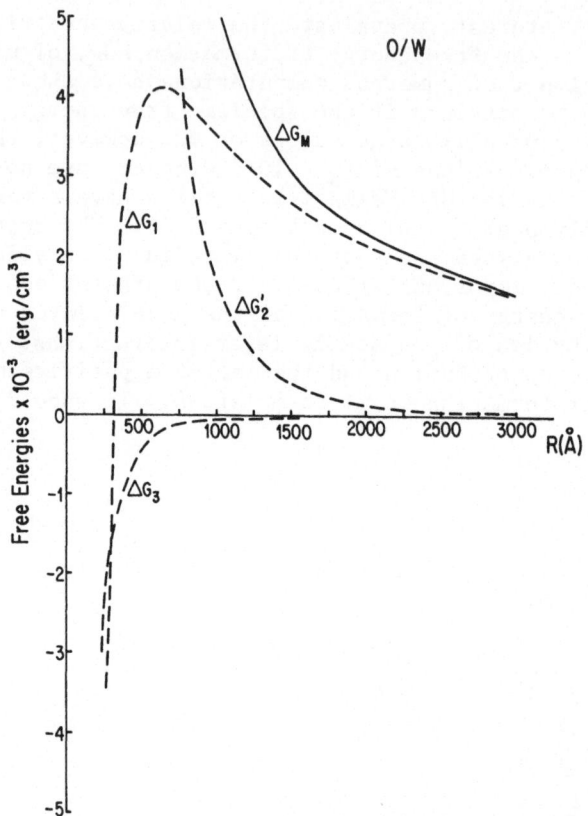

Figure 6. Comparison Between Various Contributions to the Free Energy of Formation of the O/W Microemulsions. The values of the parameters are: ψ_0 = 70 mV; ϕ_{oil} = 0.3; f_S = 10^{-1} dyne/cm; A = 10^{-13} erg; K_w^{-1} = 4 x 10^{-6} cm; K_{oil}^{-1} = 10^{-4} cm; T = 300° K. Microemulsions cannot form in these conditions.

that, for the sake of simplicity, the free energy of formation of the double layer was computed using the Debye-Hückel approximation, although the values considered for ψ_0 are outside its range of validity. Figures 5 and 6 give an idea about the contribution of various effects to ΔG_M for some specific cases.

We note that the kinetic stability of conventional emulsions is determined by double layer repulsion and van der Waals attraction. In contrast, the thermodynamic stability of microemulsions is controlled by the free energy of formation of the double layer, by the entropic effect (only for $R^* < 200$ Å) and by the repulsive double layer forces. It seems that the van der Waals interactions between globules are of secondary importance, at least for the range of parameters studied.

INTERFACIAL FREE ENERGY

The interfacial free energy, σ, between two microemulsion phases in equilibrium can be very small. This fact makes microemulsions of interest in oil recovery. Since σ can be expressed in terms of the free energy of formation of a microemulsion in the metastable and unstable regions, some comments concerning the interfacial free energy are in order in the present context. There is a rich literature concerning demixing of polymer solutions which shows both experimentally and theoretically that the interfacial free energy between the demixed phases, σ, is very low[22-26]. The theoretical treatment[20,25,26] is based upon Cahn and Hilliard's approach to the thermodynamics of non-uniform systems and the result can be used without modification to evaluate σ for microemulsions. In this theory, the interface is assumed diffuse. Accounting for the effect of the non-uniformity in concentration upon the interaction potential in a given point one obtains

$$\sigma = \int_{\phi'}^{\phi''} (L^2 \Delta G^*_d)^{1/2} \, d\phi \simeq L \int_{\phi'}^{\phi''} (\Delta G^*_d)^{1/2} \, d\phi \, , \tag{7}$$

where

$$L^2 = -\frac{4\pi}{3} \int_{R'}^{\infty} r^4 \left(\frac{u_{11}(r)}{v_1^2} + \frac{u_{22}(r)}{v_2^2} - \frac{2 \, u_{12}(r)}{v_1 v_2}\right) g(r) \, dr \, , \tag{8}$$

$$\Delta G^*_d = \Delta G^*_M - \Delta G^{**}_M \tag{9}$$

and ΔG^{**}_M is the free energy of formation corresponding to a point on the straight line connecting the equilibrium phases for the same volume fraction as ΔG^*_M. The integral in Equation (7) contains the free energy ΔG^*_d between ϕ' and ϕ'', therefore for the metastable and unstable regions. Although L depends on ϕ, this dependence is generally weak so that L is assumed constant. L is a consequence of non-uniformity and is related to the interaction potential u_{11} between two molecules of the continuous phase, u_{22} between two globules and u_{12} between one molecule and one globule. It contains also the volume per molecule of the continuous phase v_1 and the volume of the globule v_2. The lower limit of integration is not the same for all three terms in Equation (8). For each of them it is the average distance between the two kinds of particles. The radial distribution function g is also not the same for all three terms.

Equation (7) assumes that the globules are spherical, of uniform size, and that the surface properties (in particular the surface potential) are independent of position. These conditions are not generally met. In the metastable and unstable regions the radius depends strongly upon the volume fraction ϕ_{oil} (Figure 3). As noted before, conventional thermodynamics is not strictly valid in the diffuse non-uniform region; therefore, the radius has to be obtained accounting for non-uniformity. Besides, for the situation depicted in Figure 4b the topology of dispersion is very complex and difficult to predict. For these reasons, only a simple evaluation of the order of magnitude of σ will be carried out on the basis of Equation (7) assuming a constant R*. Equation (7) contains two factors: L and $(\Delta G*_d)^{1/2}$. $(\Delta G*_d)^{1/2}$ is of the order of 30 $(erg/cm^3)^{1/2}$ (see Figure 2). Concerning L^2 an evaluation will be made starting from a result obtained from spinodal decomposition measurements in binary alloys. In that case a value for L^2 of the order of 10^{-5} erg/cm, was found[27]. It is plausible to consider that this value is an upper bound for microemulsions. Indeed, in the present case the value of L^2 can be smaller if the repulsive double layer interactions between globules introduce a sufficiently large positive contribution to u_{22} and/or the attraction forces between the globule and the molecules of the continuous phase are strong enough. Since, in many cases, double layer forces dominate the attractive van der Waals interactions (see previous section), $\dfrac{u_{22}}{v_2^2}$ has for those cases a negative contribution to the value L^2 instead of the positive contribution of the same term in alloys. L^2 may become very small, perhaps even zero when the double layer repulsion and/or the attraction force between globule and molecules of the continuous phase are strong enough. Note that while for molecular mixtures σ becomes zero at the critical point because there $\phi' = \phi''$, for microemulsions σ may also become very small and perhaps zero because of L^2.

Using for $(\Delta G*_d)^{1/2}$ a value of 30 $(erg/cm^3)^{1/2}$ and for L^2 a value of 10^{-5} erg/cm, one obtains $\sigma = 10^{-1}$ dyne/cm. For reasons just discussed, σ can be much smaller than this value. In general, the low values of the interfacial free energy σ between the demixed microemulsion phases can be explained in terms of the strong repulsive double layer forces between globules, strong attraction forces between globules and the molecules of the continuous phase, and small values of $\Delta G*_d$.

For the thickness δ of the diffuse layer, dimensional considerations yield

$$\delta \simeq \frac{L}{\phi'' \int\limits_{\phi'} (\Delta G^*_d)^{1/2} \, d\phi} \simeq \frac{L^2}{\sigma} \quad . \tag{10}$$

The thickness δ can be very large. Indeed for $L^2 \simeq 10^{-6}$ erg/cm and $\sigma \simeq 10^{-3}$ erg/cm^2, $\delta = 10^{-3}$ cm. It should be noted that when σ becomes zero because $L \simeq 0$, then $\delta = 0$. This suggests that, depending upon conditions, the interface can be diffuse or sharp. However, the above treatment of the surface free energy is valid only when δ is sufficiently large, hence only when L is not too small.

DYNAMIC TREATMENT BY LINEAR STABILITY

Some insight about the differences between stable and meta-stable states on one side and unstable states on the other side can be gained by examining the dynamic response of the colloidal system to small perturbations in concentration. The treatment which follows is similar to that employed by Cahn[28] in the kinetics of spinodal decomposition. We modified[29] this for unequal volumes of particles to examine the problem of stability of concentrated colloidal systems. Consider a liquid containing a large concentration of spherical particles of uniform size interacting between them via London and double layer forces. A small perturbation is applied to the initially uniform concentration. If the response of the system is such that the perturbation decays in time, the system is stable to small perturbations in concentration; in the opposite situation the system is unstable. Both the decay as well as the growth of the perturbation involve displacements of the liquid and particles and, therefore, the starting point of such an analysis should be a transport equation. To write such an equation an expression for the diffusional fluxes is needed. An approximate expression for the present case is

$$- J_2 = J_1 = - \frac{1}{kT} \left(\frac{D_1}{v_1} \nabla \mu_1 - \frac{D_2}{v_2} \nabla \mu_2 \right) \quad , \tag{11}$$

where J_1 and J_2 are the diffusional fluxes of the liquid molecules and of the particles, and μ_1 and μ_2 are the chemical potentials per molecule and particle respectively. Because in perturbed conditions the concentration at the region of interest differs from the concentrations in the surrounding region, the non-uniformity in concentration acts as an additional field Ψ_i on a particular particle. Therefore, the chemical potential μ_i (i=1,2) can be written as

$$\mu_i = \mu_{io} + \psi_i \quad , \tag{12}$$

where μ_{io} is the chemical potential corresponding to a uniform system having the concentration of the considered position. The field ψ_i due to the non-uniformity in concentration is the difference between the interaction potential caused by the non-uniform system and that caused by a uniform system having the concentration of the considered point:

$$\psi_i(\vec{r}_1) = \sum_{j=1}^{2} \int u_{ij}(r) \{c_j(\vec{r}_2,t) - c_j(\vec{r}_1,t)\}g \, d\vec{r}_2 \quad ,$$

$$(i=1,2). \tag{13}$$

Here $c_j(\vec{r}_2,t)$ is the concentration of species j in the non-uniform system and $c_j(\vec{r}_1,t)$ is the concentration of the uniform system having the properties of the point of interest. The diffusional fluxes $\vec{J}_i$ are defined by[30]

$$\vec{J}_i = c_i(\vec{w}_i - \vec{w}) \quad , \qquad (i=1,2) \tag{14}$$

where $\vec{w}_i$ is the local velocity of species i relative to stationary coordinates, and $\vec{w}$ is the local average velocity. The fluxes of molecules and particles with respect to stationary coordinates are

$$\vec{N}_i = c_i \vec{w}_i \quad . \qquad (i=1,2) \tag{15}$$

For the unidimensional case, assuming volume conservation

$$N_1 v_1 + N_2 v_2 = 0 \quad , \tag{16}$$

one obtains

$$N_i = J_i(1 + c_i(v_j - v_i)) \quad . \qquad j \neq i \tag{17}$$

Mass conservation leads to

$$\frac{\partial c_2}{\partial t} + \frac{\partial N_2}{\partial x} = 0 \quad . \tag{18}$$

The transport equation implies an ensemble of particles which can agglomerate without coalescing and/or splitting. No convection is imposed from outside. For the linearized form of the transport equation (18) one obtains after a lengthy calculation

$$\frac{\partial c_2}{\partial t} = \frac{1 + c_{20}(v_1-v_2)}{kT} [v_2(D_2\phi_{10}+D_1\phi_{20})(\frac{\partial^2 f}{\partial\phi_2^2})_{\phi_{20}} \frac{\partial^2 c_2}{\partial x^2} +$$

$$2\pi \frac{\partial^2}{\partial x^2} \int_0^\pi \sin\theta d\theta \int_0^\infty \{c_2(x+r\cos\theta,t) - c_2(x,t)\}U(r)r^2 g(r)dr],$$

$$(19)$$

where

$$U(r) \equiv \frac{D_2}{v_2} u_{22} + \frac{D_1 v_2}{v_1^2} u_{11} - \frac{D_1 + D_2}{v_1} u_{12} \qquad . \qquad (20)$$

Applying the small perturbation δc to the colloidal system of uniform concentration

$$\delta c \equiv c_2(x,t) - c_{20} = A \exp[\alpha(\beta)t + i\beta x] \quad , \quad (i=\sqrt{-1}) \qquad (21)$$

one obtains for the growth coefficient

$$\alpha(\beta) = -\frac{1 + c_{20}(v_1-v_2)}{kT} \beta^2[v_2(D_2+(D_1-D_2)\phi_{20})(\frac{\partial^2 f}{\partial\phi_2^2})_{\phi_{20}} +$$

$$4\pi \int_0^\infty r^2(\frac{\sin\beta r}{\beta r} - 1)U(r)g(r)dr] \qquad . \qquad (22)$$

In the present case $v_2 >> v_1$ and $D_1 >> D_2$ and therefore

$$\alpha(\beta) = -\frac{\phi_{10}\beta^2}{kT} [v_2 D_1\phi_{20}(\frac{\partial^2 f}{\partial\phi_2^2})_{\phi_{20}} + 4\pi \int_0^\infty r^2(\frac{\sin\beta r}{\beta r} - 1)U(r)g(r)dr].$$

$$(23)$$

Note that $1 - \frac{\sin\beta r}{\beta r} \geq 0$ and has a maximum of about 1.218 for $\beta r \approx 1.352$. Since one expects that $U(r) < 0$,

$$0 < \int_0^\infty r^2(\frac{\sin\beta r}{\beta r} - 1)U(r)g(r)dr < - 1.218 \int_0^\infty r^2 U(r)g(r)dr \quad . \quad (24)$$

When $(\frac{\partial^2 f}{\partial\phi_2^2})_{\phi_{20}} > 0$, the growth coefficient α is negative for all values of β. In such cases the perturbation decays in time and the system is stable to small perturbations. Let us consider now the case when $\left(\dfrac{\partial^2 f}{\partial \phi_2^2}\right)_{\phi_{20}} < 0$. This time the growth coefficient α is positive for sufficiently small values of β because the integral is negligibly small for $\beta \to 0$. For large values of β the integral tends to the finite positive value $P_\infty = -4\pi \int_0^\infty r^2 U(r) g(r) dr$. If $\left|\left(\dfrac{\partial^2 f}{\partial \phi_2^2}\right)_{\phi_{20}} v_2 D_1 \phi_{20}\right| > P_\infty$, the growth coefficient α will be positive for all wave numbers β. If, however, $\left|\left(\dfrac{\partial^2 f}{\partial \phi_2^2}\right)_{\phi_{20}} v_2 D_1 \phi_{20}\right| < P_\infty$

$$\alpha(\beta) > 0 \qquad \text{for sufficiently small values of } \beta$$

and

$$\alpha(\beta) < 0 \qquad \text{for sufficiently large values of } \beta.$$

This means that the growth coefficient α grows from zero, passes through a maximum for $\beta = \beta_d$ (the dominant wave number), becomes zero for $\beta = \beta_c$ (the critical wave number) and negative for $\beta > \beta_c$.

In the stable and metastable regions of the thermodynamic phase diagram the derivative $\dfrac{\partial^2 f}{\partial \phi_2^2} > 0$. The dynamic treatment shows that in these conditions the system is indeed stable to small perturbations since the perturbation decays in time. However, linear stability cannot distinguish between stable and metastable states. Only a much more complex non-linear treatment can do this. We note, however, that the stable states are stable to any kind of perturbations either small or large, while the metastable states are unstable to perturbations larger than a critical one (a nucleus is needed for the perturbation to grow). In the region unstable from a thermodynamic point of view $\dfrac{\partial^2 f}{\partial \phi_2^2} < 0$ and the system is unstable to small perturbations whose wave-numbers are smaller than a critical value. The existence of a finite critical wave number assures the existence of a dominant wave number for which the growth coefficient has its maximum value. These dynamic considerations give some understanding of the meaning of stable and unstable systems. Besides they generate length and time scales which give further information about the system. Indeed, the reciprocal of the growth coefficient for the dominant wave number provides the time scale and the reciprocal of the dominant wave number the length scale. To obtain a simple expression for the time scale we expand $\sin\beta r$ in series and retain only the first two terms. Equation (23) becomes

$$\alpha(\beta) = -\frac{\phi_{10}\beta^2}{kT}\left[v_2 D_1 \phi_{20}\left(\frac{\partial^2 f}{\partial \phi_2^2}\right)_{\phi_{20}} - \frac{4\pi}{6}\beta^2 \int_0^\infty r^4 U(r) g(r) dr\right]. \qquad (25)$$

For the dominant wave number and the corresponding growth coefficient one obtains

$$\beta_d^2 = \frac{6}{8\pi} \, v_2 D_1 \phi_{20} \left(\frac{\partial^2 f}{\partial \phi_2^2}\right)_{\phi_{20}} \bigg/ \int_0^\infty r^4 U(r) g(r) dr \tag{26}$$

and

$$\alpha_d \equiv \alpha(\beta_d) = - \frac{3}{8\pi} \frac{\phi_{10}}{kT} \left(v_2 D_1 \phi_{20} \left(\frac{\partial^2 f}{\partial \phi_2^2}\right)_{\phi_{20}}\right)^2 \bigg/ \int_0^\infty r^4 U(r) g(r) dr. \tag{27}$$

The reciprocal of the growth coefficient α_d gives some indication of the time required for an appreciable change in the system to occur. Since not enough information is available to compute $\int_0^\infty r^4 U(r) g(r) dr$, the evaluation made in a previous section for L^2 is used to write

$$\left| 4 \int_0^\infty r^4 \frac{U(r)}{v_2 D_1} g(r) dr \right| \approx 10^{-5} \text{ erg/cm} \quad .$$

Putting $kT \simeq 4 \times 10^{-14}$ erg, $D_1 \simeq 10^{-6}$ cm^2/s, $\phi_{10} = \phi_{20} = 0.5$, R = 0.25×10^{-5} cm and $\left| 4 \int_0^\infty r^4 \frac{U(r)}{v_2 D_1} g(r) dr \right| = 10^{-5}$, Equation (27) becomes

$$\alpha_d \simeq 10^{-5} \left(\left(\frac{\partial^2 f}{\partial \phi_2^2}\right)_{\phi_{20}}\right)^2 s^{-1} \quad . \tag{28}$$

If the diffusion coefficient of the liquid molecules and the radius of the globule are smaller, the constant could be much smaller. The last expression shows that if $\frac{\partial^2 f}{\partial \phi_2^2}$ is small enough, an unstable system can survive a small perturbation a long time. This happens near the spinodal but can happen also in those colloidal systems which are always unstable thermodynamically. In the latter case L^2 can be larger and hence the numerical factor in Equation (28) smaller.

CONCLUSIONS

Microemulsions occur if the surfactant and co-surfactant are able to generate a positive surface free energy that is sufficiently small upon the surface of the globules. This positive contribution to the free energy of formation of microemulsions is compensated by the negative contributions of the entropy of dispersion and, in particular, of the free energy of formation of the double layers. These effects together with the positive free energy contribution of the repulsive double layer forces assure the existence of a finite radius for a negative minimum of ΔG_M. Note that while the kinetic stability of the ordinary emulsions is determined by the attractive van der Waals forces and repulsive double layer forces between globules, in the thermodynamic stability of microemulsions van der Waals forces seems to play a secondary role. The main role belongs now to the free energy of formation of the double layer and to the repulsive double layer forces, at least in the range of variables for which the present computations have been performed.

Phase inversion as well as phase separation can occur in microemulsions. Several basic types of multiphase systems have been identified similar to those found experimentally by Winsor and by Healy et al. The low values of the interfacial free energy between two microemulsion phases in equilibrium are explained.

A dynamic treatment is included to gain insight about the stable and unstable thermodynamic states from the response of the system to small perturbations. The reciprocal of the growth coefficient for the dominant wave number is a measure of the time needed for significant changes to occur in an unstable system.

ACKNOWLEDGEMENT

This work was supported by the NSF grant "Thermodynamic Stability of Microemulsions," 1975. I am indebted to Dr. A. Marmur for performing the numerical computations.

NOTATIONS

A   Hamaker's constant

c_1, c_2 concentrations of the molecules of the continuous phase and globules

c_{10}, c_{20} values of c_1 and c_2 in the initial, uniform system

$D_1 = D_1' \phi_1$, $D_2 = D_2' \phi_2$ Starting with Equation (19) $D_j = D_j' \phi_{j0}$

D_1', D_2' diffusion coefficients of molecules of the continuous phase and globules

f   free energy per unit volume

f_s specific surface free energy between globule and continuous phase

g   radial distribution function

ΔG_M	free energy of formation of one cm^3 of microemulsion
ΔG^*_M	the minimum of ΔG_M
ΔG^*_d	the difference between the points on the ΔG^*_M curve and on the straight line connecting the equilibrium phases, for a given volume fraction
ΔG_1	the free energy of formation of the interface between globules and the continuous phase and of the double layer
ΔG_2	the free energy contribution due to the interaction forces between globules
$\Delta G'_2$	the free energy contribution due to the double layer forces between globules
$\Delta G''_2$	the free energy contribution due to van der Waals forces between globules
ΔG_3	the free energy contribution due to the entropy of dispersion of the globules
J_2, J_1	diffusional fluxes of particles and molecules
k	Boltzmann's constant
K_1, K_2	reciprocal Debye lengths in the continuous and dispersed phases
K_w, K_{oil}	reciprocal Debye lengths in water and in oil
L	quantity defined by Equation (8)
N_1, N_2	fluxes of the molecules of the continuous phase and of the globules with respect to stationary coordinates
m	total number of droplets
r	distance between the centers of two particles
$\vec{r}_1, \vec{r}_2$	position coordinates
R	radius of a droplet
R^*	radius corresponding to the minimum of ΔG_M
R_1	average distance between droplets defined by Equation (4)

t   time

T   absolute temperature

u_{11}, u_{22}, u_{12} interaction potential between two molecules of the
                        continuous phase, between two globules and between
                        one globule and one molecule of the continuous phase

U   quantity defined by Equation (20)

v   the volume of one molecule of water

v_1, v_2 the volume of one molecule of the continuous phase
                        and the volume of a globule

$\vec{w}$ average local velocity

$\vec{w}_1$, $\vec{w}_2$ local velocities of the molecules of the continuous
                        phase and of the globules relative to stationary
                        coordinates

x   position coordinate

α growth coefficient

α_d value of α for $\beta = \beta_d$

β wave number of the perturbation

β_c critical wave number

β_d dominant wave number

δ thickness of the diffuse layer

ε_1, ε_2 dielectric constants of the continuous and dispersed
                        phases

Ψ_1, Ψ_2 fields created by non-uniformity in concentration on
                        a molecule of the continuous phase and on a globule

ϕ_1, ϕ_2 volume fraction of the continuous and dispersed phases

ϕ_{10}, ϕ_{20} values of ϕ_1 and ϕ_2 in the initial unperturbed state

ϕ_{oil} volume fraction of oil

ϕ' and ϕ'' values of the volume fractions in the two phases in
                        equilibrium

ϕ_i volume fraction at the inversion point

ψ_o absolute value of surface potential

σ interfacial specific free energy between two micro-emulsion phases in equilibrium

REFERENCES

1. B. V. Derjaguin and L.D. Landau, Acta Physico-chimica 14, 633 (1941).

2. E. W. J. Verwey and J. Th. G. Overbeek, "Theory of the Stability of Lyophobic Colloids," Elsevier, New York, Amsterdam, 1948.

3. W. Albers and J. Th. G. Overbeek, J. Colloid Sci. 14, 501 (1959).

4. W. Albers and J. Th. G. Overbeek, J. Colloid Sci. 14, 510 (1959).

5. T. P. Hoar and J. H. Schulman, Nature 152, 103 (1943).

6. J. H. Schulman and T. S. McRoberts, Trans. Faraday Soc. 42B, 165 (1946).

7. J. H. Schulman and D. P. Riley, J. Colloid Sci. 3, 383 (1948).

8. J. H. Schulman, W. Stoeckenius, and L. M. Prince, J. Phys. Chem. 63, 1677 (1959).

9. W. Stoeckenius, J. H. Schulman, and L. M. Prince, Kolloid Z. 169, 170 (1960).

10. J. H. Schulman and J. A. Friend, J. Colloid Interface Sci. 4, 1988 (1971).

11. L. M. Prince, J. Colloid Interface Sci. 23, 165 (1967).

12. D. O. Shah, A. Tamjeedi, J. W. Falco, and R. D. Walker, Jr., AIChE J. 18, 1116 (1972).

13. G. Gillberg, H. Lehtinen, and S. Friberg, J. Colloid Interface Sci. 33, 215 (1970).

14. S. I. Ahmed, K. Shinoda, and S. Friberg, J. Colloid Interface Sci. 47, 32 (1974).

15. J. W. Falco, R. D. Walker, Jr., and D. O. Shah, AIChE J. 20, 510 (1974).

16. P. A. Winsor, "Solvent Properties of Amphiphilic Compounds," Butterworth's Scientific Publications, London, 1954.

17. R. N. Healy, R. L. Reed, and D. G. Stenmark, Soc. Pet. Eng. J. 147 (June 1976).

18. K. Shinoda and S. Friberg, Adv. Colloid Interface Sci. 4, 281 (1975).

19. E. Ruckenstein and J. C. Chi, J. Chem. Soc. Faraday Trans. II 71, 1690 (1975).

20. J. W. Cahn and J. E. Hilliard, J. Chem. Phys. 28, 258 (1958).

21. H. Metiu and E. Ruckenstein, J. Colloid Interface Sci. 46, 394 (1974).

22. A. Silberberg and W. Kuhn, Nature 170, 450 (1952).

23. W. Kuhn, H. Majer, and F. Burckhart, Helv. Chim. Acta. $\underline{43}$,
     1208 (1960).
24. G. Langhammer and L. Nestler, Makromoleculare Chemie $\underline{88}$, 179
     (1965).
25. A. Vrij, J. Polym. Sci. $\underline{A-2}$, $\underline{6}$, 1919 (1968).
26. T. Nose, Polymer Journal (Japan) $\underline{8}$, 96 (1976).
27. S. Agarwal and H. Herman, Scripta Met. $\underline{7}$, 503 (1973).
28. J. W. Cahn, J. Chem. Phys. $\underline{42}$, 93 (1965).
29. E. Ruckenstein, Stability of Concentrated Colloidal Systems
     (unpublished 1972).
30. R. B. Bird, W. E. Stewart and E. N. Lightfoot, Transport
     Phenomena, John Wiley & Sons, New York (1960).

LIGHT SCATTERING OF A CONCENTRATED W/O MICRO EMULSION;

APPLICATION OF MODERN FLUID THEORIES

A.A. Caljé, W.G.M. Agterof and A. Vrij*

Van 't Hoff Laboratory for Physical and Colloid

Chemistry, University of Utrecht, Padualaan 8, Utrecht,

the Netherlands

The osmotic compressibility of a concentrated dispersion of water droplets in an organic solvent (micro emulsion) derived from light scattering experiments, is interpreted with modern fluid theories.

It is shown that the major contribution to the osmotic compressibility stems from hard repulsive forces between the particle cores. Further it is shown that as a possible source for the attraction forces, introduced to allow for deviations from the hard sphere repulsion, short range van der Waals interaction forces are responsible and that long range van der Waals interaction forces between the water droplets are not important here.

*to whom requests for reprints should be addressed.

INTRODUCTION

In 1943 Hoar and Schulman[1] found that coarse, milky
emulsions, stabilized with soap, became transparant upon addition
of a cosurfactant (e.g. an alcohol) and transformed into what
they called: micro emulsions. The system that Schulman et al[2]
investigated extensively consisted of water, benzene, potassium
oleate and hexanol. They concluded that in this case very small,
spherical particles were present with an aqueous core and a
surface layer containing a mixture of hexanol and oleate
molecules. The continuous phase contained benzene with a small
amount of hexanol. From titration studies it was found[3] that
these systems could be described satisfactory by assuming that
all soap was present in the particle surface, that the continuous
phase has a constant composition and that the dispersed particles
were monodisperse. We studied the same system but with toluene
instead of benzene.

Since then a great number of systems have been investigated.
One review was published recently[4]. Not much is known, however,
about light scattering of concentrated micro emulsions. In this
paper some data will be reported and it will be attempted to
interpret them in terms of interactions between the particles with
the help of modern theories of simple fluids. The basis for such
a theoretical treatment was given by McMillan and Mayer[5] who
showed that the grand canonical partition function for a
multicomponent system, in osmotic equilibrium, can be transformed
into an effective partition function involving the solute species
(dispersed particles) only. The interaction potential between
molecules (in vacuo) must then be replaced by a potential of
mean force between the dispersed particles. In this way an
(osmotic) equation of state can be found in the representation
of a virial series. For concentrated systems, however, many
(often unknown) terms of the series are required.

It is therefore of much interest that in the last decennium
much progress has been made in the understanding of dense fluids
by studying simple model fluids as hard spheres by means of
computer simulation. Good analytical approximations for concentrated hard sphere systems were obtained with the Percus-Yevick
(PY) and the Scaled Particle (SPT) theories. Unfortunately the
PY-theory is much less satisfactory when also attractive forces
are present. The most successful methods at the moment are the
perturbation approaches which treat the interaction forces as a
perturbation on the hard sphere potential. For reviews on these
developments see the references[6,7,8]. Formal treatments of
multicomponent systems were given by Kirkwood and Buff[9] and
Adelman[10].

One of the thermodynamic results of these theories is the (osmotic) compressibility. In practice this quantity can conveniently be obtained from light scattering.

THEORY

Light scattering is caused by local fluctuations in the particle densities. In systems containing dispersed particles (micro emulsion droplets) in a low-molecular solution, the light scattering is mainly caused by the dispersed particles. Debye[11] derived the following equation based on Einstein's fluctuation theory.

$$R_\theta = K(1 + \cos^2\theta)(N_{AV}kT)c(\partial\Pi/\partial c)^{-1} \tag{1}$$

$$K = 2\pi^2 n^2(\partial n/\partial c)^2/N_{AV}\lambda^4$$

Here R_θ is the excess scattering of the particle dispersion (micro emulsion) over that of the low-molecular solution in equilibrium with it, θ is the scattering angle, λ is the wavelength (in vacuo) of the unpolarized light, n is the refractive index of the micro emulsion, Π is the Donnan osmotic pressure, c is the particle concentration (mass/volume), k is Boltzmann's constant, T is the absolute temperature and N_{AV} is Avogadro's number.

From Equation (1) it follows that R_θ is proportional to the osmotic compressibility $\partial c/\partial \Pi$.

For hard spheres the potential of mean force, V(r), is given by

$$V(r) = \begin{cases} \infty & \text{for } r < \sigma_{hs} \\ 0 & \text{for } r \geq \sigma_{hs} \end{cases}$$

where r is the distance between centers and σ_{hs} the hard sphere diameter. Thiele[12] could solve the PY-equation for this case and found two somewhat different analytical expressions for the pressure depending on the route that was followed to connect the statistical theory with thermodynamics. This inconsistency is caused by the approximate character of the PY-theory. One of the equations was found earlier in the SP-theory[13]. A very accurate (semi-emperical) expression is found from a weight average of both PY-equations as proposed by Carnahan and Starling[14].

$$\Pi = (cRT/M)f_1(c) \tag{2}$$

$$f_1(c) = (1 + qc + q^2c^2 - q^3c^3)(1 - qc)^{-3} \tag{3}$$

where $R = N_{AV}k$ and $q = \pi\sigma_{hs}^3 N_{AV}/6M$.

Remark that qc is the volume fraction of spheres.

Other interaction potentials can be treated as perturbations of the hard sphere potential when the repulsive part is steep enough and the attractive part not too strong (see e.g. Zwanzig[15] and Barker and Henderson[16,17]). In this way Equation (2) is supplemented with a perturbation term $f_2(c)$,

$$f_2(c) = (cN_{AV}/M)^2 \partial Q/\partial c \qquad (4)$$

$$Q = 2\pi c \int_o^\infty V_2(r) g_{hs}(r) r^2 dr + \text{higher order terms}$$

Here $g_{hs}(r)$ is the radial distribution function of hard spheres and $V_2(r)$ the "attractive" part of the interaction potential. The hard sphere diameter, σ_{hs}, and $V_2(r)$ can be formulated when the form of the interaction potential is known. In situations where this is not the case it is reasonable to treat the hard sphere diameter as an adjustable parameter[8]. This will also be done by us.

EXPERIMENTAL

a. Dispersions

Toluene (Baker Ltd), hexanol (B.D.H. Liverpool) and potassium hydroxide (B.D.H. Liverpool) were Analyzed Reagents (reagent grade). The oleic acid (Merck Rein) was recrystallized twice before use. The water was freshly distilled twice. The oleic acid was neutralized with potassium hydroxide (1% excess). The stock emulsions, just titrated to transparence, contained a number of water molecules, N_w, and soap molecules, N_s, in a ratio $N_w/N_s = 25.0$. It was found from titrations[18] that the number of hexanol molecules divided by the number of toluene molecules was 6.48×10^{-2} in the continuous phase and that the number of hexanol molecules divided by the number of soap molecules was 1.12 in the mixed monolayer. Other dispersion concentrations were obtained by diluting the stock emulsions with the continuous phase. The particle concentration, c, is given in mass of discontinuous phase per unit volume of solution.

b. Methods

The light scattering intensities were measured in a Fica -50 photometer at $\lambda = 546.1$ nm. Only results at $\theta = 90°$ are reported because the systems did not show dissymmetry. The solutions were filtered through a millipore filter (0.45 μm).

It was sometimes necessary to make a correction for the attenuation of the primary and scattered beams. This was done with the help of the equation: $R_{90} = R_{90}' \exp(\tau d)$ with $\tau = (16\pi/3)R_{90}$ (see reference[19] where R_{90}' is the observed intensity and d is the cell diameter). Then R_{90} is found by iteration. The correction could sometimes amount to 5%.

Values of $\partial n/\partial c$ (see Table I) were measured in a Pulfrich refractometer (at $\lambda = 546.1$ nm) and were found to be constant over the whole concentration range.

Diffusion coefficients, D, of the particles were determined by means of light scattering fluctuation spectroscopy using a 5W Spectra Physics Model 165 Ar^+ ion laser. The autocorrelation function of the scattered light was determined as a function of θ, with a Saicor SA I-42 correlator analyzer and a Nova 1200 computer (Data General Corporation) to calculate the relaxation time t. Then D was obtained, with the help of the equation $t^{-1} = (4\pi n \sin(\theta/2)/\lambda)^2 D$, from a (linear) plot of t^{-1} versus $\sin^2 (\theta/2)$. It was found that, within 5%, D was constant over the whole concentration range. From D, a "hydrodynamic" diameter, σ_{hy}, was calculated from the equation: $\sigma_{hy} = kT/(3\pi n D)$ where n is the viscosity of the continuous medium (see Table I).

All experiments were performed at $(20 \pm 1)^\circ C$.

RESULTS

Figure 1 gives a plot of R_{90} versus c. It shows a maximum which implies that $\partial \Pi/\partial c$ increases faster with increasing c than c itself. Below a limiting concentration of 0.004 g cm^{-3}, R_{90} was vanishing small and no micro emulsion was present; this number was further subtracted in calculating c. More insight gives Figure 2 where $\partial \Pi/\partial c$, as calculated from Equation (1), divided by RT is plotted versus c. For small c, the plot is linear, and from its intercept and slope the particle molar mass and the second virial coefficient were found (see Table I) with the help of the following equation

$$\partial \Pi/RT\partial c = 1/M + 2N_{AV}bc/M^2 \qquad (5)$$

where $b = \frac{1}{2} \int_0^\infty \{1 - \exp(-V/kT)\}4\pi r^2 dr$.

From cN_{AV}/M, the number of particles per unit volume, and the added amount of water, the diameter of the aqueous core, σ_w, could be calculated (see Table I). At higher concentrations Figure 2 shows a steep increase of $\partial \Pi/\partial c$ clearly indicating the presence of a hard particle core.

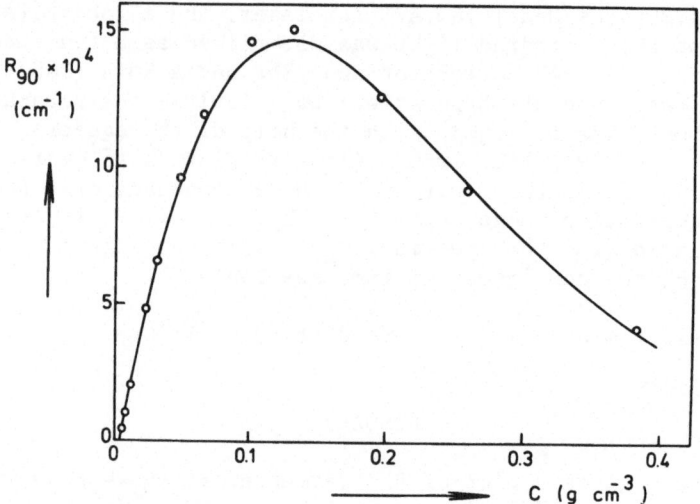

Figure 1. Light scattering ratio R_{90} as a function of the concentration. The circles (o) are measured values. The curve is calculated with Equations (1), (2), (3) and (11) with the values for σ_{hs} = 10.25 nm, ε/kT = 0.4 and λ = 1.15.

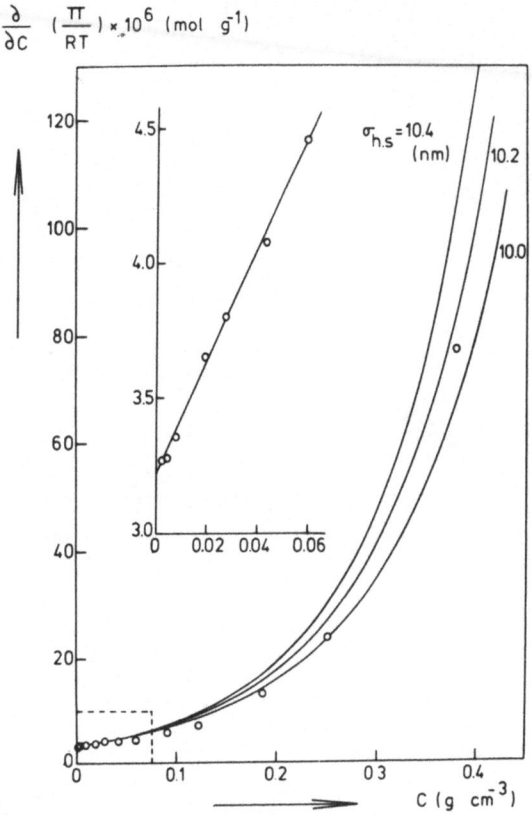

Figure 2. The osmotic compressibility $\partial\Pi/\partial c$, divided by RT, as a function of c. The circles (o) are experimental values calculated from R_{90}. The curves describe $\partial\Pi/RT\partial c$ for hard sphere systems calculated from Equations (2) and (3) with diameters σ_{hs} as indicated.

The drawn curves were calculated from the Equations (2) and (3) for some values of σ_{hs}. It is not possible, however, to make a fit over the whole concentration range. As a first attempt we tried for $f_2(c)$ a function of the van der Waals type

$$f_2(c) = - a(N_{AV}^2/M^2)c^2 \qquad (6)$$

with a constant value for a, and made plots according to the equation (see Figure 3)

$$\frac{(1+2qc)^2 - q^3c^3(4-qc)}{M(1-qc)^4} - \frac{Kc}{R_{90}} = \frac{2aN_{AV}^2}{RTM^2}c \qquad (7)$$

$$\left[\frac{\partial}{\partial c}\left(\frac{\pi}{RT}\right)_{c.s} \doteq \frac{Kc}{R_{90}}\right] \times 10^6 \,(\text{mol } \bar{g}^1)$$

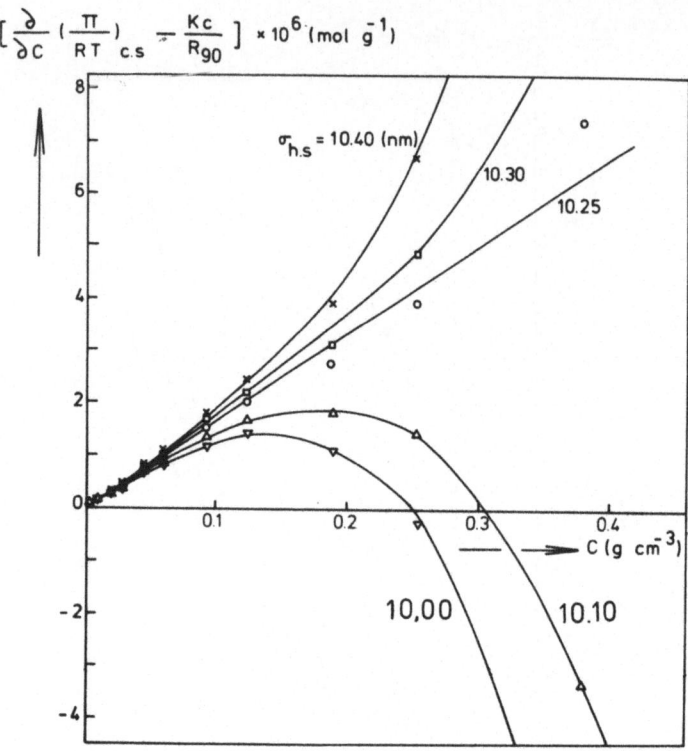

Figure 3. The osmotic compressibility $(\partial\pi/\partial c)_{c.s.}$ of a hard sphere system calculated from the equation of Carnahan and Starling (see Equations (2) and (3)) minus the measured osmotic compressibility, both divided by RT, as a function of c, for various values of the hard sphere diameter σ_{hs}.

where the first terms of the l.h.s. of Equation (7) was obtained by differentiation of Equation (3). The "best" linear plot is obtained with σ_{hs} = 10.25 nm. The slope gives $\underline{a}$ (see Table I). Observe from the Figures 2 and especially 3 that the choice of σ_{hs} is very sensitive.

Table I. Physical Constants of the Dispersion

$\partial n/\partial c$ $(\text{cm}^3\text{g}^{-1})$	$M \times 10^{-5}$ (g mole^{-1})	σ_w (nm)	σ_{hs} (nm)	σ_{hy} (nm)	$b \times 10^{18}$ (cm^3)	$(a/kT) \times 10^{18}$ (cm^3)
-0.103	3.10	8.00	10.25	12.8	1.69	1.33

DISCUSSION

From Table I one finds that the hydrodynamic diameter, σ_{hy}, is 4.8 nm larger than the diameter of the aqueous core, σ_w. This difference is somewhat larger than twice the length of an extended oleate chain (2.2 nm) which seems reasonable. The hard sphere diameter, σ_{hs}, is much smaller than σ_{hy}. The difference: $\sigma_{hs} - \sigma_w = 2.25$ nm is the minimum distance of separation between the surfaces of the aqueous cores which is much smaller than twice the length of an oleate chain. This implies that the surface layers of the opposing particle surfaces may interpenetrate each other considerably.

The results found in Figure 3 with a positive value of a imply that the hard sphere repulsion should be supplemented with an attraction. Looking for a possible origin of such a term we first assessed the contributions of long range London - van der Waals forces between the aqueous cores. According to Hamaker[20]

$$V_2(r) = -\,(A/12)\{\frac{\sigma_w^2}{r^2-\sigma_w^2} + \frac{\sigma_w^2}{r^2} + 2\,\ln\,(\frac{r^2-\sigma_w^2}{r^2})\} \tag{8}$$

where A is the Hamaker constant for water in an oil-medium. As a first approximation one can take $g_{hs}(r) = 0$ at $r < \sigma_{hs}$ and $g_{hs}(r) = 1$ at $r \geq \sigma_{hs}$ and neglect the higher order terms in Equation (4). Then using Equation (8) one obtains after integration

$$a = (\pi/9)A\sigma_w^3\{\tfrac{1}{4}\,\ln\,(\frac{s-1}{s+1}) - s - s^3\,\ln\,(\frac{s^2-1}{s^2})\} \tag{9}$$

with $s = \sigma_{hs}/\sigma_w$.

From the numbers in Table I one finds $A = 1.75 \times 10^{-12}$ erg which is about 35 times larger then values from other sources (see reference[21]) which is 5×10^{-14} erg. Although a more realistic g(r) would make the calculated A somewhat smaller, we conclude that long range attractions between the water cores cannot explain the value of a in Table I. Long range electrostatic repulsions are not active in these hydrocarbon media[22].

One could further think of contributions of van der Waals interactions on shorter distances due to the exchange of solvent molecules by segments of the oleate chains when the particle surfaces approach each other. These would have to be supplemented by exchange terms of entropic origin. The solubility parameters of aromatic and aliphatic groups seem sufficiently different to make cyclohexane a thêta solvent for polystyrene (at 34°C) and benzene a thêta solvent for polyisobutylene (at 24°C). At the moment we will not elaborate on these mechanisms but

direct the attention on the approximate character of Equation (6).
This equation must be correct for very small c where only pair
interactions are important. Often it turns out (see e.g. ref. 8)
that it is also a good approximation for large c, but not
necessarily with the same value of $a^{8)}$. This is indeed the case
with our system. We have investigated this in more detail with
the help of a simple model fluid having a square well potential,
i.e.

$$V(r) = \begin{cases} \infty \text{ for } r < \sigma_{hs} \\ -\varepsilon \text{ for } \sigma_{hs} < r < \lambda\sigma_{hs} \\ 0 \text{ for } r > \lambda\sigma_{hs} \end{cases} \qquad (10)$$

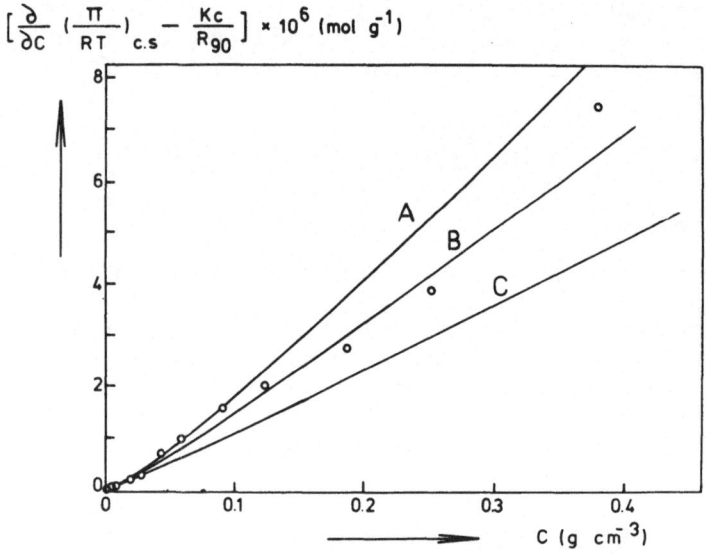

$$\left[\frac{\partial}{\partial c} \left(\frac{\pi}{RT} \right)_{c.s} - \frac{Kc}{R_{90}} \right] \times 10^6 \text{ (mol } g^{-1})$$

Figure 4. The l.h.s. of Equation (7) is plotted versus c for
$\sigma_{hs} = 10.25$ nm (see the circles (o). The curves are with the same
hard sphere diameter calculated with Equation (11) for several
combinations of ε and λ as indicated. A($\varepsilon = 0.6$ kT, $\lambda = 1.09$);
B($\varepsilon = 0.4$ kT, $\lambda = 1.15$); C($\varepsilon = 0.2$ kT, $\lambda = 1.29$).

Here ϵ is the depth and $(\lambda-1)\sigma_{hs}$ the width of the square well. Recently Ponce and Renon[23] derived an analytic, approximate expression for the perturbation $f_2(c)$

$$f_2(c) = - (cN_{AV}/M)^2 kT[(2/3)\pi\sigma_{hs}^3 \epsilon/kT\{\lambda^3 - \frac{qc+2}{2}(\frac{1-qc}{2qc+1})^3\}$$

$$- \frac{\pi}{3}\sigma_{hs}^3 (\epsilon/kT)^2 (\frac{1-qc}{2qc+1})^3 \{(\lambda^3 - (\frac{qc+2}{2})(\frac{1-qc}{2qc+1})^3)$$

$$(6q^2c^2+7qc-1) - \frac{qc}{2}(\frac{1-qc}{2qc+1})^3 (2q^2c^2+8qc+17)\}] \qquad (11)$$

The expression for the attractive part of the osmotic compressibility $\partial f_2(c)/\partial c$, which is very lengthy and will not be given here, contains three adjustable parameters, σ_{hs}, ϵ and λ. Barker and Henderson[6] have pointed out that $\partial f_2(c)/\partial c$ is very nearly a linear function of c at high concentrations for a hard-core fluid like the square well fluid.

In Figure 3 we obtained the "best" straight line at $\sigma_{hs} = 10.25$ nm, but this line (through the origin) is an overall fit over all concentrations and its slope does not fit well with the slope at low concentrations. In Figure 4, the same data points as in Figure 3 are plotted (with $\sigma_{hs} = 10.25$ nm) but also calculated lines for $\partial f_2/\partial c$ at several combinations of ϵ and λ are given. These combinations of ϵ and λ are so chosen that the value calculated for the second virial coefficient, b, for a square well fluid, Equation (12), is in agreement with the measured value in Table I.

$$b = b_{hs}\{1 + (\lambda^3-1)(1 - \exp \epsilon/kT)\} \qquad (12)$$

where $b_{hs} = 2\pi\sigma_{hs}^3/3$, is the hard core second virial coefficient. Ponce and Renon calculated in their approximation for b:

$$b = b_{hs}\{1 - (\lambda^3-1)\epsilon/kT(1 + \epsilon/2kT)\} \qquad (13)$$

Equations (12) and (13) are equivalent when $(\epsilon/kT)^2 < 1$.

In Figure 4 one can see that the calculated curve for the combination $\epsilon/kT = 0.4$ and $\lambda = 1.15$ gives the best fit. The width of the square well is in this case $(\lambda-1)\sigma_{hs} \approx 1.5$ nm, which is indeed comparable with the width of the overlapping surface layers (≈ 2.1 nm).

Summarizing one can say that the compressibility of the system is described well by particles having a hard core repulsion and a small short range attraction. This is finally shown in Figure 1, where the drawn curve was calculated from the middle curve of Figure 4.

ACKNOWLEDGEMENTS

For one of us (W.G.M.A.) this work was part of the research programme of the "Stichting voor Fundamenteel Onderzoek der Materie" (FOM) with financial support from the "Nederlandse Organisatie voor Zuiver-Wetenschappelijk Onderzoek (ZWO)". We wish to thank Dr. H. Fijnaut and Mr. H. Mos for their help with light scattering fluctuation spectroscopy. We also wish to thank Mr. B. Schalk for programming the lengthy calculations. Finally we thank Miss H. Miltenburg for typing the manuscript and Mr. W. den Hartog for drawing the illustrations.

REFERENCES

1. T.P. Hoar, J.H. Schulman, Nature (London) 102, 152 (1943).
2. J.H. Schulman, J.A. Friend, Koll. Z. 115, 67 (1949).
3. J.H. Schulman, W. Stoeckenius, L. Prince, J. Phys. Chem. 63, 1677 (1959).
4. K. Shinoda, S. Friberg, Adv. Coll. Interface Sci. 4, 281 (1975).
5. W.G. Mc Millan, J.E. Mayer, J. Chem. Phys. 13, 276 (1945).
6. J.A. Barker, D. Henderson, Ann. Rev. Phys. Chem. 23, 439 (1972).
7. G.A. Neece, B. Widom, Ann. Rev. Phys. Chem. 20, 167 (1969).
8. W. R. Smith, in "Statistical Mechanics", A Specialist Periodical Report, Vol. 1, K. Singer, Editor, The Chemical Society, London, 1973.
9. J.G. Kirkwood, F.P. Buff, J. Chem. Phys. 19, 774 (1954).
10. S.A. Adelman, J. Chem. Phys. 64, 724 (1976).
11. P.J.W. Debye, J. Phys. Chem. 51, 18 (1947).
12. E.J. Thiele, J. Chem. Phys. 39, 474 (1963).
13. H. Reiss, H. Frisch, J.L. Lebowitz, J. Chem. Phys. 31, 369 (1959).
14. N.F. Carnahan, K.E. Starling, J. Chem. Phys. 51, 635 (1969).
15. R.W. Zwanzig, J. Chem. Phys. 22, 1420 (1954).
16. J.A. Barker, D. Henderson, J. Chem. Phys. 47, 2856 (1967).
17. J.A. Barker, D. Henderson, J. Chem. Ed. 45, 2(1968).
18. J.E. Bowcott, J.H. Schulman, Z. Electrochemie, 59, 283 (1955).
19. P. Putzeys, E. Dory, Ann. Soc. Sci. Brux. Ser. I 60, 37 (1940).
20. H.C. Hamaker, Physica 4, 1058 (1937).
21. R.H. Ottewill, T. Walker, Kolloid-Z. Z. Polym. 227, 108 (1968).
22. W. Albers, J.T.G. Overbeek, J. Coll. Sci. 14, 510 (1959).
23. L. Ponce, H. Renon, J. Chem. Phys. 64, 638 (1976).

MICROEMULSIONS CONTAINING IONIC SURFACTANTS

Stig Friberg

Department of Chemistry

University of Missouri-Rolla, Rolla, MO 65401

Irena Buraczewska

The Swedish Institute for Surface Chemistry

Stockholm, Sweden

W/O microemulsions were shown to be inverse
micellar solutions and the influence of added
electrolyte on their stability was related to
their structure.

INTRODUCTION

Microemulsions have been used in practice for a long time[1];
the scientific introduction came through the contribution from
J. Schulman[2]. Schulman prepared the microemulsions by addition
of a medium chain length alcohol such as hexanol to an emulsion
of water and oil stabilized by an ionic surfactant, such as a
soap. At a certain concentration of alcohol a transition spontaneously takes place from a coarse turbid emulsion to a transparent microemulsion.

With this approach a concentration on the interfacial aspects
of the problem was a natural and expected development and
J. Schulman and his school[3-10] have also emphasized the role of

interfacial free energy for the stability of microemulsions,
thereby neglecting other factors[11].

The main thoughts of the Schulman school may be summarized
according to Prince[12]. The addition of the alcohol to the
emulsion stabilized by an ionic surfactant will further reduce the
interfacial tension between water and oil, and at a certain concentration the interfacial tension will become negative. This
means that a negative free energy γdA is available to break the
emulsion droplets into smaller units. Sufficient amount of
alcohol leads to droplets of a diameter too small to be optically
observed. The approach is well illustrated by Figure 1 from one
of Schulman's contributions[7].

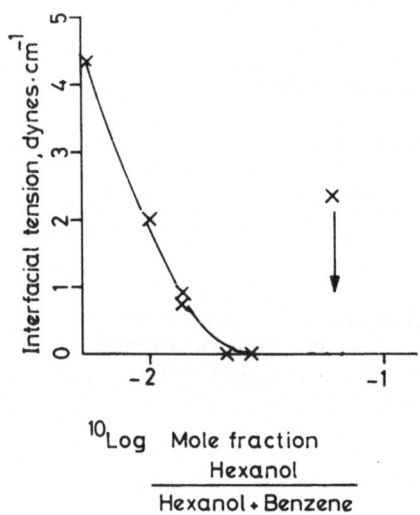

Figure 1. Interfacial tension between an aqueous solution of
potassium oleate (10^{-2}M) containing 0.7M KCl and a benzene
solution as function of concentration[7].

In addition, Prince[9] has pointed out the importance of the
ratio between interfacial tensions towards water and oil. A
higher interfacial tension towards water leads to W/O systems and
vice versa. The important point is, however, the negative interfacial tension, according to Prince[12], "a generally accepted condition" for the microemulsion state.

The opinion of interfacial tension as the dominating factor is obviously an oversimplification[11]; other factors play equally important roles for the stability of microemulsion systems. The attention has recently been drawn to the free energy contributions from entropy[14], from the Van der Waals interaction and from the compression of the diffuse electric double layer[13]. Although the theories developed do not yet contain all factors needed for a complete explanation of the microemulsion phenomena, the present results[13,14] demonstrate the possibility of a positive interfacial tension.

The present investigations were mainly undertaken in order to obtain reliable information on the phase regions in a model system and to obtain a preliminary information on the influence of the presence of an electrolyte on the phase regions in W/O micro-emulsions.

EXPERIMENTAL

The oleic acid, the benzene, the pentanol and the sodium chloride were all p.a. qualities and used without further purification. Water was twice distilled. The potassium oleate was prepared by titration with oleic acid of an ethanolic potassium ethoxide solution[15].

Solubility regions were determined by titration and the limits of solubility areas were checked by long time storage of compositions close to phase separation.

RESULTS

The investigation was focussed on the W/O microemulsion system; an O/W system appeared at high electrolyte content and will be briefly mentioned. The W/O system will first be described for an aqueous phase without electrolyte. The results on the influence of electrolyte will follow.

W/O Microemulsions with Water as the Aqueous Phase

The W/O microemulsions are with advantage connected to the association conditions of the three "structure forming elements"; water (H_2O), potassium oleate (KO1) and pentanol (C_5OH). Figure 2 shows the increase of the KO1 solubility from 2.5% (by weight) to 57% by addition of 11% water. This corresponds to three molecules of water per soap molecule to dissolve the surfactant. A maximum solubilization of water, 57% by weight, was observed for a pentanol/soap molecular ratio of 12.

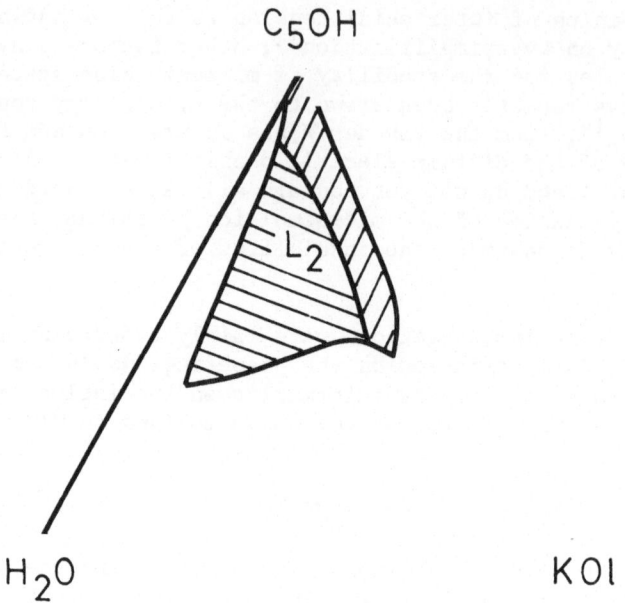

Figure 2.   Isotropic liquid pentanol solution of potassium oleate
and water   ⧄⧄ inverse micelles   ⧄⧄   ion pairs.

It is essential to understand the different mechanisms for
the increase of solubility of the soap and the water. The soap is
dissolved as small aggregates, presumably ion pairs[16,17], whilst
the high solubilization of water is due to the formation of inverse
micelles. The two areas are approximately marked in Fig. 2, the
ion pairs are to the right in the figure the inverse micelles to
the left.

The microemulsions contain a hydrocarbon in addition to the
three structure forming compounds. They should have a high
hydrocarbon content in order to deserve the notation microemulsion
and should be found at high levels in the complete system built as
a tetrahedron with the hydrocarbon forming the fourth corner (Fig.
3A). The solubility areas for W/O systems are marked in black on
the planes for 25, 50, 75 and 90% of benzene. The maximum
solubility of water on the plane for 50% benzene gives a microemulsion containing 50% benzene, 10% soap, 12% pentanol and 38%
water; a reasonable composition. The diagram clearly demonstrates
the direct connection between the microemulsion areas and the
inverse micellar solutions described by Ekwall and coworkers[16,17].

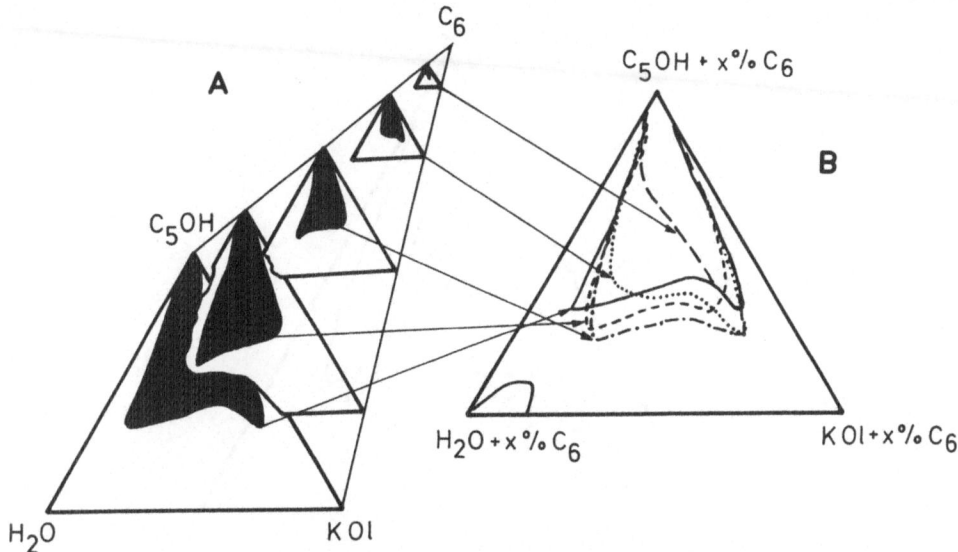

Figure 3. The W/O microemulsions are a direct continuation of the
isotropic pentanol solution at zero percent hydrocarbon and
contain corresponding ion pair and inverse micellar areas (A). The
area changes little with increased amounts of hydrocarbon (B).
x = 0   20   50   75   and 90% (W) benzene

In fact the W/O emulsions belong to the same solubility
region; microemulsions may well be prepared by addition of hydrocarbon to the inverse micellar solutions in Fig. 2.

Fig. 3B shows the solubility areas projected on the base
diagram of the three structure forming elements. The ratios
between the three structure forming elements remained approximately constant for the solubility region independent of the
hydrocarbon content for less than 50% benzene. The solubility
area was changed only to a small extent at 75% benzene (dotted
line in Fig. 3B). The minimum amount of water to obtain solubility
was exactly constant for all hydrocarbon contents, the maximum
amount remained constant to 50% benzene, when calculated as weight
function $H_2O/(H_2O+KOl+C_5OH)$. The pentanol/soap ratio for obtaining
the maximum water solubilization was decreased with increasing
benzene content to 50% benzene but reduced for higher benzene
contents. The variation (Fig. 3B) was small, however.

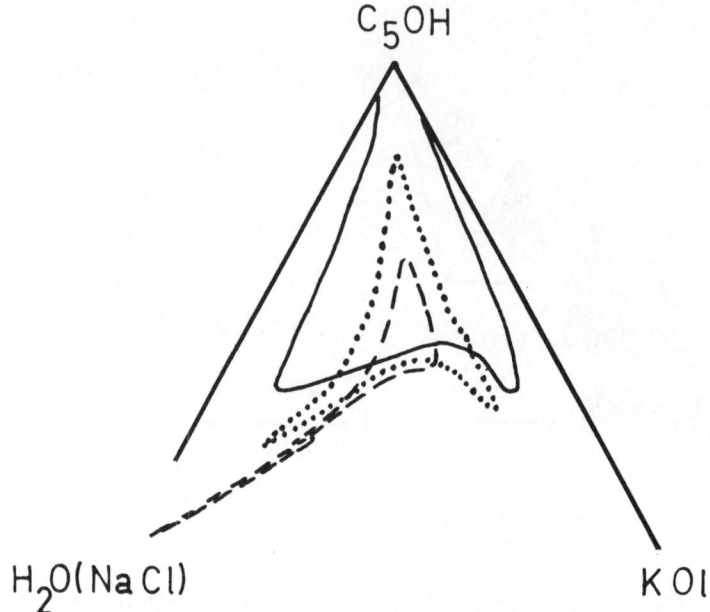

Figure 4. Solubility regions in the system of the three structure-
forming elements for different concentrations of sodium chloride
in the water.
   [NaCl] =   0   0.5M   1M

          The influence of electrolyte content is displayed by Fig. 4.
Enhanced amounts of electrolyte gave solubility regions with
increased amounts of minimum water content to dissolve the soap
(Fig. 4, left solubility borders). The small solubility of the
soap in alcohol remained, but the solubility of water in that
solubility range was too small to be marked on the diagram. The
minimum of soap content to obtain solubility was also increased
(Fig. 4, right borders). The <u>maximum</u> water solubility was
<u>increased</u> with electrolyte content with a drastic reduction of the
soap/pentanol ratio range for the solubility region. The narrow
solubility region reaching towards the water corner for 1M sodium
chloride solution was a <u>separate</u> area; a discontinuity was observed
at 56% water.

DISCUSSION

The results were pertinent to illustrate the relation between inverse micellar solutions and W/O microemulsions. As is well known a prolonged discussion has taken place as to whether W/O microemulsions should be considered as inverse micellar solutions[18,21] or not[22]. The latest contribution from the anti-micellar school[22] "The interactions that are needed to form the phases of this diagram (showing microemulsion regions) are of a higher order of complexity than those required for the phases of Refs. 1-7 (micellar solutions)" deserves a further discussion.

The present results clearly demonstrated the microemulsions to be identical to inverse micellar solutions. Fig. 2A is in itself a convincing illustration of the identity between the inverse micellar solutions and microemulsions, but the following considerations may be useful further to clarify the matter. A microemulsion containing 50% hydrocarbon may be obtained by emulsification of the hydrocarbon and water using the soap as stabilizer. Addition of the alcohol will cause clarification of the system and the microemulsion is obtained. This is the traditional Schulman-Prince approach that has focussed the attention on interfacial properties delaying a proper evaluation of the stability factors.

The microemulsion could as well be prepared by addition of the hydrocarbon to an inverse micellar solution of soap and water in pentanol according to Fig. 2B. As is easily realized the hydrocarbon will serve only as a dilutant for the pentanol solution taking little part in the structure of the inverse micelles. The solution of water and soap in the pentanol is obviously a solution containing inverse micelles, the addition of hydrocarbon means no change of the structure and it is obvious that W/O microemulsions are identical to inverse micellar solutions.

There is no advantage of considering the W/O microemulsions as different from inverse micellar solutions and the stability of inverse micelles is sufficiently complex[23-26] without introduction of "higher order of complexity".

The present results also illustrated the importance of the association structures in the water poor part of the inverse micellar solution. The common use of nomenclature such as "inverted micellar"[22] is misleading. The extensive investigations by Ekwall and coworkers[15-17] have shown the aggregates to be small; presumably monomeric specia. The present results indirectly supported the later interpretation. Addition of electrolyte to

the solutions enhanced the minimum water in the water poor part of
the system where no inverse micelles were present, whilst the
water rich part with its micellar actually could accommodate more
water when electrolyte was added. This would be the result if ion
pairs with a minimum amount of water to change the polarity of the
soap ion pairs were present. Addition of electrolyte would require
further water to dissolve the added electrolyte. An estimation of
the minimum requirements of water gave 33±2 moles of water per mole
sodium chloride for the solubility limits to the right in Fig. 4.

The reason for enhanced nonionic water solubilization
capacity with added electrolyte is not yet clarified. At present
it may suffice to point out the different factors for stabilization of an inverse micelle[22-26]; the first moment of interfacial
tension[23] may play a more important role, when the presence of
salt has reduced the energy of the minor diffuse electric double
layer.

It is our hope that the present result may serve to end the
lengthy discussion of the nature of W/O microemulsions. The
phenomena involved in the stability of an inverse micelle[23-26],
in their mutual interactions[13], in the solubility/solubilization
phenomena and in the transition between inverse micelles and
liquid crystalline phases are sufficiently complex without adding
artificial boundaries.

ACKNOWLEDGEMENT

Financial support from The Swedish Board for Technical
Development and Délégation Générale pour Recherche Scientifique
et Technologique, France is gratefully acknowledged.

REFERENCES

1. V.R. Kokatnur, U.S. Pat. 2.111.100 (1935).
2. T.P. Hoar and J.H. Schulman, Nature (London) 152, 102 (1943).
3. J.H. Schulman, R. Matalon and M. Cohen, Discuss. Faraday Soc.
    11, 117 (1951).
4. J.H. Schulman and D.P. Riley, J. Colloid Sci. 3, 383 (1948).
5. J.H. Schulman and J.A. Friend, Ibid. 4, 497 (1949).
6. J.E.L. Bowcott and J.H. Schulman, Z. Electrochem. 59, 283
    (1955).
7. C.E. Cooke and J.H. Schulman in "Surface Chemistry" Munksgaard,
    Copenhagen, 1965.
8. W. Stockenius, J.H. Schulman and L. Prince, Kolloid-Z. 169,
    170 (1960).
9. L.M. Prince, J. Colloid Interface Sci. 23, 165 (1967).

10. D.O. Shah and R.M. Hamlin, Science 171, 483 (1971).
11. C.A. Miller and L.E. Scriven, J. Colloid Interface Sci. 33, 360 (1970).
12. L.M. Prince, J. Colloid Interface Sci. 29, 216 (1969).
13. E. Ruckenstein and J.H. Chi, Chem. Soc. Faraday Trans. 71, 1690 (1975).
14. H. Reiss, J. Colloid Interface Sci. 53, 61 (1975).
15. L. Mandell and P. Ekwall, Acta Polytechn. Scand. Chem. 74, 1 (1968).
16. P. Ekwall and L. Mandell, Acta Chem. Scand. 21, 1612 (1967).
17. G. Gillberg and P. Ekwall, Ibid. 21, 1630 (1967).
18. G. Gillberg, H. Lehtinen and S. Friberg, J. Colloid Interface Sci. 33, 40 (1970).
19. K. Shinoda and H. Kunieda, J. Colloid Interface Sci. 42, 381 (1973).
20. K. Shinoda and S. Friberg, Adv. Colloid Interface Sci. 4, 281 (1975).
21. D.H. Rance and S. Friberg, J. Colloid Interface Sci. (In Press).
22. L. M. Prince, J. Colloid Interface Sci. 52, 182 (1975).
23. C. Murphy, Thesis, University of Minnesota, Minneapolis, 1966.
24. A.W. Adamson, J. Colloid Interface Sci. 29, 261 (1969).
25. S. Levine and K. Robinson, J. Phys. Chem. 76, 876 (1969).
26. H.F. Eicke and H. Christen, J. Colloid Interface Sci. 48, 417 (1974).

INTERACTIONS AND REACTIONS IN MICROEMULSIONS

R. A. Mackay, K. Letts, and C. Jones

Department of Chemistry, Drexel University

32nd and Chestnut Streets, Philadelphia, Penna. 19104

A microemulsion is a clear, stable fluid consisting of essentially monodisperse oil in water (o/w) or water in oil (w/o) droplets with diameters generally in the range 10-60 nm. The intermediate size of the microdroplets (between classical emulsion drops and spherical micelles) and the high volume fraction they occupy (20-80%) render micellar emulsions of great potential value for the study of reactions and interactions at microscopic oil-water interfaces. We have conducted a number of physical and chemical investigations in principally ionic systems of the o/w type in order to survey the properties of these media. The studies include acid-base equilibrium in both the continuous and disperse phases, and dye adsorption and reaction. The results are correlated with the physical studies and are discussed in terms of the roles of the continuous phase and the droplet surface (interphase) and interior.

INTRODUCTION

A microemulsion[1] differs from a classical emulsion in several significant respects. As the name implies, the droplet size is smaller, in the range 5-150 nm and more often 10-60 nm. In this range, the microemulsion is monodisperse.[2,3] Due to the small droplet size, microemulsions are transparent but do exhibit the Tyndall effect with visible light. A microemulsion generally consists of four components, water, oil, ionic surfactant, and and alcohol, [1,3] although a suitable nonionic surfactant may be effective.[4,5] The formation of the microemulsion is spontaneous, suggesting that it is thermodynamically stable.[6,7] In any event,

it is mechanically stable. For oil in water (o/w) systems of
the type examined here, the internal structure may thus be
described as a stable collection of "oil" microdroplets in an
aqueous continuous phase. Each droplet consists of a 60-600
$\overset{\circ}{A}$ diameter "bulk" oil drop surrounded by a 20-30 $\overset{\circ}{A}$ thick surface
phase (interphase) consisting mainly of alcohol molecules and
cationic or anionic detergent ions.

The volume fraction of the disperse phase can generally be
varied over a fairly wide range (e.g., 20-80%). The molecular
weight of the disperse phase is greater than that of most micelles
containing solubilized materials, and high ratios of solubilized
to solubilizing substances can be obtained.[8]

During the past fifteen years, there have been an increasing number of investigations of the effect of surfactants on the
courses and rates of chemical reactions.[9,10] These studies
have been primarily concerned with the influence of normal
micelles in aqueous solutions, although the action of reversed
micelles[11] and lyotropic liquid crystalline phases[12] have
recently been examined. Reactions in monolayers and in the
presence of polyelectrolytes, which have many features related
to those of reactions in micellar systems, have also been investigated.[13,14] Common to all of these systems is the presence
of microscopic "oil-water" interfaces.

Virtually no comparable studies have been performed in microemulsions. Stoeckenius, Schulman and Prince[3] reacted unsaturated
acid with OsO_4 in an o/w microemulsion to stain it for electron
microscopy. Recently, a study of metalloporphyrin formation in
o/w microemulsions was reported.[15] Microemulsions possess some
unique differences compared with micelles, monolayers or polyelectrolytes. An aqueous micelle may incorporate one, or perhaps
a few, solute molecules. Since a microdroplet in a micellar
emulsion has a sizeable surface and bulk interior phase of
varying polarity, a large number of solute molecules may be
accommodated. A coarse emulsion might be somewhat similar to
a microemulsion in this regard, but has lower surface charge,
is polydisperse, inherently unstable and turbid. The metal ion
incorporation studies mentioned above[15] showed that an understanding of the interaction of ionic species with the surface,
and the distribution of ionic and neutral species between the
internal phases was essential. We present here the results of
some studies designed to survey the behavior of solutes in o/w
microemulsions as well as the properties of the systems themselves.

EXPERIMENTAL

The oil in water microemulsions were prepared by mixing the
oil, alcohol and surfactant in the proportions given in Table I,
and then diluting with water or aqueous solutions of dye and
buffer. Agitation was not required, but the mixture was normally
stirred to speed the formation of the transparent microemulsion.

The spectral measurements were performed on a Cary 14 spectrophotometer. For the Kinetic studies, a 1 cm quartz cell was placed in
a hollow metal cell holder through which water was circulated from
a constant temperature bath. All reactants were equilibrated in
the bath before mixing. A dip cell with platinized platinum
electrodes in conjunction with a Serfass conductance bridge oper-
ating at 1 KHz was employed for the conductivity measurements. The
pH of the buffer solutions was measured with a Beckman Expando-
matic pH meter and a Corning Tripurpose glass electrode. The Tween
and Span surfactants were supplied courtesy of ICI United States,
Inc.

SYSTEMS

Composition and Properties

A number of ionic and nonionic systems were employed in these
studies. A few of the microemulsions were used for somewhat
detailed investigation, while others were utilized only for purposes of comparison. In all cases, the micellar emulsions were
of the oil in water (o/w) type, using either pentanol or cyclohexanol as co-surfactant. Pentanol was the co-surfactant with
hexadecane or mineral oil, and cyclohexanol with benzene. The
initial composition of surfactant, co-surfactant and oil for each
system is given in Table I. These mixtures were then diluted with
water to give o/w microemulsions of appropriate composition.

Table I.   Initial Composition of Microemulsion Systems

Surfactant	%Surfactant	%Alcohol	%Oil
Tween 40[a]	72.3	21.0	6.7
Tween 60[b]	48.2	25.2	26.6
Tween 60/Tween 81[a]	38.6/17.1	30.7	13.6
Tween 60/Span 80[a]	52.0/17.3	21.3	9.4
SCS[a,c]	30.6	47.9	21.5
CPB/CTAB[a,d]	18.0/16.1	50.9	15.0
SCS[c,e]	34.9	51.7	13.4
CPB[d,e]	40.0	47.4	12.6
PO[e,f]	27.6	54.3	18.1

a) pentanol/mineral oil b) pentanol/hexadecane c) sodium cetyl
sulfate (SCS) d) cetylpyridinium bromide (CPB) and cetyltri-
methyl-ammonium bromide (CTAB). e) cyclohexanol/benzene
f) potassium oleate (PO).

All of the systems could be diluted with water to give a
clear fluid over a certain range of percent water, and some of
these will be discussed below. Some systems could be diluted
indefinitely with water with no apparent signs of breaking, while
others would turn milky at a given amount of water added, within
a drop or two. In general, the microemulsions containing about
30-40% water by weight were stable with respect to acid or base
over the range 1 < pH < 12, with the exception of the potassium
oleate system which precipitated at pH < 7.5, and to added
electrolyte in the vicinity of 0.1M depending upon the salt.
The micellar emulsions could accomodate larger amounts of 1:1
than 2:1 electrolyte. The ionic systems are stable over a fairly
wide temperature range. The upper limit is near the boiling
point of the oil or water, whichever is lower. Benzene microemulsions freeze at about 0°C, while those with mineral oil
solidify at about 15°C. In both cases, the microemulsion reforms
on thawing. The nonionic systems become turbid at lower
temperatures, the Tween 40 system at about 50°C.

Stability

One of the questions which frequently arises is whether the
system is thermodynamically or kinetically stable. Contrary
evidence can prove the latter, but it is difficult if not
impossible to prove the former. It is our feeling that some
of the clear, isotropic systems are true equilibrium phases,
while others are indeed kinetically stable. Strictly speaking,
the term "solubilized micellar solutions" might more appro-
priately apply to the former, while the latter are truly "micro"-
emulsions.

One aspect of this type of behavior is illustrated by
dilution studies. For example, the benzene in water systems
may be diluted with water indefinitely and remain clear and
isotropic. However, an indication of the internal structural
changes is provided by conductivity measurements. The conductance curve of a water/benzene/cyclohexanol/cetyl-pyridinium
bromide (CPB) microemulsion is shown in Figure 1. The undiluted emulsion is given a relative concentration of 1.0.
The equivalent conductance (Λ) based on the amount of surfactant
present, is about one-third that expected for a 1:1 electrolyte.
The value of Λ begins to rise sharply at a relative concentration
of about 0.02, corresponding to a 50-fold dilution with water.
At very high dilution, the curve corresponds closely to that
of aqueous CPB. It thus seems likely that at relative concentrations below 0.02, aqueous micelles are present. However,
the fact that Λ decreases and goes through a minimum with
increasing dilution indicates that other changes are taking
place prior to this point. It may be mentioned that all of
the ionic systems exhibit qualitatively the same behavior.
The sodium cetyl sulfate (SCS) system which is clearly unstable

below a relative concentration of about 0.45, is also shown
in Figure 1. A phase diagram of this system is shown in
Figure 2. At present, we feel that this is a true equilibrium
phase since it is independent of the order of mixing or titration
of components and it exhibits consistent and reversible
behavior with respect to perturbations such as temperature and
salt addition. Low angle X-ray studies indicate a droplet
diameter on the order of 10 nm (100Å), which would correspond
to a droplet concentration of about 10^{-3}M for the system with
a phase volume of 40%.

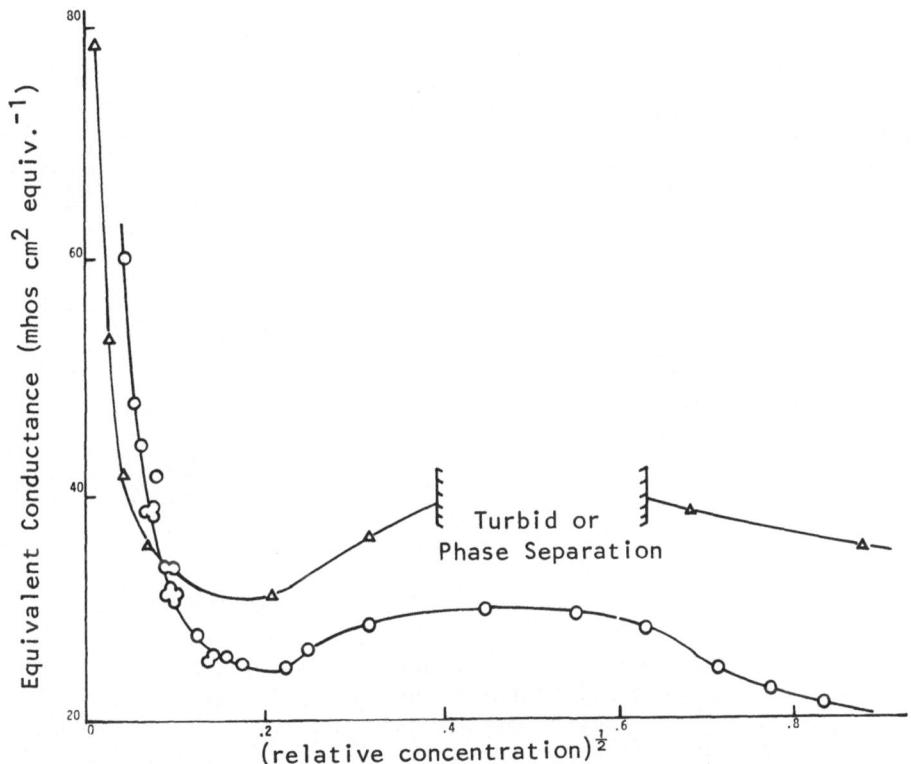

Figure 1. The equivalent conductance of a water/benzene/CPB/
cyclohexanol microemulsion (-O-) and a water/mineral oil/
SCS/n-pentanol microemusion (-△-). The undiluted microemulsion,
which is .5M in CPB or in the case of SCS, .38M, is given a
relative concentration of 1.0.

    Returning to the type of structural changes which are taking
place, we feel that in the SCS system the minimum in Λ, which is
clearly in a region of kinetic stability, may be due to the
presence of larger droplets, or even w/o droplets within the o/w

droplets. Ultracentrifugation in excess of 100,000 g fails to
cause phase separation in the SCS micellar emulsions of relative
concentration 0.1-0.2, but they will eventually become turbid on
standing for an extended period of time. In any event, the
similar conductance curves indicate that all the ionic systems
become unstable even if no visible turbidity appears on dilution
with water.

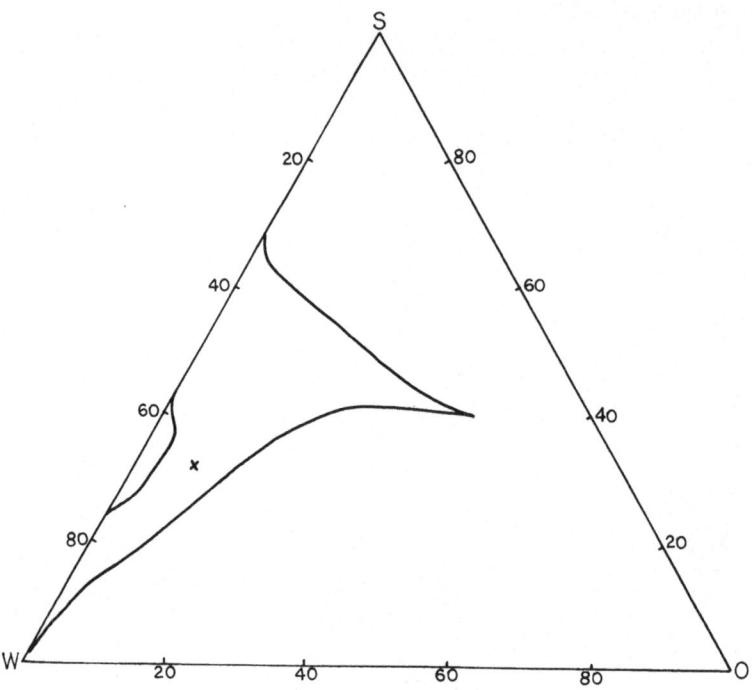

Figure 2. Pseudo-three component phase diagram of water/
mineral oil/SCS/n-pentanol system. The outlined area shown is
a clear, single phase region corresponsing to o/w microemulsions.
Composition is given in weight percent. W=water, O=mineral oil
and S=surfactant (40 wt. % SCS + 60 wt. % n-pentanol). W+O+S=
100 wt. %. The cross labeled a corresponds to the relative concentration of 1.0. (vide text).

ACID-BASE EQUILIBRIUM

Continuous Phase

    The compound 1-methyl-4-cyanoformylpyridinium oximate (CPO)
is a water soluble molecule which possesses solvent sensitive
intramolecular charge transfer bands[16]. The molecule exhibits
no tendency to be adsorbed by micelles, and thus should provide

Table II. Short Wavelength Band Maximum of CPO in Various Solvents
and Microemulsion Media

Solvent	approx. Aq. range (%)	HLB[a]	λ_{max} (nm)
Pentanol	_____	_____	376
Methanol	_____	_____	363
Tween 40	25-85	15.6	364
Tween 60	25-80	14.9	360
Tween 60/ Tween 81[b]	30-45	13.3	346
Tween 60/ Span 80[c]	25-65	12.3	350
SCS (nujol)	30-80	_____	344
CPB/CTAB	25-60	_____	350
H_2O[d]	_____	_____	341

a) Atlas Hydrophile-Lipophile Balance. b) 69% Tween 60, 31%
Tween 81. c) 75% Tween 60, 25% Span 80. d) Same in SCS, CPB
and PO benzene/cyclohexanol systems.

a true indication of the nature of the continuous phase in o/w
emulsions. The data shown in Table II tends to bear out this
hypothesis. Although there is a rough correlation of λ max with
HLB for the nonionic microemulsions, the systems containing
tween 81 and Span 80 appear to be reversed. However, the per-
centage of water soluble Tween 60 is greater in the Span 80
system, and thus the value of λ_{max} appears to be a better gauge
of the composition of the aqueous phase than the (overall) HLB,
which in any event does not take the alcohol cosurfactant into
account. The value of 344 nm for the SCS microemulsion indicates
that the composition of the continuous phase is very close to
that of pure water. It was, therefore, decided to use the CPO
acid-base equilibrium in the SCS microemulsion to obtain information on the behavior of buffers in ionic systems. Two
different buffer systems were examined, oxalate and acetate.
The oxalate buffer contained mono- and divalent negative ions
and should remain in the aqueous phase and not be adsorbed by
the (negative) droplet interface. On the other hand, the (neutral)
acetic acid component of the acetate buffer system could be
partitioned between the droplet and continuous phase. The
equilibrium examined is shown in Equation (1), and the results
are summarized in Table III.

Table III. Values of pK_a for CPO in water and the SCS Micro-emulsion at 25°C

buffer[a]	Oxalate			Acetate		
pH	4.09	4.56	5.00	4.26	4.69	5.07
water[b]	4.56	4.70	4.56	4.55	4.57	4.63
SCS[c]	5.39	5.45	5.45	5.14	5.15	5.14

a) Oxalate, 0.025M; Acetate, 0.050M (sodium salts). b) reported pK_a in water is 4.6 (ref.[16]). c) 60% water.

$$CPOH^+ = CPO + H^+, \qquad (1),$$

The values of pK_a in the SCS microemulsion were calculated using the pH of the buffer as measured in water. Only the CPO concentration was measured spectrally, the $CPOH^+$ concentration being determined by difference from the initial amount of CPO added. Since it is unlikely that either anion of the oxalate buffer is adsorbed, the difference in pK_a is ascribed entirely to binding of the species $CPOH^+$ by the droplet. It may be noted that the surface concentration of H^+ will of course be higher than the bulk. However, hydrogen ion should not be more strongly bound than the sodium counter ion, and the buffer capacity should not be overtaxed. Since the CPO concentration is ~ 5 x 10^{-5}M while the droplet concentration is ~ 10^{-3}M, we may write,

$$(CPOH^+)_{drop} = k_\alpha (CPOH^+) (1-\alpha), \qquad (2)$$

where α is the fraction of free counterion (generally ~ .3), and k_α is a "selectivity" constant which gives the binding of $CPOH^+$ relative to the Na^+ counterion. All concentrations in Equation (2) refer to overall concentrations. Then, the partitioning of $CPOH^+$ between the droplet surface and bulk (continuous) phases is given by equation (3),

$$K_\alpha = \frac{(CPOH^+) \, drop}{(CPOH^+) \, bulk} = \frac{k_\alpha (1 - \alpha)}{1 - k_\alpha (1 - \alpha)} \qquad (3)$$

and the difference between the measured pK_a's in the SCS microemulsion and water (Table III) are given by Equation (4),

$$pK_{SCS} = pK_{H_2O} + \log (1 + K_\alpha). \qquad (4)$$

Using the average oxalate and water values, $K_\alpha = 5.6$. For values of $\alpha = 0.2 - 0.4$. $k_\alpha = 1.1 - 1.4$, and we may conclude that the $CPOH^+$ ion is bound more strongly than Na^+ to the Stern layer of the droplet.

Turning to the acetate buffer system, if K_D is the distribution coefficient for acetic acid between the phases (droplet/continuous) then,

$$pK_{SCS} = pK_{H_2O} + \log (1 + K_\alpha) - \log \left[1 + \frac{\phi K_D}{1 - \phi} \right] \quad (5)$$

where ϕ is the volume fraction of the disperse (droplet) phase. Using a value of $K_\alpha = 5.6$ and $\phi = 0.40$, $K_D = 2.73$. This means that acetic acid is more soluble in the droplet than in the aqueous phase. This result is somewhat surprising in view of the fact that acetic acid is miscible with water in all proportions, and indicates that the interphase (surface) region of the droplet is playing a major role here. In any event, the data shows that appropriate acid-base equilibrium studies can provide corrections for use of buffers in o/w microemulsions and shed some light on ion binding.

Surface Phase

We now turn our attention to the examination of an acid-base equilibrium which must occur in the surface phase. The molecule employed is tetraphenylporphin (TPPH$_2$), an oil soluble molecule which has been previously employed in studies of metalloporphyrin formation in microemulsions[15]. As opposed to CPO, the visible absorption spectrum of TPPH$_2$ is not very solvent sensitive, although the extinction coefficients do exhibit some variation with medium as shown in Table IV.

Table IV. The Visible Absorption Spectrum of Tetraphenylporphin in Benzene and Benzene in Water Microemulsions Stabilized by Cyclohexanol and Sodium Cetyl Sulfate (SCS) or Potassium Oleate (PO)

benzene	0.6 SCS[b]	SCS[c]	PO
513 (22.6)	513 (20.2)	513 (18.8)	513 (18.2)
547 (9.5)	547 (8.6)	547 (7.9)	548 (8.1)
590 (6.7)	590 (6.0)	590 (5.1)	590 (4.5)
645 (4.6)	647 (3.7)	647 (3.2)	647 (3.4)

a) wavelength (extinction coefficient) in nm (M^{-1} cm^{-1} x 10^{-3}).
b) $\phi = 0.18$ c) $\phi = 0.45$

The changes in extinction coefficient may reflect both the influence of the interphase on the TPPH$_2$ and the presence of alcohol in the "benzene" droplet interior. That the latter effect should be important is indicated by the fact that the vapor pressure of benzene at 15 °C in the SCS ($\phi = 0.27$) system is only 0.55 of the vapor pressure of pure benzene. For a nearly

ideal solution, this would mean an alcohol/benzene mole ratio
on the order of unity, and an alcohol/SCS ratio of about 5:1
in the interphase.

The protonation of the free base porphyrin ($TPPH_2$) with HCl
can lead to the mono and dication species, $TPPH_3^+$ and $TPPH_4^{2+}$,
respectively. The spectra, upon addition of varying amounts of
HCl to the SCS/benzene microemulsions containing the porphyrin,
showed two sharp isobestic points indicating the presence of
only two species (the free base and dication). If the mono-
cation were present, a "floating" isobestic point would be obtained[17]. The data are given in Table V.

Table V. Tetraphenylporphin Dication to Free Base Ratio at
Various Overall HCl concentrations in the SCS/Benzene Microemulsion
System[a]

(H^+) $(M \times 10^2)$	$\dfrac{(TPPH_4^{2+})}{(TPPH_2)}$ [b]	K^c $(M^{-2} \times 10^{-3})$
0.2	0.056	14
0.4	0.124	7.8
0.6	0.257	7.1
0.8	0.462	7.2
1.0	0.687	6.9
1.2	0.950	6.6
1.4	1.36	6.9
3.5	7.36	6.0

a) $\phi = 0.27$ b) Total porphyrin concentration = 8.69×10^{-5} M.
c) Measured equilibrium constant $K = (TPPH_4^{2+}) / (TPPH_2) (H^+)^2$.

The values of K vary more than would be expected, but this
may reflect a change in surface potential due to H^+ since the
HCl concentration is relatively large and is changing by over
an order of magnitude. In other words, if Equation (2) also
holds for H^+, then,

$$K = K' \frac{(1 - k')^2}{1 - \phi} \exp (+2e\Psi_o/kT). \qquad (6)$$

where $k' = k_\alpha (1-\alpha)$, Ψ_o is the surface potential, K the measured
value (Table V) and K' the actual value.

This assumes that the ratio of free base to dication at
the surface defined by Ψ_o is the same as the overall ratio. This
factor may also be responsible for the variation in the values of K.
Using a Ψ_o of 50 mV and reasonable values for the other parameters
in equation (6), $K \approx 10^2 K'$. Here K is ~10^4, while for $TPPH_2$ in organic

solvents such as nitrobenzene[17] the value is ~10^6. This indicates that the dication/free base ratio at the Ψ_o surface is orders of magnitude greater than the overall ratio. A major obstacle in this type of study is the difficulty in obtaining a reliable value of K', and no account was taken of the contribution of bound hydrogen ion. A combination of CPO studies using HCl in place of buffer and the use of $TPPH_2$ and buffer in a suitable nonionic system may be able to help further delineate the contributions of these factors.

DYE ADSORPTION AND REACTION

Adsorption

We have examined an oil soluble substance and a water soluble substance which remains in the aqueous phase. In this section we will discuss the behavior of a water soluble substance which is adsorbed by the droplet surface.

Crystal violet is a stable triphenylmethane carbonium ion, and like virtually all indicator dyes appears to be strongly adsorbed at a surfactant stabilized oil-water interface. The spectrum is independent of pH, except at very low pH. The visible absorption maximum of this dye in various media is given in Table VI. The wavelength (λ_{max}) in aqueous micelles varies,

Table VI. Visible Absorption Maxima of Crystal Violet[a] in Various Media

SOLVENT	λ_{max} (nm)
water	591
n pentanol	591
cyclohexanol	593
aqueous SDS[b]	593
aqueous CPB[b]	595
O/W microemulsion[c]	600
benzene	605
mineral oil	605[d]

a) crystal violet: $(p\text{-}(CH_3)_2NC_6H_4)_3C^+Cl^-$. b) water containing ca. 0.01 M sodium dodecyl sulfate (SDS) and cetylpridinium bromide (CPB). c) same λ_{max} (±1) for both water/benzene/cyclohexanol/CPB or SCS and water/mineral oil/n-pentanol/SCS or tween 40 microemulsions. d) extrapolated

depending upon the particular surfactant. In microemulsion
media, λ_{max} appears to be essentially independent of the
nature of both the oil and the surfactant. Dilution studies
provide additional insight into the location of the dye in the
microdroplet. The absorption spectrum of crystal violet in the
SCS/mineral oil system as a function of dilution is shown in
Figure 3.   The spectrum in water, dilute aqueous SCS, or pentanol

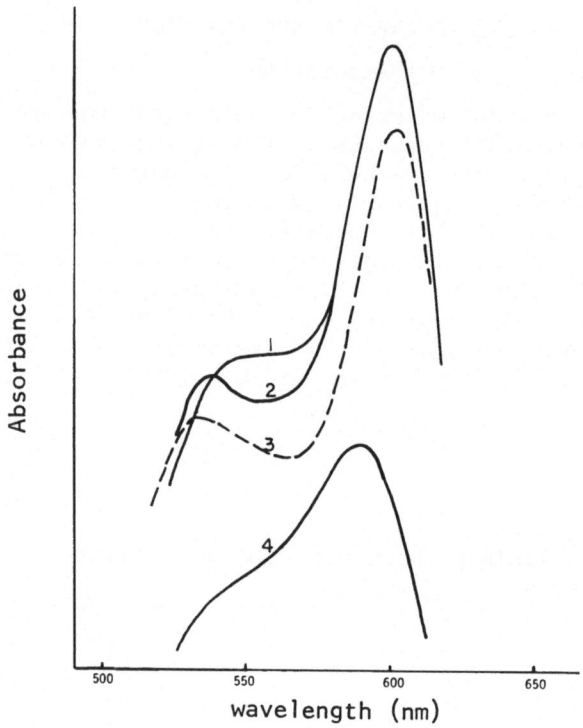

Figure 3. Visible absorption spectrum of crystal violet in
various media.   1) SCS/mineral oil microemulsion, $\phi=0.4$ (C_{rel} =
1.0);   2) SCS/mineral oil, $C_{rel}=0.1$;   3)   90% (v/v)
mineral oil/pentanol;   4)   water.

consists of a peak at 591 nm and shoulder on the short wavelength
side of this band. In the microemulsion the peak is at 601 nm
and the shoulder is almost flat. In a diluted system of relative
concentration (C_{rel}) of 0.1 the shoulder becomes a peak at 535 nm,
similar to the spectrum in mineral oil containing a little pentanol. These data are given in Table VII. The dilutions on the
phase diagram (Figure 2) are along the line connecting the cross

Table VII. Absorption Spectrum of Crystal Violet in SCS/Mineral Oil Microemulsions and Mineral Oil/Pentanol Mixtures[a]

SCS Mineral Oil Microemulsion			Mineral Oil/Pentanol		
C_{rel}[b]	λ_{max}	Sh/p[c]	%Oil[d]	λ_{max}	sh/p[c]
0	591	560 sh	10	591	566 sh
.01	593	560 sh	20	590	567 sh
.05	590	560 sh	30	590	570 sh
.10	600	535 p	60	595	563 sh
.15	599	535 p	70	597	560 sh
.45	597	545 sh	80	600	546 p
.50	600	545 sh	90	603	535 p
.70	599	555 sh	92.5	603	533 p
.90	599	555 sh	95	604	530 p

a) wavelengths in nm. b) C_{rel} = 1.0 is for ϕ = 0.40 c) sh = shoulder, p = peak. d) volume percent

and the water apex. Our interpretation is that, in accord with the phase diagram, a thermodynamically stable microemulsion does not exist below a C_{rel} of 0.45. However, a kinetically stable system can be produced by rapid dilution below a C_{rel} of 0.2.

This is accompanied by the conversion of the shoulder to a new peak at 535 nm, and the beginning of a drop in conductance (Figure 1). At a C_{rel} of 0.05, which corresponds to the minimum in the conductance curve, the spectrum shifts abruptly to that characteristic of aqueous (micellar) solutions. We may also note, by comparison of the data in Tables VI and VII, that the principal absorption maximum is indicative of the general polarity of the medium, while the shoulder/peak is more indicative of the specific composition. For example, λ_{max} is at 599 and 600 nm for C_{rel} of 0.8 and 0.1, respectively, while the positions of the shoulder correspond to 25% and 15% pentanol respectively. We thus tentatively conclude that all of the dye is adsorbed, and likely has an average location in the palisade layer. In the kinetically stable region, there are fewer but larger drops resulting in a decrease in conductivity and the transfer of some pentanol from the interphase to the bulk droplet interior, accounting for the concommitant changes in the crystal violet spectrum.

Reaction

Crystal violet reacts with hydroxide ion to form a colorless carbinol. In the classical work of Duynstee and Grunwald[18], the second order rate constant k_2 was found to be increased by cationic micelles and decreased by anionic ones. The results were attributed largely to electrostatic effects. A latter study by Ritchie, Skinner and Badding[19] in various pure solvents has stressed the role of solvent reorganization as an important contribution to the activation free enthalpy. A recent study by Cordes and coworkers[20] using alkyltrimethylammonium bromide has shown, among other things, that k_2 is larger for longer chain surfactants. It was considered that this may be due to a change in medium of the micellar surface and/or an increased electrostatic field. It was suggested that direct contribution of hydrophobic interactions may not be very important.

We have measured the second order rate constant in water, aqueous CPB, and the CPB/benzene/cyclohexanol microemulsion. In the microemulsion, k_2 did not vary by more than 15% for phase volumes of 0.09, 0.23 and 0.36. Judging by the conductance data (Figure 1), the most dilute system is just within the region of kinetic stability. The other data are given in Table VIII. The surface charge density for the aqueous micelle should be about the same or greater than for the micro-droplet, so that

Table VIII. Second Order Rate Constant for the Reaction of Crystal Violet with Sodium Hydroxide

Solvent	$k_2(M^{-1} min^{-1})$	$k_2/k_2(H_2O)$	λ (nm)
water	12	1.0	591
aq. CPB[a]	59	4.9	595
CPB/benzene[b]	135	11.3	601

a) 0.015M   b) $\phi = 0.23$

nonelectrostatic effects must also be important. There appears to be a correlation with the absorption frequency, which should be related to the polarity of the medium (vide supra). Further investigation of this effect seems warranted.

SUMMARY

1. Both thermodynamically and kinetically stable o/w "microemulsions" may be formed. For the ionic systems, conductance data can aid in the interpretation of the changes in internal structure with composition.

2. The compound CPO may be used as a probe of the composition of the continuous phase, as well as to determine the effect of

the oil microdroplets on added buffer. In addition, the binding
of the counter ion relative to protonated CPO can be obtained.

3. Although more complicated, equilibria involving oil soluble
probes such as porphyrins are capable of yielding information
on solute distribution and reaction surface potential.

4. The absorption of water soluble dyes such as crystal violet
can be used to obtain structural information complimentary to the
conductance and phase studies. Initial results indicate that the
dye is principally located in the hydrocarbon tails region of
the interphase. The reaction with base indicates a large contribution by medium effects as well as electrostatic effects
and should be investigated in more detail.

ACKNOWLEDGEMENT

The authors wish to thank Ms. Kathy Mazaika, and the late
Mr. Mike Del Vacchio for technical assistance with some of the
measurements, and Dr. Rameshwar Agarwal for many helpful
discussions. This work was partially supported by a grant from
the U. S. Army Research Office.

REFERENCES

1. P. Sherman, Editor, "Emulsion Science", Academic Press, New
    York, N.Y. (1968), p. 205.
2. J. H. Schulman, W. Stoeckenius, and L. M. Prince, J. Phys.
    Chem. 63, 1677 (1959).
3. W. Stoeckenius, J. H. Schulman, and L. M. Prince, Kolloid-
    Z., 169, 170 (1960).
4. J. H. Schulman, R. Matalon, and M. Cohen, Discuss. Faraday
    Soc. 11, 117 (1951).
5. K. Shinoda and H. Kunieda, J. Colloid Interface Sci. 42, 381
    (1973).
6. W. C. Tosch, S. C. Jones, and A. W. Adamson, J. Colloid Inter-
    face Sci. 31, 297 (1969).
7. S. Levine and K. Robinson, J. Phys. Chem. 76, 876 (1972).
8. C. E. Cooke, Jr., and J. H. Schulman in "Surface Chemistry",
    P. Elwa, K. Groth, and V. Runnstrom-Reio, Editor, p. 231,
    Academic Press, New York, N.Y. (1965).
9. E. H. Cordes and R. B. Dunlap, Acc. Chem. Res. 2, 329 (1969).
10. E. J. Fendler and J. H. Fendler, Adv. Phys. Org. Chem. 8,
    271 (1970).
11. S. Friberg and S. I. Ahmad, J. Phys. Chem. 75, 2001 (1971).
12. S. I. Ahmad and S. Friberg, J. Amer. Chem. Soc. 94, 5196 (1972).
13. J. T. Davis, Advan. Catalysis 6, 1 (1954).
14. H. Morawetz, ibid. 20, 341 (1969).
15. K. Letts and R. A. Mackay, Inorg. Chem. 14, 2990, 2993 (1975).
16. R. A. Mackay and E. J. Poziomek, J. Amer. Chem. Soc. 94, 6107
    (1972).

17. S. Atonoff, J. Phys. Chem. <u>62</u>, 428 (1958).

18. E. F. J. Duynstee and E. Grunwald, J. Amer. Chem. Soc. <u>81</u>, 4540, 4542 (1950).

19. C. Ritchie, G. Skinner and U. Badding, J. Amer. Chem. Soc. <u>89</u>, 2065 (1967).

20. J. Albrizzio, J. Archila, T. Rudolfo and E. H. Cordes, J. Org. Chem. <u>37</u>, 871 (1972).

DISCUSSION

On the paper by S. Friberg and I. Buraczewska

H. Chun, *General Mills*: (1) During my preparation of microemulsions with non-ionic surfactant I have found, sometimes, that an originally clear microemulsion – in either gel or liquid form – developed some "white" color on the surface upon standing at 45°C for a prolonged period. Sometimes this "white" colored stuff very slowly expanded and then made the originally clear microemulsion turn hazy. Sometimes this "white" stuff also developed on the surface around the beaker during the mixing process (the beaker was covered). My system contained: H₂O ~ 20%-30%

$$H_2O \sim 20\%\text{-}30\%$$
$$\text{Mineral oil} \sim 40\%$$
$$25 \text{ mole ethoxylated phytosterol} \sim 15\%$$
$$3 \text{ mole ethoxylated oleyl alcohol} \sim 10\%$$

"Oil" and "water" phases were both heated to ~ 80°C, then the water was added to the oil with mixing action.
(2) In your opinion, what is the best way to judge whether a microemulsion is o/w or w/o?

S. Friberg: (1) It is very common that a liquid crystalline phase is dispersed in the microemulsion via particles so small that they are not observed. With time, they separate. <u>Short</u> time ultracentrifugation is a good test to avoid this occurrence.
(2) Electric conductance is good; so is the rate at which a water droplet and an oil droplet are dissolved. A water droplet might take days to dissolve in a w/o microemulsion.

Part VII
General Papers

Part VII

Bacterial Parasites

MIXED MICELLES OF METHYL ORANGE DYE AND CATIONIC SURFACTANTS

Richard L. Reeves and Shelley A. Harkaway

Research Laboratories

Eastman Kodak Company, Rochester, New York 14650

Interaction of the anionic dye with cationic surfactants gives a sharp new absorption band at surfactant concentrations below the cmc for surfactant
homomicelles. The band is shown to result from a
dye-dye stacking interaction rather than from a
change in dye geometry. Variation of the surfactant:
dye (S:D) ratio gives distinct absorption bands characteristic of three disperse states: (1) a band similar
to that of free dye at S:D ratios near the equivalence
point that is due to a microcrystalline suspension of
the insoluble salt; (2) the dye aggregate band at
larger S:D ratios characteristic of mixed micelles
having a significant population of dye molecules
occurring side-by-side with their molecular axes
nearly parallel; and (3) a band similar to that of
dye dissolved in an organic medium at surfactant
concentrations near the cmc. The intermediate mixed
micelles are unstable and slowly reorganize to the
insoluble salt. The dye is weakly surface active.
Dye-CTAB mixtures are more surface active than CTAB
alone at surfactant concentrations near the cmc. The
results are discussed in terms of possible structures
of the mixed micelles.

INTRODUCTION

Amphiphilic dyes constitute a major class of colloidal electrolytes. A common class, the dye sulfonates, consists of ionic
head groups with localized charge attached to a hydrophobic moiety
consisting of aryl groups and heteroatoms. The tendency of ionic
dyes to aggregate is well recognized, but detailed knowledge of
such dye aggregates lags far behind our present knowledge about
the aggregation of the other major class of amphiphiles, the ionic
surfactants. Part of this deficiency lies with the formidable task
of obtaining highly purified dye preparations, the tendency of many
dyes to aggregate to metastable, slowly equilibrating aggregates,
and the multitude of available structures that has tended to make
the study of colloidal properties of ionic dyes haphazard. In
general, the nonionic moieties of dye sulfonates are more solvated
in water than the hydrocarbon chains of surfactants so that the
hydrophilic-hydrophobic balance is rather different in the two
classes of amphiphiles. This causes the dyes to be weakly surface
active[1-3] and to show little tendency to form micelles in the
absence of added salts unless they contain pendent alkyl groups.

Several studies have been made of mixed micelles of surfactants with similar head groups[4,5] and with oppositely charged head
groups.[6-10] Most of these studies involved mixtures of surfactants
that differed little in their hydrophobic-hydrophilic balance.
Studies of mixed amphiphiles having oppositely charged head groups
have usually been carried out by varying the total concentration
at a fixed 1:1 stoichiometric ratio. Very few studies have been
made of mixed micelles of amphiphiles which differ as much in
their hydrophobic-hydrophilic balance as do dyes and surfactants.
Although there is considerable evidence for interaction of dye
sulfonates with cationic surfactants,[11-17] existing studies have
not shed much light on the detailed nature of the resulting
colloids.

We have shown through spectral[18] and kinetic studies[19,20] that
anionic dye sulfonates and cationic surfactants form a continuum of
mixed aggregates in water depending on the total concentration and
the surfactant:dye ratio (S:D). These range from microcrystalline
suspensions of the insoluble salts at low, near-equivalent S:D
ratios through a series of mixed micelles of varying composition,
up to surfactant micelles of normal structure containing small
fractions of solubilized dye. The latter micellar structures form
at surfactant concentrations near the critical micelle concentration (cmc) for surfactant homomicelles, but the intermediate mixed
micelles form at surfactant concentrations far below the cmc's.
Where insoluble 1:1 crystalline salts of the dyes and surfactants
exist, mixed micelles formed at low S/D ratios are unstable and
slowly reorganize to microcrystalline suspensions. The present

study is concerned with the mixed colloidal aggregates formed from the azo dye sulfonate, methyl orange (MO,I) and three cationic surfactants, hexadecyltrimethylammonium bromide (CTAB), hexadecylammonium chloride (HAC), and surfactant II, and with the interaction

$$(CH_3)_2N \text{—} C_6H_4 \text{—} N{=}N \text{—} C_6H_4 \text{—} SO_3^-$$

$$\xleftarrow{\hspace{1.5cm}} 1.5 \text{ nm} \xrightarrow{\hspace{1.5cm}}$$

I

$$\overset{+}{C_{15}H_{31}CONH(CH_2)_3N(CH_3)_2C_2H_4OH} \ Cl^-$$

II

of the dye with a polycation, poly(3-methyl-1-vinylimidazolium) methosulfate (PMVI).

Klotz and his coworkers found that the binding of MO to cationic poly(ethyleneimine) containing apolar side chains gave rise to an intense new band at ~375 nm that is not found in aqueous solutions of dye alone under any conditions.[21] They attributed the band to dye stacking on the polyion. Quadrifoglio and Crescenzi found the new absorption band in colloids formed from interaction of the dye with cationic surfactants and with additional polycations.[15] The fact that a sharp, well separated band is not found on interaction with all polycations led these workers to question whether the band arises from dye stacking on the polyion. They suggested that the band might arise from the thermally unstable cis-isomer of the dye which is somehow stabilized in the bound state.

In aqueous solutions, the cis- to trans-conversion of MO in aqueous solutions is fast.[22] When the dye is bound to proteins such as serum albumin, however, the acid catalyzed cis- to trans-isomerization is slowed considerably[22] so that the proposal of Quadrifoglio and Crescenzi could not be ruled out. We felt that the existence of a unique absorption band characteristic only of the bound state could yield useful structural information on mixed micelles containing the dye if the origin of the band were understood. Irradiation experiments have enabled us to reject the hypothesis of Quadrifoglio and Crescenzi. By use of molecular exciton theory[23] for paired dye interactions of bound dye and by comparison of spectra of dye bound in mixed micelles with that bound to a polycation, we can draw some qualitative but comparatively detailed conclusions about the nature of the micelles. The present results provide additional confirmation of conclusions drawn from our earlier studies.

EXPERIMENTAL

The dye was a commercial sample that was recrystallized repeatedly to constant absorptivity (2.70×10^4 in water). The CTAB was a sample from an earlier study.[19] HAC was obtained by treating hexadecyclamine in ethanol with 37% aqueous hydrochloric acid and recrystallizing the amine salt from ethanol-ethyl acetate. Surfactant II was synthesized by first condensing equimolar amounts of palmitic acid and dimethylamino propylamine in refluxing toluene and then quaternizing the resulting N-(3-dimethylaminopropyl) palmitamide, m.p. 57-8°, with 2-chloroethanol in refluxing aceto-nitrile. The surfactant was recrystallized repeatedly from 2-buta-none and from acetonitrile to a constant transition temperature (solid-mesophase) of 95°. The cmc determined by the dye method (Rhodamine 6G) was 6×10^{-4} mol dm^{-3}.

Methyl orange solutions in 0.01 $\underline{M}$ sodium hydroxide were irradiated in the infrared mode of a Cary Model 14 spectrophoto-meter with white light from the tungsten source. The absorption curve of the photostationary state was scanned monochromatically during the irradiation. A different photostationary level of the cis-isomer was obtained by varying the intensity of the irradiation with a neutral density filter. The original absorption curve was obtained after the irradiated sample had re-equilibrated in the dark.

Solutions of the dye with the surfactants and polyion were prepared by mixing equal volumes of a dye stock solution with solutions of surfactant or polyion from separate motor-driven syringes through impinging nozzles. Absorption curves of the resulting mixtures were recorded within five minutes of mixing on a Beckman DK-2A or a Cary 118C spectrophotometer.

Ultrafiltrations were done through Nuclepore membranes with uniform pore sizes of 0.6 μm. Light scattering observations were made on filtered solutions in dust-free vials using the focused beam of a high intensity microscope lamp.

Surface tensions were measured by the Wilhelmy plate method on solutions at 25° in a Teflon plastic dish. The platinum plate was held in the surface until constant readings were obtained.

RESULTS AND DISCUSSION

Absorption curves of the dye in water, in ethanol, and in the bound state on poly(3-methyl-1-vinylimidazolium)methosulfate (PMVI) are shown in Figure 1. The dye is completely bound at a polymer site:dye ratio (P:D) of 4. Two important facts should be

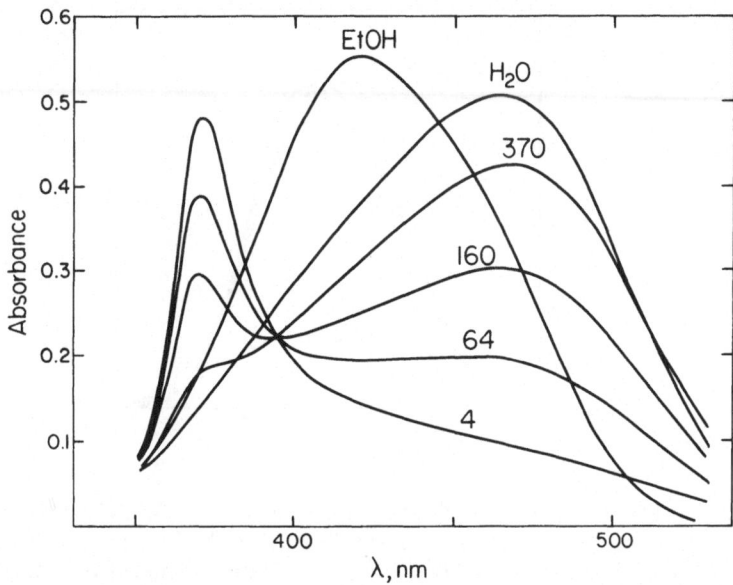

Figure 1.   Absorption curves of a 2.0 x 10^{-5} mol dm^{-3} solution of MO in water, in a nonaqueous environment, and bound to PMVI at the indicated P:D ratios.

noted: (1) The absorption band at about 370 nm for bound dye is not characteristic of the dye in a hydrophobic or nonaqueous binding site but is unique to the bound state on the polyion or in micelles with cationic surfactants.   The band is not obtained by aggregation of the dye in solution.[15]   (2) The molar absorptivity of the 370 nm band is similar to that of the <u>trans</u>-isomer of the dye in solution.   Thus, if the 370 nm band is characteristic of the <u>cis</u>-isomer, irradiation of a solution of the <u>trans</u>-form should give an increase in absorbance at 370 nm in the photostationary state that is nearly equal to the decrease in absorbance of the <u>trans</u>-isomer at 462 nm.

The results of such an irradiation experiment are shown in Figure 2.   Although only about 20% conversion of the dye was achieved in the photostationary level, it is clear that this degree of conversion does not produce a corresponding increase in absorbance at 370 nm.   The absorption curve of the <u>cis</u>-isomer was constructed from the curves of the pure trans-isomer and the two photostationary levels using the method of Fischer.[24]   The constructed curve shows that the <u>cis</u>-isomer does have an absorption band centered near 370 nm, but it is too weak to account for the intensity of the observed absorption of the bound dye.

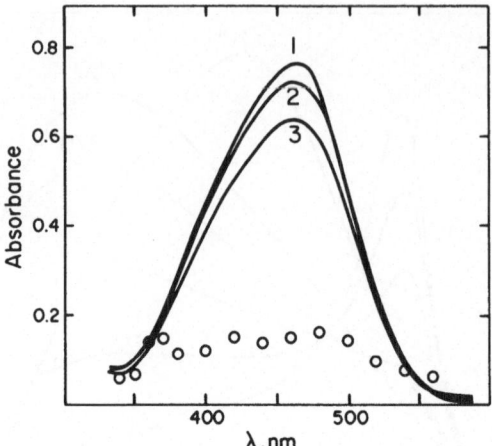

Figure 2. Effect of white-light irradiation on a 3.0 x 10^{-5} mol
dm^{-3} solution of MO in 0.01 mol dm^{-3} sodium hydroxide. Curve 1,
pure trans-dye before irradiation and after irradiation and dark
equilibration. Curves 2 and 3, photostationary levels of cis- and
trans- mixtures. Open circles are absorbances of pure cis-dye
estimated from curves 1-3.

Resonance Raman spectra of MO also show that MO retains its
trans- structure on interaction with CTAB.[25]

The 370 nm band is accounted for by application of molecular
exiton theory for paired interactions of dyes stacked in sandwich
structures with parallel transition dipoles. The energy level diagram for such a parallel dimer of identical molecules, A and B, is
shown in Figure 3. If the wave function for the dimer ground
state is $\Psi_G = \psi_A \psi_B$, the first excited state is described equally
well by two wave functions, $\phi_1 = \psi_A \psi_B^*$ and $\phi_2 = \psi_A^* \psi_B$. Thus, the
excited state level is degenerate and is split into a high energy
and a low energy level. The physical interpretation is that
excited state splitting arises from electrostatic interaction of
transition dipoles on neighboring molecules.[23] The low level
corresponds to interaction of antiparallel transition dipoles and
the high level to parallel dipoles. The interaction energy is
represented by E. Excitation from the ground state to the low
energy level is forbidden, whereas a transition to the higher level

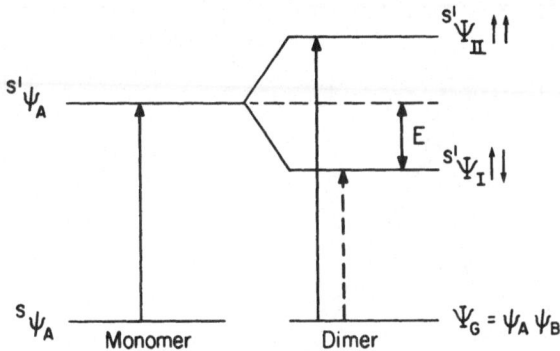

Figure 3. Schematic-energy level diagram of exciton splitting in the first excited state of parallel dimers. Solid arrows represent allowed transitions and the dashed arrow a forbidden transition.

is allowed. Thus, a sandwich dimer or a card pack n-mer gives an absorption band that is blue-shifted relative to that of the monomer.

The interaction energy for dye in a sandwich dimer may be estimated by the exciton model from the spectral shift of the 370 nm band relative to the dye monomer band. The absorption curve of dye monomer in water has been shown[26] to consist of two unresolved bands centered respectively at about 21000 and 24000 cm^{-1}. Thus, interaction energies of 3000 or 6000 cm^{-1} are computed, depending on which monomer band is taken as reference. These energies fall within the range calculated by exciton theory for parallel dimers of dyes with unit oscillator strengths at separations of 0.5-0.7 nm.

Deviations from parallelism of the transition dipoles of the dimer decreases the interaction energy, narrows the splitting of the excited state energy levels, and moves the more allowed transition to lower energies. Since the line of the transition dipole in MO corresponds to the long molecular axis, a distribution of angular orientations of paired dye molecules in a mixed micelle or on the polyion should lead to a distribution of exciton splittings. This will broaden the aggregate band and cause it to be skewed toward longer wavelengths. Thus, the molecular exciton model permits us to draw qualitative conclusions regarding molecular arrangement, structural tightness, and angular orientation in mixed micelles containing MO.

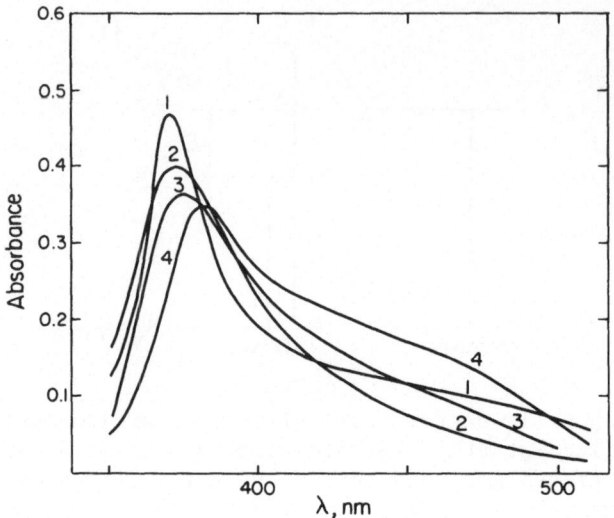

Figure 4. Absorption curves of 2.0×10^{-5} mol dm^{-3} MO bound to
PMVI at a P:D ratio of 4 and in mixed aggregates with HAC (curve 2),
CTAB (curve 3), and surfactant II (curve 4). Total surfactant con-
centrations are 2.0×10^{-4} mol dm^{-3}.

Figure 4 shows a comparison of the absorption curves of MO
bound to PMVI and in mixed micelles with CTAB, HAC, and surfactant
II at a S:D ratio of ten. The narrower aggregate band on the polyion is consistent with a narrow distribution of molecular orientations on the polyion. The broadening and increased skewness of the
band in the mixed micelles suggest a looser structure and a broader
distribution of orientations or separations of paired dye molecules.
It is interesting that the C16 surfactant with an ammonium head
group gives a mixed micelle with slightly stronger dye pairing and
lower absorbance due to unpaired dye molecules than the surfactant
with the trimethylammonium head group. The more hydrophilic surfactant (II) gives even lower dye-dye interaction energies and a
higher ratio of unpaired dye molecules, consistent with a looser
structure.

The nature of the disperse species, as judged by the absorp-
tion curves and light scattering, depends on the stoichiometric
S:D ratio and on the absolute concentrations of both amphiphiles.
Figure 5a shows the spectral changes observed when increasing con-

centrations of CTAB are added to a constant dye concentration of 4×10^{-6} mol dm^{-3}. Figure 5b shows the changes for increasing concentrations of CTAB and MO at a constant S:D ratio of ten. Qualitatively, similar effects are found for both variations. At low concentrations and/or low S:D ratios, the absorption curves are similar in shape to that of dye in molecular dispersion, but with reduced absorptivity. Scattered light from these dispersions is due almost entirely to reflections from microcrystalline suspension of the insoluble salt. The particles can be removed by ultrafiltration (0.6 µm pore sizes) to give initially nonscattering filtrates. On standing, additional crystalline material forms in the filtrates. The spectra show that the suspensions of the insoluble crystalline salts do not give rise to the band centered near 370 nm and therefore do not have dye molecules oriented with parallel molecular axes.

At higher concentrations of MO and CTAB but at concentrations considerably below the cmc for CTAB homomicelles, the colloidal aggregates are characterized by principal absorption due to the

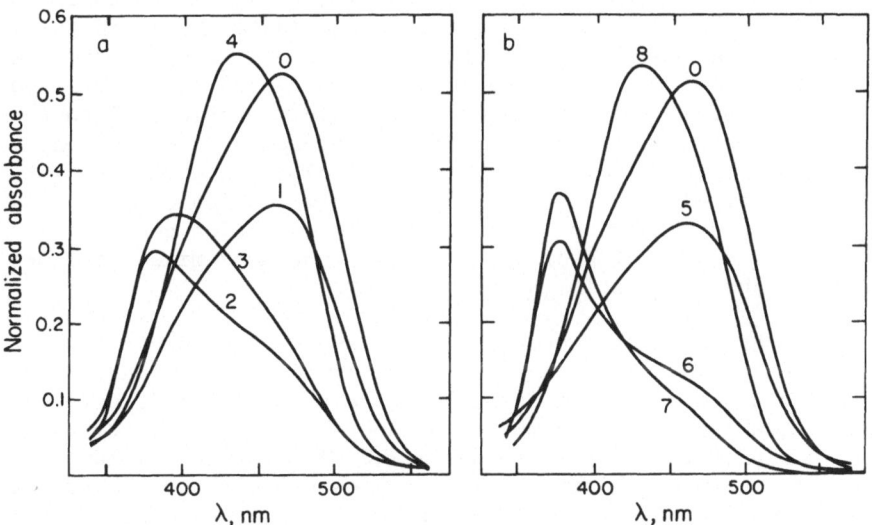

Figure 5.   (a) Absorption curves of 4.0×10^{-6} mol dm^{-3} MO in the presence of the following concentrations of CTAB: (0) 0; (1) 2.0×10^{-5} mol dm^{-3}; (2) 2.0×10^{-4} mol dm^{-3}; (3) 4.0×10^{-4} mol dm^{-3}; (4) 2.0×10^{-3} mol dm^{-3}.   (b) Effect of the total concentration of MO and CTAB on the absorption curves of MO at a constant S:D ratio of ten.   [CTAB] = (5) 4.0×10^{-5} mol dm^{-3}; (6) 1.0×10^{-4} mol dm^{-3}; (7) 4.0×10^{-4} mol dm^{-3}; (8) 2.0×10^{-3} mol dm^{-3}.

dye stacking interaction. Light scattering from these solutions
is primarily from amorphous particles with some contribution at low
S:D ratios from crystallites. Curve 6 of Figure 5b is typical of
a suspension of both types of particles. The absorption centered
near 460 nm is characteristic of the crystallites and absorption
near 370 nm is from the amorphous mixed micelles. Ultrafiltration
removes the crystallites but not the micelles. The filtrate shows
scattering only from the amorphous particles and absorption from
the 370 nm band but with reduced absorption at 460 nm.

At concentrations of CTAB near its cmc, a third dye absorption
band with a maximum near 430 nm is found for both high and low
concentrations of dye. This band is characteristic of MO in a nonaqueous microenvironment and is obviously due to monomeric dye
solubilized in CTAB micelles consisting mainly of surfactant and
with structures approaching that of a CTAB homomicelle.

The spectra in Figure 5 show that the general nature of the
mixed aggregates depends largely, but not entirely, on the surfactant concentration. Colloids giving similar absorption curves
tend to form at similar surfactant concentrations. This is because
the dye is a very weak micelle former compared to the surfactant.
The curves show, however, that the structure of the colloids depends
to some extent on the dye concentration. Curves 3 and 7 were measured on solutions containing the same stoichiometric CTAB concentration but which contained a ten-fold difference in MO concentration. The lower dye concentration (curve 3) gives a micelle with
diminished paired dye interactions and increased amounts of isolated
dye. It is reasonable to conclude that the solution containing the
stoichiometric S:D ratio of 100 (curve 3) contains mixed micelles
having a higher S:D ratio in the micelle.

The present results provide direct spectral evidence for our
earlier conclusions based on less direct evidence. Anionic dyes
and cationic surfactants interact to give a continuum of colloidal
aggregates which vary in composition and structure from crystallites
at one end of the spectrum, though mixed micelles containing varying
ratios of the two amphiphiles, to micelles closely resembling surfactant homomicelles. In such a system, there is no single micellar
species and no cmc.

The mixed micelles with CTAB and HAC are unstable at low S:D
ratios owing to the low solubility of the crystalline 1:1 salts.
The micelles reorganize to crystallites on aging, even when the
crystallites formed initially during mixing have been removed. The
more hydrophilic surfactant (II) gives more soluble salts so that
the mixed micelles with this surfactant were stable at the concentrations of our experiments.

Figure 6 shows possible partial structures of the mixed micelles. In all the structures, a high degree of charge neutralization by the oppositely charged head groups permits the mixed micelles to attain large size. In 6A, the core of the micelle is formed by the more hydrophobic alkyl chains of the surfactant and the dye molecules are absorbed as oriented counterions. When it is remembered, however, that the bulk of the hydrophobic end of the dye molecule is nearly as great as that of the surfactant alkyl group, structure A would give a micelle with a hydrophobic surface and a Stern layer in a region intermediate between the surface and the hydrocarbon core. Any structure having the oppositely charged head groups in close proximity and the molecular axes of paired dyes reasonably parallel require that the hydrophobic moiety of the dye be turned completely outward into bulk water as in A, or completely inward as in B and C.

Regions of the mixed micelles having unpaired dyes (structure B) probably exist in micelles containing a large excess of surfactant, but structure C is the most probable one for the micelles giving the absorption band near 370 nm. In this structure, dye is not solubilized randomly but prefers to occupy a position adjacent to another dye. This preference for dye pairing is seen dramatically with MO bound to PMVI. Figure 1 shows that even at large excesses of binding sites to dye, the absorption from paired dye interactions persists. Dye is not bound in completely independent sites until the P:D ratio exceeds about 400. If dye were bound randomly on independent sites, the absorption at 370 nm should have diminished at much lower P:D ratios because of the r^{-3} dependence of the interaction energy on the separation of transition dipoles.

In an earlier paper,[20] we gave kinetic evidence that the addition of low concentrations of electrolyte to solutions of dye-surfactant mixed aggregates at low S:D ratios gave significant changes in the aggregate structures. The changes appeared to be

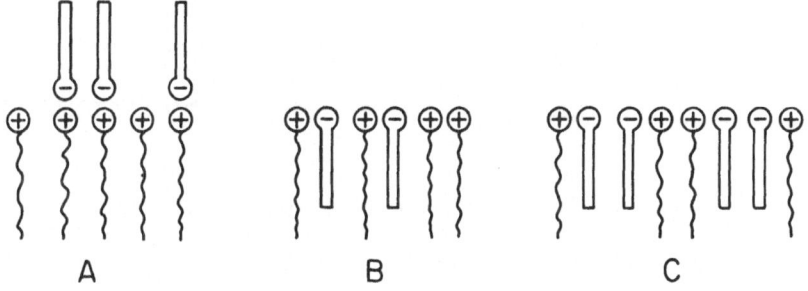

Figure 6. Possible structures of dye-surfactant mixed aggregates.

in the direction of stabilizing mixed micelles with a higher surfactant content than was possible in the absence of added salt.
We have confirmed these conclusions in the present study with direct
spectral evidence. Figure 7 shows the effect of potassium chloride
on the spectra of mixtures of MO and surfactant II at a S:D ratio
of 4. In the absence of added salt, the absorption curve is that
of a mixture consisting of micellar aggregates displaying the sharp
band near 385 nm and the 1:1 salt with maximum absorption near 460
nm. Addition of only 2.5×10^{-3} mol dm^{-3} of potassium chloride
alters the structure of the aggregate to give enhanced dye stacking.
A salt concentration of $0.01 \cdot$ mol dm^{-3} alters the structure still
further to give micelles containing sufficient excess surfactant
to eliminate paired dye interactions and to isolate individual dye
molecules in the micellar environment (Figure 5). Thus, at the
dye and surfactant concentration used in the experiment, 0.01 mol
dm^{-3} chloride ion can induce a micellar structure that is only
obtained at 7-8 times the surfactant concentration in the absence
of added chloride. These and our earlier results show that the
various structures of the mixed aggregates differ very little in
free energy and show the necessity of carefully specifying all
concentrations in describing the aggregates.

It seems unlikely that the degree of stacking of MO in the
mixed micelles or on the polyion exceeds that of dimerization.
Space filling models of PMVI show that considerable steric strain

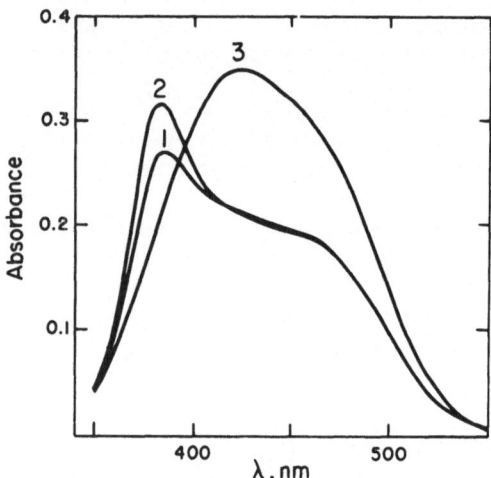

Figure 7. Effect of KCl on the structures of mixed aggregates of
MO with surfactant II. [MO] = 2.0×10^{-5} mol dm^{-3}; [II] = 8.0×10^{-5} mol dm^{-3}; [KCl] = (1) 0; (2) 0.0025 mol dm^{-3}; (3) 0.001
mol dm^{-3}.

is introduced when neighboring imidazolium groups occupy parallel positions as required to align the molecular axes of bound dye. However, syndiotactic triads permit next nearest-neighbor imidazolium groups to occupy perfectly parallel positions without strain. In such conformations, the separation of cationic sites is approximately 0.5 nm. This separation is in excellent agreement with the separation of aligned dyes computed by exciton theory from the spectral shift. We do not know what the overall tacticity of the PMVI is, but it seems reasonable that the exciton band arises most probably from the summation of numerous paired interactions in small syndiotactic regions rather than from large n-mers requiring a large number of imidazolium groups to be aligned on the same side of the polymer backbone. Similarly, in the mixed micelles, formation of dimers as in structure C permits charge neutralization by the surfactant head group. Formation of dye n-mers in the micelles would create regions of high negative charge in the Stern layer that would have to be shielded by an uneven surface accumulation of counterions. The fact that the absorption maximum for the dye aggregates is the same in the micelles and on the polyion requires that the average degree of aggregation be the same in both cases. Exciton theory predicts that an increase in the degree of aggregation increases the splitting of the exciton band and gives a corresponding shift of the more allowed transition to higher energies.

The surface tensions of solutions of CTAB, of MO, and of mixtures of the two amphiphiles were measured to gain additional insight into their mutual aggregation. The results are shown in Figure 8. The measurements on mixtures were made at a constant MO concentration of 1.0×10^{-5} mol dm^{-3} and varying CTAB concentrations. The very low hydrophobic rejection of the nonionic portion of the dye is seen from the absence of appreciable surface activity of the dye solutions. The results show, however, that in mixtures with CTAB, the dye is strongly adsorbed at the air-water interface, giving greater lowering of the surface tension than with CTAB alone. Measurements could not be made at lower S:D ratios than those shown because aging effects associated with micellar reorganization and crystallization prevented the achievement of a surface equilibrium. The results show that in the dissociated dye anions, the hydrophobic-hydrophilic balance in the amphiphile produces little hydrophobic rejection at concentrations up to near saturation. When the charge on the dye is neutralized, however, the hydrophobicity of the non-ionic portion of the dye is apparent. Whereas orientations A, B, or C (Figure 6) at the interface could account for the increased surface activity in the mixtures of MO with CTAB, orientations B and C seem most probable. Recent measurements of monolayer coverage of MO-CTAB salts at the water-CCl4 interface[27] show that MO and CTAB contribute equally to the coverage as required by 6B or 6C (Figure 6).

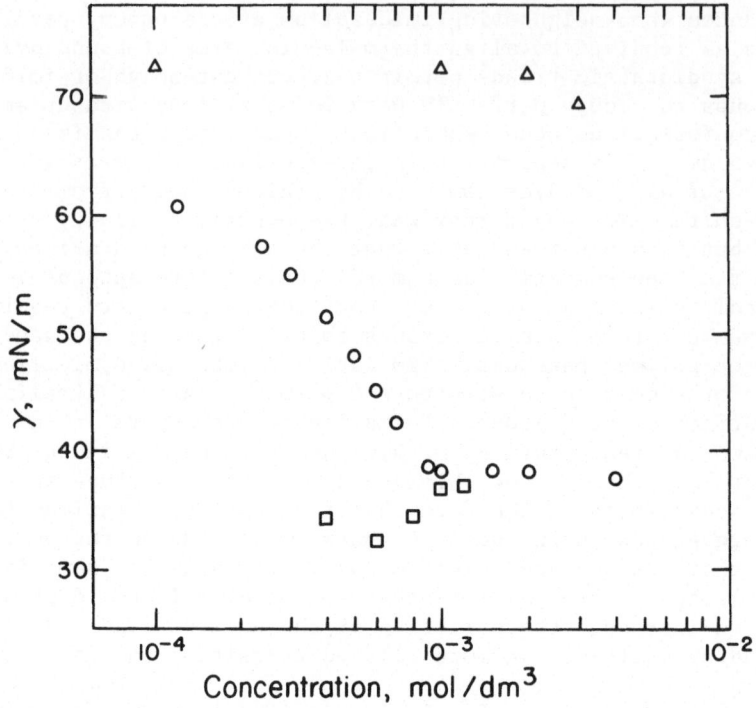

Figure 8.   Surface tension of aqueous solutions of MO and CTAB at
25° as a function of MO concentration (triangles), CTAB concentration (circles), and CTAB concentration in the presence of 1.0 x
10^{-5} mol dm^{-3} MO (squares).

     The idea of paired dye interactions as the origin of the band
centered near 370 nm is not inconsistent with the results of
Quadrifoglio and Crescenzi[15] which prompted them to seek an alternate interpretation. Although they only found the sharp, blue-
shifted band when MO was bound on certain polycations, they did
find new short wavelength absorption with all their polyions. In
some cases, this was very broad and centered at wavelengths intermediate between 370 nm (parallel dimers) and 460 nm (isolated monomers). By the exciton interpretation, the broad intermediate high
energy absorption is from dyes bound with weaker interaction of
transition dipoles than with those cases where the sharp 370 nm
band is found, i.e., with greater separations and/or less parallel
orientations. In view of the differences in polymer conformations
and steric interactions among polyions with different bulky pendent
groups, differences in the abilities of various polycations to
orient the bound dye with parallel molecular axes are not surprising. The dye stacking interpretation also explains the fact that
an increase in the length of alkyl groups on the terminal tertiary

nitrogen of the dye (ethyl orange, butyl orange) causes severe degradation of the 370 nm band on a polyion where this band is sharp with bound MO. Molecular models show that when the sulfonate groups of parallel dye molecules are held at intermolecular distances of 0.5-0.6 nm, increasing size of alkyl groups on the terminal amine causes increased tipping of the molecular planes of adjacent dyes. Such tipping causes a more rapid decrease in interaction energies of transition dipoles than rotational twisting through the same angle.[23,28]

ACKNOWLEDGMENTS

We wish to thank Mr. L. Costa and Dr. M. Long of our laboratories for carrying out the irradiation experiments, and Drs. W. Cooper and R. Klingbiel for helpful discussions.

REFERENCES

1. C. W. Gibby and C. C. Addison, J. Chem. Soc., 1306 (1936).
2. P. Alexander and D. A. Charman, Textile Res. J., 20, 761 (1950).
3. C. H. Giles and A. H. Soutar, J. Soc. Dyers Colourists, 87, 301 (1971).
4. H. W. Hoyer and A. Marmo, J. Phys. Chem., 65, 1807 (1961).
5. J. H. Clint, J. Chem. Soc., Faraday I, 71, 1327 (1975).
6. A. B. Scott, H. V. Tartar, and E. C. Lingafelter, J. Amer. Chem. Soc., 65, 698 (1943).
7. E. W. Anacker, J. Colloid Sci., 8, 402 (1953).
8. H. W. Hoyer, A. Marmo, and M. Zoellner, J. Phys. Chem., 65, 1804 (1961).
9. H. W. Hoyer and I. L. Doerr, J. Phys. Chem., 68, 3494 (1964).
10. B. W. Barry and G. M. T. Gray, J. Colloid Interface Sci., 52, 314 (1975).
11. C. F. Hiskey and T. A. Downey, J. Phys. Chem., 58, 835 (1954).
12. L. K. J. Tong, R. L. Reeves, and R. W. Andrus, J. Phys. Chem., 69, 2357 (1965).
13. M. L. Corrin and W. D. Harkins, J. Amer. Chem. Soc., 69, 679 (1947).
14. C. E. Williamson and A. H. Corwin, J. Colloid Interface Sci., 38, 567 (1972).
15. F. Quadrifoglio and V. Crescenzi, J. Colloid Interface Sci., 35, 447 (1971).
16. R. Hague and W. U. Malik, J. Phys. Chem., 67, 2082 (1963).
17. W. U. Malik and S. P. Verma, J. Phys. Chem., 70, 26 (1966).
18. R. L. Reeves, R. S. Kaiser, and H. W. Mark, J. Colloid Interface Sci., 45, 396 (1973).
19. R. L. Reeves, J. Amer. Chem. Soc., 97, 6019 (1975).

20. R. L. Reeves, J. Amer. Chem. Soc., 97, 6025 (1975).

21. I. M. Klotz, G. P. Royer, and A. R. Sloniewsky, Biochemistry, 8, 4752 (1969).

22. R. Lovrien, P. Pesheck, and W. Tisel, J. Amer. Chem. Soc., 96, 244 (1974).

23. E. G. McRae and M. Kasha, in "Physical Processes in Radiation Biology," L. Augenstein, R. Mason, and B. Rosenberg, Editors, pp. 23-42, Academic Press, New York, 1964.

24. E. Fischer, J. Phys. Chem., 71, 3704 (1967).

25. B-K. Kim, A. Kagayama, Y. Saito, K. Machida, and T. Uno, Bull. Chem. Soc. Japan, 48, 1396 (1975).

26. R. L. Reeves, R. S. Kaiser, M. S. Maggio, E. A. Sylvestre, and W. H. Lawton, Can. J. Chem., 51, 628 (1973).

27. T. Takenaka and T. Nakanaga, J. Phys. Chem., 80, 475 (1976).

28. W. Kauzman, "Quantum Chemistry," p. 507, Academic Press, New York, 1957.

ANIONIC SURFACTANT COMPLEXES WITH CHARGED AND UNCHARGED
CELLULOSE ETHERS

E. D. Goddard and R. B. Hannan

Union Carbide Corporation

Tarrytown Technical Center, Tarrytown, New York 10591

The interaction of sodium dodecyl sulfate (SDS) with
a cationically charged cellulose ether has been examined
by measurements of dye solubilization and viscosity.
The results support our previously developed picture of
a partial first layer of surfactant on the polymer, a
complete layer (precipitation) and a developing second
layer: thus, at constant polymer concentration, solubilization takes place at very low ratios of added SDS,
drops to zero in the precipitation range, and in the
post-precipitation range occurs with an efficiency
greater than that in the micellar SDS range, encountered
at higher concentrations. Solubilization by methyl
cellulose/SDS systems resembles that in the third range
given above, while there is little evidence of this
range for hydroxyethyl cellulose/SDS or of polymer/sur-
factant interactions. Viscosity measurements, however,
provide definite evidence of the latter, and also provide a picture for association in the methyl cellulose/
SDS consistent with that obtained from solubilization
measurements. Addition of SDS to the cationic polymer
results in a significant drop in its solution viscosity,
especially in the region of double layer adsorption.
Surface tension measurements are provided for all of the
polymer/SDS systems.

INTRODUCTION

Several investigators have reported the solubilization of oil-
soluble dyes[1-10] and simple hydrocarbons[11,12] by association com-

plexes of water-soluble polymers, including proteins, and ionic surfactants. This phenomenon has provided improved understanding of
the nature of polymer-surfactant interactions. At constant polymer
concentration, the surfactant concentration at which solubilization
begins generally corresponds to that at which there is an increase
in viscosity[3-5,7] and the beginning of a plateau in the surface
tension of the solutions when plotted against the logarithm of the
surfactant concentration[6,8-10]. The limits of this plateau have
been considered as defining the initial involvement and eventual
saturation of the interaction sites along the polymer[6,8-10],
although the association of alkyl sulfates with polyethylene
oxide[6,9] and bovine serum albumin[1,2] actually takes place prior to
the onset of solubilization. The fact that uptake of dyes and
hydrocarbons starts below the critical micelle concentration of the
surfactant indicates that at this point the surfactant binds to the
polymer in aggregates which have solubilization properties similar
to those of ordinary micelles. Furthermore, it has been found by
varying the alkyl chain length of the surfactant that the free
energy of transfer, per methylene group, of surfactant from solution
to aggregate is similar for both the polymer association and micellization processes[2,8].

    In our previous work[13] we have shown that the interaction of a
cationic cellulose ether with sodium dodecyl sulfate occurs in two
stages. The first, corresponding to charge neutralization, produces an essentially insoluble, highly surface active species.
Addition of further surfactant is accompanied by charge reversal
and solubilization of the primary complex. It was therefore of
interest to investigate the solubilizing and viscosity behavior of
association complexes of SDS with the charged cellulose and to
include a study of the behavior of two uncharged cellulose ethers
for comparison. Surface tension measurements of the latter systems
were undertaken to augment the information.

                            EXPERIMENTAL

                             Materials

    Sodium dodecyl sulfate (SDS) was a purified specimen from BDH,
Poole, England used in our previous study[13]; its surface tension
behavior shows that it has a small amount of residual surface active
impurity. Orange OT dye, 1-o-tolylazo-2-naphthol, was prepared from
o-toluidine and 2-naphthol and purified by precipitation from benzene
solution with ethanol. Oil Blue A, 1,4-bis-(isopropylamino) anthro-
quinone, was obtained from the Dupont Company and recrystallized
three times from ethanol. Neither dye displayed detectable
solubility in water. Polymer JR-400[14], a quaternary nitrogen
substituted cellulose ether, and Cellosize QP-300, hydroxyethyl

cellulose (HEC), are products of Union Carbide Corporation.
Methocel A4C, a methyl cellulose, was obtained from the Dow Chemical
Company. All of the polymers are in the high molecular weight
range (100,000 - 500,000) and were used as received. The water used
was twice distilled, the second time from alkaline permanganate
solution in a boro-silicate glass still.

Methods

Solubilization of Orange OT and Oil Blue A was determined
spectrophotometrically at 495nm and 646nm, respectively. Solutions
were prepared by adding SDS concentrate to polymer concentrate to
which the appropriate amount of water had been added. An excess of
dye was then added and the solutions agitated for three days at
$23 \pm 2°C$. The excess dye solids were then separated by centrifuga-
tion and the absorbance of the supernatant determined with a Cary
Model 14 recording spectrophotometer. Viscosity measurements were
made with a Cannon-Fenske viscometer at $25 \pm 0.01°C$. Solutions
were prepared as above and any insolubles in the polymer concentrate
were removed by filtration. Surface tension was determined at
$25 \pm 0.1°C$ by the Wilhelmy plate method using a Rosano Tensiometer
with a 2.5cm wide platinum blade, sand blasted to achieve complete
wetting. A stock solution was prepared from a filtered polymer
concentrate by adding sufficient SDS to give a clear solution.
Aliquots of the stock solution were then diluted with a solution of
polymer of identical concentration. The final mixtures were allowed
one hour to achieve constant temperature and each surface was cleaned
by suction with a fine capillary tube 10 minutes prior to measurement.

RESULTS

Solubilization

Figure 1 presents the solubilization of Orange OT versus SDS
concentration plot for the surfactant alone and in the presence of
0.1% of the cationic hydroxyethyl cellulose and the uncharged methyl
cellulose. Similar data obtained using Oil Blue A dye appear in
Figure 2 which includes results for HEC. The critical micelle
concentration (CMC) for SDS alone, obtained by this technique,
agrees well with that by the surface tension (ST) technique. See
below. As pointed out by Schott[15], the absorbance A above the CMC
obeys a relationship of the type

$$A = (C - CMC)/b$$

where b is a constant. A is defined by

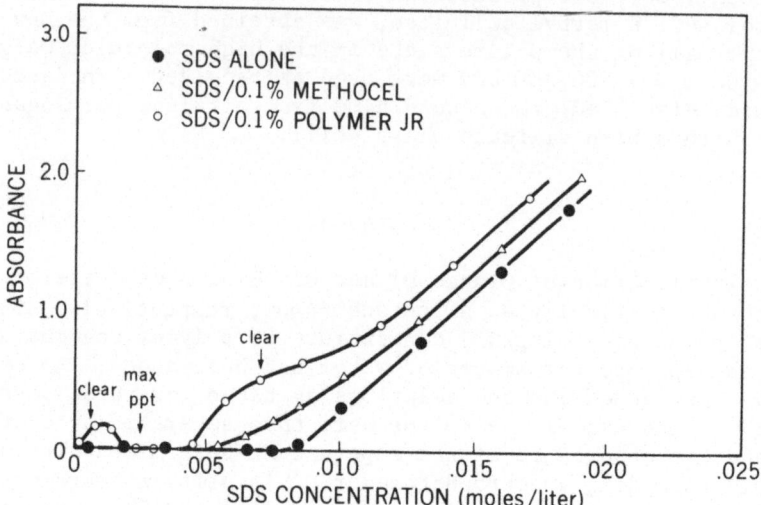

Figure 1 - Solubilization of Orange OT by SDS alone and in the presence of 0.1% methyl cellulose and 0.1% Polymer JR. The appearance
of the Polymer JR/SDS system is indicated above the solubilization
plot.

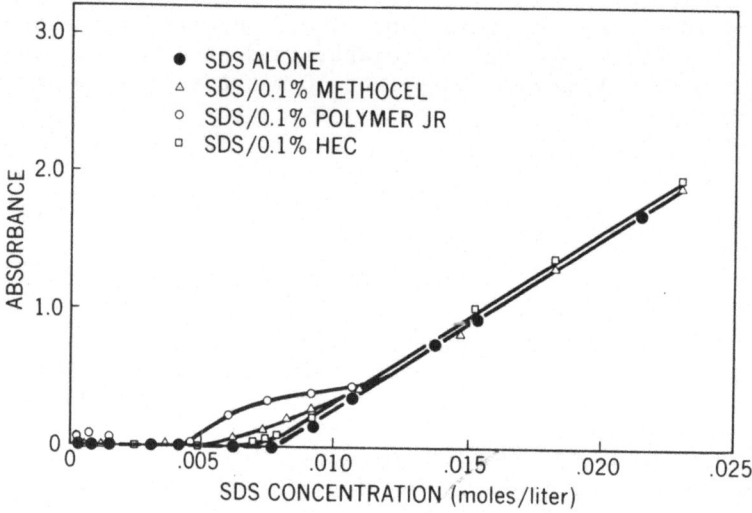

Figure 2 - Solubilization of Oil Blue A by SDS alone and in the
presence of 0.1% methyl cellulose, 0.1% Polymer JR, and 0.1%
hydroxyethyl cellulose.

$$A = C_D \times \varepsilon \times l$$

where C_D is the concentration of dye, ε the molar extinction coefficient, and l the cell path length. The ε values for Orange OT and Oil Blue A were found to be 1.99×10^4 and 1.43×10^4 1/mole cm, respectively. Schott adduced evidence that the solubilization capacity of many surfactants is one dye molecule per micelle, although this conclusion has been challenged[16] and other evidence has been obtained[17,18] that the solubilization capacity changes with surfactant structure. It is of considerable interest that use of Schott's assumption and his derived relationship for micellar molecular weight (MMW), viz.,

$$MMW = b \times l \times \varepsilon \times MW_{SDS},$$

leads to identical values for the MMW of SDS micelles, viz., 32,900, for the two different dyes. This value agrees well with published information[15].

Somewhat above the CMC, the amount of dye solubilized in the presence of the polymers becomes proportional to the SDS concentration with a slope of the absorbance versus concentration plot equal to that of the surfactant alone. At this point, the polymers are saturated with respect to the surfactant and further addition of surfactant generates ordinary micelles. Below this point, changes in solubilization behavior reflect interaction between the polymer and surfactant. Unlike the other two polymers, HEC leads to very little increased solubilization by SDS in the submicellar region, but causes a slight shift of the linear portion of the absorbance plot to lower concentrations. See Figure 2.

Surface Tension

The ST versus log surfactant concentration plots of SDS alone and in the presence of 0.1% of the polymers are given in Figure 3. The appearance of the solutions containing the cationic polymer is shown above the ST curve. The presence of this polymer results in a marked reduction in ST at low concentrations of SDS. This ST lowering has been ascribed to the polymer's being modified and rendered highly surface active by adsorption onto it of DS anions[13]. Two plateaux are evident. The first begins at concentrations of SDS lower than that at which turbidity can be detected and ends at the point corresponding to maximum precipitation. The second plateau begins shortly after the first and continues through a precipitation, and then a clarification, zone before the curve finally coincides with the ST curve of micellar SDS solutions. This occurs at ca. 1.5×10^{-2}M SDS, the "saturation" concentration. We have suggested[13] that the first plateau corresponds to the ionic binding of DS anions to the cationic sites of the polymer, and the second to

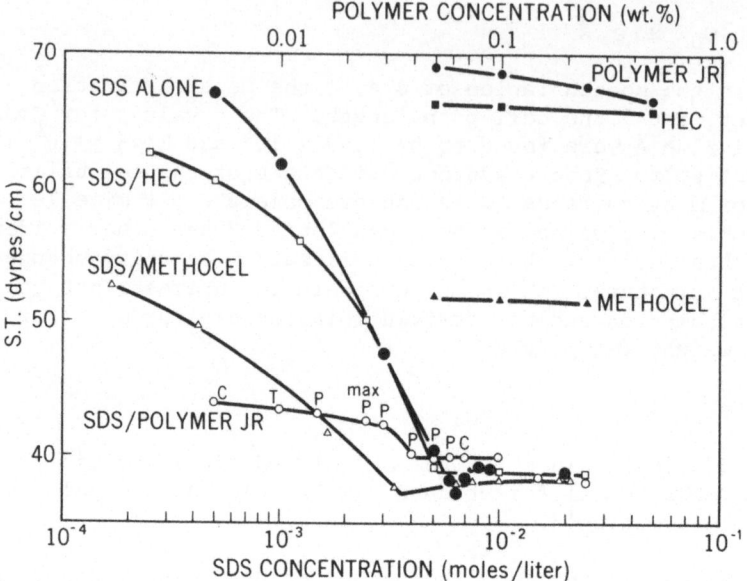

Figure 3 - Surface tension versus log concentration (mole/liter) of
SDS alone and in the presence of 0.1% of the polymers. The appearance of the Polymer JR/SDS system is indicated above the ST plot:
(C) clear, (P) precipitate. Surface tension versus log concentration (wt.%) of the polymers alone.

adsorption of a second layer of surfactant ions. In some respects
the ST depression and precipitation caused by the cationic polymer
resemble that brought about by the addition of low levels of other
cation active materials, e.g., the dye Pinacyanol, to SDS although,
because of higher hydrophobicity, this dye gave rise to a pronounced minimum in ST before eventually being solubilized[19].

We consider next the surface tension behavior of methyl
cellulose/SDS mixtures. Unlike the cationic polymer, methyl
cellulose is surface active on its own. At the concentration
employed, viz., 0.1%, it lowers the ST of water by over 20 dynes/
cm. In its presence, the ST values of SDS solutions are appreciably
lowered. See Figure 3. The generation of a ST minimum is often
encountered in mixtures of two surface active species, and provides
an indication of interaction between the two components, at least
at concentrations of SDS exceeding the minimum, viz., ca. 4×10^{-3}M.

Hydroxyethyl cellulose is weakly surface active, a 0.1% solution having a ST of 66 dynes/cm. There is little indication of interaction of this polymer and the surfactant except for a slight shift of the apparent CMC to a lower concentration, as was found in the dye solubilization experiments.

Viscosity

Viscosity results for mixtures of SDS and each of the polymers at 0.1% are presented in Figure 4 as values relative to that of water. Considering first the cationic polymer, we see there is little change in viscosity on adding the first amounts of SDS, i.e., up to ~ 2 x 10^{-4}M. However, further additions lead to a drop in viscosity, and, in the region just prior to the development of turbidity, the drop is very marked. Highly depressed viscosity persists into the region just beyond the precipitation range, and indeed values are attained little above that of pure water. On adding further SDS a slight increase in viscosity is found, followed by an arrest at 1.5 x 10^{-2}M SDS and a more gradual increase thereafter. This concentration is approximately the same as the saturation concentration observed from ST measurements. Above it the slower increase in viscosity reflects the properties of micellar solutions of SDS of increasing concentration.

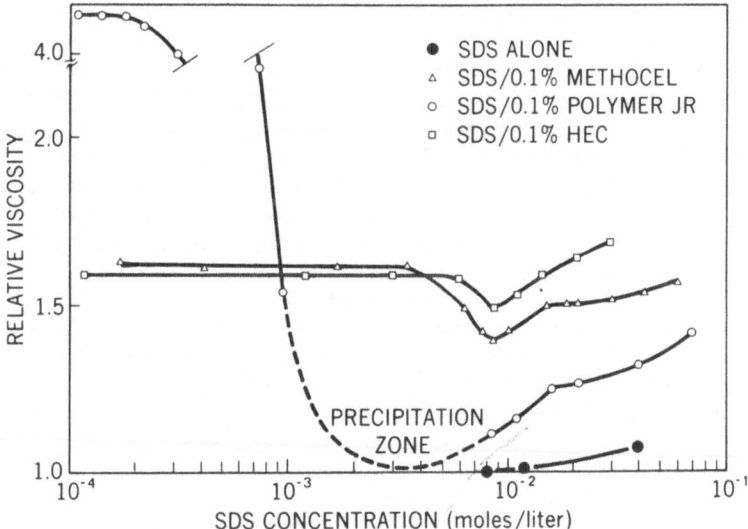

Figure 4 - Relative (to water) viscosity versus log concentration (mole/liter) of SDS alone and in the presence of 0.1% of the polymers.

In the case of methyl cellulose, little change in viscosity of the polymer solution is found until a concentration of ~4mM SDS is reached, at which level of surfactant a definite decrease is noted. This concentration is near that of the ST minimum referred to above. It is followed by a minimum and there is an arrest in the subsequent viscosity increase at a concentration of $1.5 \times 10^{-2}M$, close to the saturation concentration of the cationic polymer system. For 0.1% HEC solutions a smaller, but definite, decrease in viscosity is observed when the concentration of added SDS reaches a value of ca. 6mM. A minimum and a subsequent rise in viscosity are again encountered.

DISCUSSION

Cationic Hydroxyethyl Cellulose

It is convenient to discuss the results in terms of the three concentration regions which are defined on the ST plot by the two plateaux and the region beyond where there is coincidence with micellar SDS. Dye solubilization in the very low SDS concentration region is found only in the presence of the polycation and is attributed to the adsorption of DS anions by ionic attraction into the "primary" layer on the polymer backbone. Here, the dye uptake increases with surfactant concentration until the onset of pre-cipitation occurs as complete charge neutralization is approached. As has been mentioned, solubilization below the CMC of the sur-factant suggests that the surfactant adsorbs onto the polymer in clusters which have solubilizing power similar to that of ordinary micelles. Although little change in viscosity (in the dye-free system) was observed in this region, corresponding experiments carried out at a higher concentration (1%) of the polymer revealed a considerable increase in solution viscosity in the region just preceding the onset of precipitation. This phenomenon may reflect interaction and bonding of surfactant molecules attached to separate polymer molecules. In this respect the cationic polymer/SDS system resembles certain ionic surfactant solutions to which increasing addition of inorganic salt is made: prior to eventual precipitation, marked increases in viscosity are frequently observed which are attributed to the formation of large anisotropic, interacting micelles, which sometimes result in the formation of coacervates[20].

The second region of interaction, viz., that between the point of maximum insolubility and the onset of micellization, is ascribed to the formation of a second layer of DS anions on the ionically bound primary layer. Two points are of interest concerning solu-bilization in this region. The first is that the concentration, at which the slope of the absorbance versus surfactant concentration plot attains the same value as in the absence of polymer, is close

to the saturation concentration as shown in the ST and viscosity
plots, confirming that saturation of the polymer interaction sites
has occurred and ordinary micelles are forming. The second point
is the decreasing slope of the absorbance versus surfactant concentration plot observed throughout this region, which indicates
that the complex being formed is increasingly less efficient (per
surfactant molecule) in solubilizing dye. This may be a consequence
of cluster size or conformation. Unfortunately, it is not possible
to deduce information on the intrinsic solubilizing power of adsorbed
DS, since the concentration of the free SDS is not known. Measurements of dialysis equilibrium would allow determination of the
latter. The very low viscosity in the second interaction region is
particularly interesting. The relatively high viscosity of the
cationic polymer in solution by itself is attributable to both its
ionic charge and the acknowledged "stiffness" of its cellulose
backbone. The model we have proposed for the resolubilized complex
in the second region is of a negatively charged polyelectrolyte,
dimensionally larger than the starting polycation by virtue of an
adsorbed layer of DS anions. Since neither charge nor size can
explain the depressed viscosity, one concludes that the configuration
of the macromolecule with an adsorbed double layer of surfactant has
become more isometric, leading to less chain entanglement in
solution. This configurational change must also be involved in the
reduced adsorption affinity of the modified polyion for several
surfaces[21].

In the third region, where the polymer is saturated and
micellization of the surfactant occurs, solubilization is proportional to the concentration of SDS, as observed in the absence of
polymer. In the solubilization of Orange OT the polymer provides a
constant increment above the solubilization observed for the surfactant alone (Figure 1), while no such increment occurs in
solubilizing Oil Blue A (Figure 2). This shows that adsorbed
aggregates of surfactant may have the same or different efficiency
over regular micelles in solubilizing dye, and that this effect
depends on the dye itself.

Methyl Cellulose and Hydroxyethyl Cellulose

For methyl cellulose/SDS only two regions of solubilization
exist, and the properties of the mixed systems here are similar to
those observed for the cationic polymer/SDS system in the areas
beyond the precipitation regions. The properties of methyl
cellulose/SDS mixtures have been examined in some detail by Lewis
and Robinson[22] by viscosity and dialysis equilibrium experiments.
The latter revealed that binding of the surfactant first occurs at
a concentration of ca. 4mM SDS, and they conclude that the viscosity
decrease at this point results from deaggregation of associated
methyl cellulose molecules. We believe methyl cellulose attracts

DS anions largely through a slight residual positive charge on its
ether oxygens, which initiates a process of cluster formation,
generates solubilization ability and undoubtedly causes configurational changes of the polymer, as observed with the cationic poly
polymer. The association is probably augmented by the somewhat
hydrophobic character of the polymer, as revealed by ST.

The results show clearly that interaction with SDS is weakest
with hydroxyethyl cellulose and only the third solubilization
region apparently exists. Although little evidence is obtained of
association from ST measurements, the viscosity measurements show
unequivocally that interaction does occur just below the CMC of the
surfactant. Although little is known about the solution properties
of HEC, the explanation of Lewis and Robinson, viz., a deaggregation
effect on the polymer in adsorbing DS ions seems plausible in this
case. The displacement of the solubilization line to a lower concentration yet parallel position to that of the polymer free system,
indicates that the solubilizing efficiency for Oil Blue A of sorbed
DS is greater than that of DS in regular micelles. Such variation
in efficiency has been reported previously, and we have shown, for
example, that the opposite effect for this dye is observed in the
case of the polyvinyl alcohol/SDS system, i.e., reduced efficiency
versus that of the polymer free system.

REFERENCES

1. I. Blei, J. Colloid Sci. 14, 358 (1959).
2. I. Blei, J. Colloid Sci. 15, 370 (1960).
3. S. Saito, J. Colloid Sci. 15, 283 (1960).
4. S. Saito and M. Yukawa, J. Colloid Interface Sci. 30, 211(1969).
5. S. Saito, T. Taniguchi and K. Kitamura, J. Colloid Interface
    Sci. 37, 154 (1971).
6. M. N. Jones, J. Colloid Interface Sci. 23, 36 (1967).
7. H. Arai and S. Horin, J. Colloid Interface Sci. 30, 372 (1969).
8. H. Arai, M. Murata and K. Shinoda, J. Colloid Interface Sci.
    37, 233 (1971).
9. M. Schwuger, J. Colloid Interface Sci. 43, 491 (1973).
10. M. Murata and H. Arai, J. Colloid Interface Sci. 44, 475 (1973).
11. M. Breuer and U. P. Strauss, J. Phys. Chem. 64, 224 (1960).
12. S. Saito, J. Colloid Interface Sci. 24, 227 (1967).
13. E. D. Goddard and R. B. Hannan, J. Colloid Interface Sci. 55,
    73 (1976).
14. F. W. Stone and J. M. Rutherford, U.S. Patent 3,472,840,
    October 14, 1969.
15. H. Schott, J. Phys. Chem. 70, 2966 (1966).
16. E. W. Ancker, J. Phys. Chem. 72, 379 (1968).
17. H. Schott, J. Phys. Chem. 71, 3611 (1967).
18. F. Tokiwa, J. Phys. Chem. 72, 4331 (1968).
19. E. D. Goddard and T. G. Jones, Research 8, No. 8 (1955), A1.

20. I. Cohen and T. Vassiliades, J. Phys. Chem. 65, 1781 (1961).
21. E. D. Goddard, R. B. Hannan and J. A. Faucher, VII International Congress of Detergency, Moscow, September 1976.
22. K. E. Lewis and C. P. Robinson, J. Colloid Interface Sci. 32, 539 (1970).

PROPOSAL FOR A NEW THEORY OF MOLECULAR TRANSPORT ACROSS MEMBRANES:

IMPLICATIONS FOR LUNG GAS TRANSFERENCE

Bernard Ecanow, Bernard H. Gold,
Reuben Balagot, and R. Saul Levinson

University of Illinois at the Medical Center
833 South Wood Street
Chicago, Illinois 60612

It has been proposed that water insoluble non-polar gases are solubilized by lung surfactants. To test this hypothesis, aqueous dog lung preparations were made which contained increasing concentrations of surfactant. Solubilization was followed by placing samples in a Tonometer, admitting a nonpolar vapor, extracting gas samples and analyzing by gas chromatographic methods. Using oxygen, halothane, ether, and trichloroethylene, the results indicate that surfactant preparations in the micellar and coacervate states can readily solubilize the nonpolar gases. Little gas is dissolved by dilute aqueous lung or dilute synthetic surfactant solutions. The effect of powder particles accumulating on lung surfactant solutions was studied. The results showed that solid particles on the concentrated surfactant suspension reduced the ability to solubilize gas. From the above and from film balance studies of lung surfactant, a molecular mechanism of gas transfer across the aqueous alveolar film is proposed. On lung contraction, the alveolar surfactants form first micellar and then coacervate films which solubilize the nonpolar gases; on lung expansion the films lose their coacervate and most of their micellar structures and nonpolar gas is released to the subphase. This hypothesis is generalized to explain molecular transfer into and out of tissue through the mechanism of matrix conformational changes.

847

A physiologically important gas such as oxygen and most general anesthetic gases are nonpolar in character and only slightly
soluble in water. The limited solubility of these gases in the
water of physiologic media is reduced further because of the presence
of salts in such water. In addition, since body temperature exceeds
room temperature, the solubility of these gases in water is
decreased even further when compared to solubility at room temperature.

Simple diffusion across a water barrier membrane cannot possibly account for the quantity of nonpolar gases which are known
to be transported.[1] Given these obstacles to gas dissolution, an
additional molecular mechanism has to be proposed to account for
the transport of large volumes of gas from the lung lumen across
the aqueous membrane tissue (the alveolar lining) to the pulmonary
capillaries.

In a previous publication addressed to this topic, Ecanow et
al. advanced the idea that a molecular mechanism involving the
aggregation of lung surfactants is critical to the solubilization
and transport of oxygen and other nonpolar gases.[2,3] Solubilization, in this instance, refers to the spontaneous dissolving of a
normally water insoluble substance by an aqueous solution of surfactant. It is to be noted that where surfactant molecules aggregate
to form groups referred to as micelles or when the surfactant molecules aggregate to form an aqueous phase referred to as a coacervate changes in solvent capability occur.

Electron microscopy studies of the lung lining[4] demonstrate
that the lung is a micellar system in which the surfactant molecules,
chiefly phospholipids are aggregated into micelles.[5,6,7,8] Distributed throughout the aqueous matrix of the lung lining, micelles
are sufficiently concentrated so that expansion and contraction of
the lining during the respiratory cycle alternately disociates and
aggregates them. Thus, under appropriate physiological conditions
of electrolyte concentration, temperature, and state of contraction
of lung lining, micelles aggregate to form the coacervate phase.
As the cycle continues, the process is reversed and the coacervate
phase is converted back to the micellar state.

The presence of micelles in the lung lining can be demonstrated
by the properties of lung surfactant preparations in in vitro
experiments.[9]

The hysteresis exhibited by lung surfactant preparations on a
modified Langmuir-Wilhelmy film balance demonstrates the interre-
lationship of the various aggregated states (thermodynamic states)
which are involved in the micellization and coacervation process.

EXPERIMENTAL STUDY

Hysteresis data was obtained by the following method: Normal saline solution was introduced into the lungs of a series of dogs; after a period of ventilation, the resulting foamy solution was extracted from the lung by means of suction.[10] The concentrated solution was then poured onto a film balance. With 100 percent expansion of the film of lung surfactant preparation, surface tension values were observed to be at a maximum, ranging from 55 to 60 dynes (Figure 1). In this expanded state, the film preparation is generally polar and has minimal ability to absorb nonpolar gases such as oxygen[11] or the anesthetic gases such as halothane.[2] Carbon dioxide molecules, however, are able to diffuse into and through the lung solution during this expanded stage.[12]

Upon contraction of the surfactant film, the surfactant molecules become increasingly more aggregated. This structural change is marked by a corresponding decrease in surface tension values. As contraction continues, a point of maximum micellization is reached, and there is a relative leveling off of surface tension values from roughly 30 to 5 dynes. Examination of the curve in Figure 1 reveals that the process of contraction of the surfactant film from a maximal expansion of 50 dynes to one of 30 dynes is a gradual one. From 30 to 5 dynes, however, the drop is more precipitous. With still further contraction, surface tension values of 2 to 5 dynes can be obtained. It is our position that values in this range indicate the presence of a coacervate system. This position is supported by electrical conductivity studies which indicate that (a) reduced conductivity reflects the transition from a micellar to a coacervate phase, and (2) electrical conductivity of solution preparations with surface tension values of 1 to 2 dynes is significantly lower than that obtained from micellar solutions.[13]

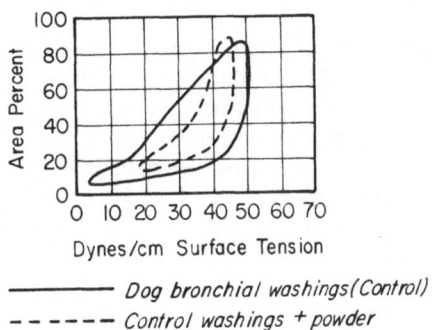

Dynes/cm Surface Tension

——————— Dog bronchial washings(Control)

— — — — — Control washings + powder

Figure 1. Typical surfactant curves showing the effects of powder present in the surface of bronchial washings (see reference 15).

The film balance model of expansion and contraction of the surfactant film just described provides the basis for the details of our analysis of the respiratory process as it occurs in the lung. In the respiratory cycle, the coacervate state constitutes the maximum state of contraction and has the maximum ability to solubilize the nonpolar gases, i.e. oxygen, etc. The re-expansion process of the lung lining is initiated with the deaggregation of the coacervate phase. Deaggregation is associated with an increase in surface tension, polarization of the lung surfactant solution, and therefore the release of the previously solubilized gases.

To initiate the inhalation of atmospheric gases, the lung cavity is expanded and a pressure less than ambient pressure is created. The pressure differential induces the atmospheric oxygen to enter the lung. The question now arises why under such conditions of negative pressure, the oxygen from the pulmonary blood stream is not also drawn into the lung. Our answer is that this does not occur because the negative pressure differential is insufficient to overcome the aqueous polar barrier present in the lung lining. In the presence of a negative pressure, this expanded membrane is no barrier to the escaping soluble carbon dioxide. When the lung lining contracts, the resulting micellar and then coacervate states provide the necessary conditions for the solubilization of oxygen from the cavity.

These events can be summarized as follows: With contraction of the lung lining, there is formation of micellar and coacervate phases, and the solubilized oxygen enters the lung lining. Upon re-expansion and deaggregation of lung surfactants, increasing polarization of the lung lining solution occurs. The nonpolar gases are no longer solubilized and are released to diffuse into the pulmonary capillaries to be absorbed by the circulating red blood cells.

THE EFFECTS OF PARTICULATE MATTER ON GAS EXCHANGE

In vitro studies indicate that the normal processes of lung gas exchange can be seriously disrupted by the presence of particulate matter.[14] Thus, if a lung solution preparation is placed on a film balance, expanded to its maximum degree, and dusted with colloidal particulate matter, lung surfactant molecules are adsorbed and concentrated on the surfaces of the particulate matter.[15] The resulting lowered surface tension values are in the range associated with concentrated surfactant preparations. When a film preparation with such a value is contracted, the resulting hysteresis encloses a much smaller area than is observed in contracted, undusted coacervate preparations.[9] Depending upon the nature and amount of the particulate matter dusted on the solution, a minimum of 20 dynes is produced rather than the expected value of 5 dynes. Expansion of the system with particulate contamination never completely deaggregates the micellar forms; and accordingly, the necessary dilute aqueous state of the lung lining is never established.

Long periods of exposure to particulate matter as in exposure to cigarette smoke exerts effects on gas exchange in the alveolar membrane of the lung identical to those described above. After prolonged contamination by the particulate matter of cigarette smoke, the alveolar lining with its adsorbed particulate matter on expansion never completely releases solubilized gas from the now permanently aggregated micelles. On contraction, the micelles which are still partly saturated with oxygen from the previous cycle cannot absorb the normal amount of oxygen. As a consequence, the polluted system is unable to pick up a sufficient quantity of oxygen at the contracted phase of the cycle and cannot release oxygen at the expanded phase of the cycle. As increasing quantities of particulate matter accumulate, the particulate-surfactant-solution-complex forms a crusty cement-like matrix. The matrix is incapable of any gas exchange function, and unencrusted portions of the respiratory system are left to carry on the gas exchange. In compensation, the alveolar sacs enlarge, and in the process, lose some of their elasticity. The respiratory pathology known as emphysema is the common consequence.

THE BIOLOGICAL UNIVERSALITY OF EXPANSION AND CONTRACTION PHENOMENA

The conformational changes in the surfaces of membranes or molecular interfaces in aqueous media which result from expansion and contraction phenomena produce solvent abilities which range from nonpolar to polar. Since all cellular cytoplasmic membranes interfacing with the aqueous media undergo conformational changes continuously, they also have continuous corresponding subtle changes in their solvent properties.

Processes involving molecular structural conformational changes are not limited to the respiratory system. An analogous process occurs as one example, with bile salts and the membrane of the intestinal tract.

It is now assumed that fats, nonpolar drugs, etc. are transported across the intestinal tract membrane by means of solubilization of the nonpolar solutes in the micelles which are formed by the bile salts.[16] This occurs in the following stages: The nonpolar solute is first solubilized by the bile salts in the lumen of the intestinal tract. The micellized bile salts-solute complex then fuses with the membrane of the intestinal wall. Structural conformational changes which are continuously occurring in the intestinal lining during peristalsis enable the transport of the solute into the lymphatic system.[16] The conformational changes associated with peristalsis are analogous to those occurring during the respiratory cycle.

The universality of the conformational change phenomena may be demonstrated at the molecular level by the functioning of the mitochondria. The mitochondria are arranged in the form of a double helix and are believed to exist in the structural form of a gel. Under Brownian motion and other influences, the helix structure lengthens, the twisting adjacent strand surfaces come closer. The increased proximity increases both the gel structuring and the nonpolarity of the mitochondria system. The increased gel structure and nonpolarity now allow nonpolar oxygen plus nonpolar metabolites such as fatty acids to diffuse readily into the mitochondria. After oxidation of the metabolites and hence their increased polarity, the oxidized metabolites are "forced out" from the nonpolar gel into adjacent, more polar aqueous protoplasm. As the helix structure shortens and the strands are loosened at the reciprocal phase of the cycle, the strand surfaces move apart and the gel structure becomes more polar. Now only water soluble metabolites such as glucose can diffuse into the mitochondria matrix to be metabolized.

SUMMARY

Existing models of molecular transport across membranes are unsatisfactory since they do not rest upon accepted scientific principles nor can they account for commonly observed physiological phenomena. The model proposed in this paper is based upon established physical chemistry concepts. It maintains that the cell membrane is composed of protein, phospholipids, other macromolecules or surfactants, electrolytes and water, all of which exist collectively in a dynamic equilibrium system, which undergoes cyclic conformational changes. At one phase of the cycle, the system acts as a nonpolar solvent, absorbing nonpolar metabolites and expelling polar solutes. At the other phase of the cycle, the process is reversed as the system becomes more polar and a corresponding change in solvent abilities occurs. Biological systems are dynamic systems characterized by polar and nonpolar phases. Physiological functioning corresponds to the phase of the cycle existing at the moment.

REFERENCES

1. J. S. Haldane and J. G. Priestley, Respiration, pp. 264, Clarendon Press, Oxford (1935).
2. B. Ecanow, R. Balagot, and V. Santelices, Nature, 215, 1400 (1967).

3. B. Ecanow and H. Klawans, in "Models of Human Neurological Diseases", H. Klawans, Editor, pp. 253–284, Excerpta Medica Amsterdam, 1974.

4. E. R. Weibel and J. Gil, Resp. Physiol., 4, 42 (1968).

5. E. S. Brown, Fed. Proc., 21, 438 (1962).

6. R. E. Pattle, Physiol. Rev., 45, 68 (1965).

7. R. DePalma, C. A. Hubay, and S. Levey, J. Amer. Med. Assoc., 195, 943 (1966).

8. D. E. Korn, Science, 153, 1491 (1966).

9. R. W. Webb, A. W. Coll, and J. M. Lanius, Amer. Rev. Resp. Dis., 95, 244 (1967).

10. S. Bondurant and D. A. Miller, J. Appl. Physiol., 17, 167 (1962).

11. W. F. Stanaszek, B. Ecanow and R. Levinson, J. Pharm. Sci., 65, 142 (1967).

12. B. Ecanow (1975), unpublished data.

13. J. Woznicki and B. Ecanow (1976), unpublished data.

14. F. C. Lowell, W. Franklin, A. L. Michelson and I. W. Schiller, Ann. Intern. Med., 45, 268 (1956).

15. B. Ecanow, R. Balagot, and V. Bandelin, Amer. Rev. Resp. Dis., 99, 106 (1969).

16. G. Borgstrom, in "Metabolism and Physiological Significance of Lipids", Dawson and Rhodes, Editors, John Wiley and Sons, New York, 1964.

INTERFACIAL TENSION MINIMA IN TWO-PHASE MICELLAR SYSTEMS

E. Franses, M. S. Bidner* and L. E. Scriven

Department of Chemical Engineering & Materials Science

University of Minnesota, Minneapolis, MN 55455

For single and mixed nonionic surfactants, the
Gibbs adsorption isotherm and Gibbs-Duhem equations
reveal two possible causes of tension minima in dilute
systems: (1) sign change in interfacial excess inventory of a surfactant, and (2) a maximum in chemical
potential of a surfactant with respect to its total
inventory. Whether either occurs depends on details
of solution, adsorption, partitioning and association
behavior as linked by the requirement of mass conservation. Specific cases are examined using the Bury-
Hartley limit of multiple association equilibria, and
also, for mixed surfactants, the more approximate
phase-separation limit. A single surfactant in two
phases, as in one, can ordinarily have no maximum in
chemical potential. With two surfactans, mixed
micelles must form in at least one phase to cause a
maximum in any of the cases studied.

INTRODUCTION

Sharp minima in interfacial tension versus surfactant concentration are reported for certain systems consisting of water,
sodium chloride, mixed hydrocarbons and mixed hydrocarbon sulfonates.[1-3] On either side of the minimum the interfacial tension
rises thirty-fold over a ten-fold range of initial surfactant

*Present address: Departamento de Ingenería Química,
 Universidad Nacional de La Plata, 47 y 1, La Plata, Argentina.

855

concentration in the aqueous phase. The minimum is also with respect
to sodium chloride content and is equally dramatic in that dimension.
Wade and coworkers[4-7] find minima versus carbon number of alkanes
and equivalent carbon numbers of other hydrocarbons, their compari-
sons being made at a fixed, low concentration of surfactant (0.2
weight percent in water usually, though Doe and Wade[6] use 0.07
weight percent in water) and a fixed salinity (1 weight percent
sodium chloride in water). Only for a commercial petroleum sulfonate and three hydrocarbons have these investigators reported
measurements of concentration dependence.[5] In all of these reports
the only mention of an attempt to determine a critical micelle concentration is by Doe and Wade,[6] who remark that with 8-phenyl
hexadecane sulfonate in water there is no *pronounced* minimum of
surface tension versus concentration.

From the available reports it is not clear that the interfacial
tensions measured are for systems in metastable or absolutely stable
thermodynamic equilibrium. The order of adding sodium chloride and
surfactant to water matters.[2,5] Time of contact between aqueous
and hydrocarbon phases is at least four hours but apparently not
more than twelve hours or so. Studies of a nonionic surfactant
which sometimes gives sharp minima in interfacial tension in the
same concentration range reveal that different results can be
obtained depending on the number of *days* the two phases are pre-
contacted before measurements are made.[8] More recent, unpublished
studies in the same laboratory indicate similar effects with a
commercial hydrocarbon sulfonate surfactant.

Nevertheless, these reports of interfacial tension minima with
respect to surfactant concentration in oil-water systems raise the
question of whether micellization phenomena can produce dramatic
minima in two-phase systems. Incidentally, there are micellar
systems of water, sodium chloride, hydrocarbon and significantly
higher concentrations of surfactant, in which very low interfacial
tensions are measured between brine and equilibrium microemulsion,
and between equilibrium microemulsion and hydrocarbon. The data
are often plotted in such a way that the casual reader may infer a
minimum.[9,10]

Surface tension (s.t.) minima at solution-air interfaces are
well known and have been reviewed by Mysels and Florence.[11] Interfacial tension (i.t.) minima in simple, well-defined systems have
been reported.[12,13] The minima are attributed to the presence of
more than one surface-active or interface-active species, whether
it is deliberately added or present as an impurity.[11] It seems
generally agreed that s.t. minima are due to micellization phenomena, and theoretical explanations have been given.[14,15]

The purpose of the present paper is to examine systematically
a sequence of simple thermodynamic models of solutions of

micellizable nonionic surfactants, to see what features are
necessary to account for a minimum in equilibrium i.t. versus surfactant concentration. There is interest in nonionic systems which
give low i.t.[8] Furthermore it is not difficult to incorporate into
a model those additional features that are peculiar to ionized surfactants.[14,16,17] The results in all cases pertain as well to
adsorption maxima at liquid-solid interfaces.[18] It proves instructive to consider s.t. minima in one-phase systems as the starting
point for examining i.t. minima in two-phase systems. Most of the
models examined involve mixed micelles. A key issue is how the
composition of a mixed micelle at equilibrium depends on the concentrations of unassociated surfactants, i.e., on surfactant
'monomers.'[19,20] Here the dependence is found for a relatively
unrestricted mass-action model of micellization, by direct application of the principle of minimum free energy, thereby providing
the basis for a more general thermodynamic treatment of mixed
micelles than Clint's,[20] which is strictly applicable only in the
phase-separation limit.[21,22]

BASIC THEORY

     Of concern are closed, equilibrium systems of two phases, A
and B, and four components. The two surfactants, 1 and 2, may
exist in solution not only as monomers but also in aggregates; the
solvents, 3 and 4, are totally immiscible and inert (the possibility
that solvent may be solubilized into the opposing phase by aggregates
there is touched on below and is the subject of a sequel). At a
specified temperature and pressure, the compositions of the phases
and of the interfacial region are governed by:

   (i)   the equation of solution state of each species
          in each phase
   (ii)   the conditions of equilibrium partitioning be-
          tween phases
   (iii) the conditions of equilibrium adsorption in the
          interface
   (iv)   the conditions of association equilibrium be-
          tween aggregates and monomers
   (v)   the four material balance requirements on the
          independent components of the closed system

To be complete a thermodynamic model must contain all of these
elements. Surface and interfacial tensions each require an
additional equation:

   (vi)   the equation of interfacial state.

     Neighboring equilibrium states are related by three equations:
the Gibbs adsorption isotherm (which is a Gibbs-Duhem equation for
the interface) and the Gibbs-Duhem equation for each of the bulk

phases. Under certain mild conditions these can be combined, as shown by Defay et al.[23] to yield

$$(d\sigma)_{T,p} = -[\Gamma_{1(3)} - \Gamma_{4(3)}\frac{c_{3A}c_{1B} - c_{1A}c_{3B}}{c_{3A}c_{4B} - c_{3B}c_{4A}}]d\mu_1$$

$$-[\Gamma_{2(3)} - \Gamma_{4(3)}\frac{c_{3A}c_{2B} - c_{2A}c_{3B}}{c_{3A}c_{4B} - c_{3B}c_{4A}}]d\mu_2$$

$$-\sum_{i=5}^{\infty} [\Gamma_{i(3)} - \Gamma_{4(3)}\frac{c_{3A}c_{iB} - c_{iA}c_{3B}}{c_{3A}c_{4B} - c_{3B}c_{4A}}]d\mu_i \tag{1}$$

Here the Γ's are surface excesses relative to that of component 3, a solvent, and the suffix i denotes aggregates or micellar species. c's are molar concentrations, μ's are chemical potentials, and σ is tension. This is the equation with which local extrema of tension are identified, for at them $(d\sigma)_{T,p} = 0$. On the grounds that the relative surface excess of 4 with respect to 3 is negligible, the cumbersome terms in Equation (1) are suppressed, leaving the conventional form

$$(d\sigma)_{T,p} = -\Gamma_{1(3)}d\mu_1 - \Gamma_{2(3)}d\mu_2 - \sum_{i=5}^{\infty}\Gamma_{i(3)}d\mu_i \tag{2}$$

For premicellar and micellar association equilibria

$$n_1 S_1 + n_2 S_2 \rightleftharpoons S_{1,n_1 S_2, n_2} \tag{3}$$

The equilibrium condition is of course

$$\mu_{n_1 n_2} = n_1\mu_1 + n_2\mu_2 \tag{4}$$

For this reason it is convenient to define total interfacial excesses of aggregated surfactants (with respect to solvent 3)

$$\Gamma_{1\Sigma} \equiv \sum_{n_1}\sum_{n_2} n_1\Gamma_{n_1 n_2} \quad , \quad \Gamma_{2\Sigma} \equiv \sum_{n_2}\sum_{n_1} n_2\Gamma_{n_1 n_2} \tag{5}$$

Monomers are excluded from the summations. With obvious notational identifications (e.g. $\Gamma_{5(3)} \equiv \Gamma_{2,0}$), with c_T the total molality of both surfactants, and with c_i the molality of surfactant monomer in either phase, Equations (2) and (5) give (cf. Equation 16.168)[15] and Equation (8.48)[23])

$$\left(\frac{\partial\sigma}{\partial c_T}\right)_{T,p} = -(\Gamma_1 + \Gamma_{1\Sigma})\left(\frac{\partial\mu_1}{\partial c_1}\right)\left(\frac{\partial c_1}{\partial c_T}\right) - (\Gamma_2 + \Gamma_{2\Sigma})\left(\frac{\partial\mu_2}{\partial c_2}\right)\left(\frac{\partial c_2}{\partial c_T}\right) \qquad (6)$$

To have a local minimum of tension with respect to total surfactant content it is necessary that $\partial\sigma/\partial c_T$ change sign, regardless of the equation of interfacial state. Thermodynamic stability of the bulk phases demands that $\partial\mu_1/\partial c_1$ and $\partial\mu_2/\partial c_2$ be positive. Therefore to have a minimum it is necessary (yet not sufficient) that an interfacial excess inventory, $\Gamma_1 + \Gamma_{1\Sigma}$ or $\Gamma_2 + \Gamma_{2\Sigma}$, change sign, or that a rate of change of monomer molality, $\partial c_1/\partial c_T$ or $\partial c_2/\partial c_T$, do so.

Monomers of dilute surfactant are generally positively adsorbed into the interface, and so $\Gamma_1 > 0$ and $\Gamma_2 > 0$. Premicellar aggregates, depending on their size, arrangement and flexibility, can be expected to adsorb more or less strongly. But nearly complete micelles as well as complete ones can be expected to desorb, that is, to be present in lower concentration than in the contiguous bulk phase, and therefore to have a negative surface excess. So if micelles come to preponderate as total surfactant content increases, there is a possibility that one of the interfacial excess inventories, $\Gamma_1 + \Gamma_{1\Sigma}$ or $\Gamma_2 + \Gamma_{2\Sigma}$, may change sign.

If surfactants are added in fixed ratio to a system, then the monomer concentrations can be expected to increase monotonically with total surfactant content, unless *mixed* micellation intervenes. An accepted hypothesis is the "regression in chemical potential of the more surface-active species," i.e. the idea that concentration of monomers of a more surface-active surfactant falls when micelles of a less surface-active one form.[14,16,17,24] In other words, it is supposed that $\partial c_2/\partial c_T$ turns negative while $\partial c_1/\partial c_T$, $\Gamma_1 + \Gamma_{1\Sigma}$ and $\Gamma_2 + \Gamma_{2\Sigma}$ all remain positive and the ratio c_{1T}/c_{2T} is held fixed. However, there are other possibilities, as discussed below. Furthermore, there is no record of the hypothesis being validated with a complete thermodynamic model.

For the models examined in this paper the equation of solution state which is chosen is the quasi-ideal solution that corresponds to Denbigh's Convention III:[25]

For solvent s: $\mu_s = \mu_s^o + RT\ell n x_s$

$$(7)$$

For solutes j: $\mu_j = \mu_j^o + RT\ell n c_j$

where x_s is the mole fraction of solvent and c_j is the solute molality (gm-mol per 1000g of solvent); this accounts for a certain type of nonideal mixing entropy which is related to the molar volumes. For equilibrium partitioning the equation chosen is the

Nernst distribution law with constant distribution or partitioning coefficient:

$$\frac{c_{jB}}{c_{jA}} = \exp\left(-\frac{\mu^o_{jB} - \mu^o_{jA}}{RT}\right) \equiv P_j \tag{8}$$

The standard free energy change is that of solute transfer at infinite dilution. For equilibrium adsorption the simplest choice is the linear relation

$$\frac{\Gamma_j}{c_{jA}} = \exp\left(-\frac{\mu^o_{jI} - \mu^o_{jA}}{RT}\right) \equiv H_j \tag{9}$$

The standard free-energy change is that of adsorption into the interface from infinitely dilute solution. Langmuir-type isotherms are perhaps next simplest and can account for competitive adsorption. In this paper Equation (9) suffices where a specific equation is needed. Otherwise all that is assumed is a monotone relation between relative surface excess Γ_j and bulk phase concentration c_{jA} or c_{jB}. Surfactant inventory in the interface and at solid surfaces is disregarded in material balances.

With the quasi-ideal solution model, Equation (7), the mass-action equilibrium expressions for premicellar and micellar association equilibria in either phase are

$$\frac{c_{jn}}{c_j^n} = \exp\left(-\frac{\mu^o_{jn} - n\mu^o_j}{RT}\right) \equiv K_{jn} \tag{10}$$

for an aggregate of n monomers. The standard free energy change is that of formation from monomers at infinite dilution. For mixed micelles consisting of two surfactants in the molar ratio $x_n : (1 - x_n)$ the analogous equilibrium expression is (see Equation (36) below):

$$\frac{c_{mn}}{c_1^{x_n n} c_2^{(1-x_n) n}} = \exp\left(-\frac{\mu^o_{mn} - x_n n\mu^o_1 - (1-x_n)n\mu^o_2}{RT}\right) \equiv K_{xn} \tag{11}$$

The notation is appropriate when mixed micelles of size n have a unique composition x_n. There must of course be a distribution of sizes, but in this paper the premicellar and micellar distributions are replaced, for algebraic simplicity, by the Bury-Hartley limit, in which micelles of only a single aggregation number N are present.[26-27] The limit is singular but nevertheless accounts for the features of micellization of greatest importance here.[28]

A typical material balance is that for a micellizable surfactant 1 added in the amount c_{1T} gm-mol per 1000g of solvent 3 (phase A). If N-mers are the largest aggregates that form, the equation is

$$c_{1T} = [c_1 + \sum_{n=2}^{N} n(c_{1n} + x_n c_{mn})]_{\text{in A}} + \phi[c_1 + \sum_{n=2}^{N} n(c_{1n} + x_n c_{mn})]_{\text{in B}} \tag{12}$$

$$+ \psi[\Gamma_1 + \sum_{n=2}^{N} n(\Gamma_{1n} + x_n \Gamma_{mn})]_{\text{in I}} + \chi[\Gamma_1 + \sum_{n=2}^{N} n(\Gamma_{1n} + x_n \Gamma_{mn})]_{\text{in S}}$$

Here ϕ is the mass ratio of solvent 4 (phase B) to solvent 3; ψ is the area of A/B interface I per 1000g of 3; and χ is the area of A/Wall surface S per 1000g of 3. These last two terms represent inventories in adsorbed states; there may be another set at the B/Wall surface, which would require another term, as would the presence of yet other solid surfaces. In this paper only the first two terms, which represent surfactant inventory in two bulk phases, are retained.

SINGLE SURFACTANT

In this case Equations (6) and (7) reduce to

$$\frac{\partial \sigma}{\partial c_T} = - (\Gamma_1 + \Gamma_{1\Sigma})(\frac{RT}{c_1})(\frac{\partial c_1}{\partial c_T}) \tag{13}$$

If the surfactant is restricted to a single phase, Equations (10) and (12) give

$$c_T = c_1 + \sum_{n=2}^{N} n K_n c_1^n \tag{14}$$

(superfluous subscripts are suppressed). Obviously c_1 is monotonic in c_T; therefore

$$\frac{\partial c_1}{\partial c_T} > 0 \tag{15}$$

and the only possibility of a local minimum of surface tension is for the net interfacial excess inventory, $\Gamma_1 + \Gamma_{1\Sigma}$, to vanish. This conclusion is altered neither by the presence of surfactant in a second phase nor by its adsorption into the interface and onto surfaces, at least as long as (8) and (9) apply, because these guarantee monotonicity of c_{1A} in c_T.

The possibility of a local minimum is easily examined in the Bury-Hartley limit with micelles of aggregation number N. Equation

(14) becomes

$$c_T = c_1 + NK\, c_1^N \ , \quad K \equiv \exp\left(-\ \frac{\mu_{1N}^o - N\mu_1^o}{RT}\right) \tag{16}$$

and

$$\frac{\partial c_1}{\partial c_T} = (1 + N^2 K\, c_1^{N-1})^{-1} > 0 \tag{17}$$

From the many definitions of a 'critical micelle concentration' —
which is in fact a concentration range — the one chosen here is
c_1^* such that surfactant inventories in monomers and micelles are
equal[29]

$$c_1^* = NK\, c_1^{*N} = c_T/2 \tag{18}$$

Thus

$$c_1^*(N) = \left(\frac{1}{NK}\right)^{1/(N-1)}, \quad \left(\frac{\partial c_1}{\partial c_T}\right)_{CMC} = \frac{1}{1+N} \tag{19}$$

and

$$-\frac{1}{N}\,\ell nK = \frac{1}{N}\,\ell nN + \frac{N-1}{N}\,\ell nc_1^* = \frac{1}{RT}\left(\frac{\mu_{1N}^o}{N} - \mu_1^o\right) \tag{20}$$

In the limit as micellar size increases without bound, $N \to \infty$ and

$$c_1^*(\infty) = \exp\left(\frac{\mu_{1m}^o - \mu_1^o}{RT}\right) \tag{21}$$

where

$$\mu_{1m}^o \equiv \lim_{N \to \infty} \frac{\mu_{1N}^o(N)}{N} \tag{22}$$

is the standard chemical potential per mole of surfactant in what
may be regarded as a micellar 'phase.' In this case the so-called
phase-separation model, or pseudo-phase-separation limit, Equation
(21), is the limit of the Bury-Hartley limit as aggregation number
N becomes very large.[22,30] This is illustrated in Figure 1.

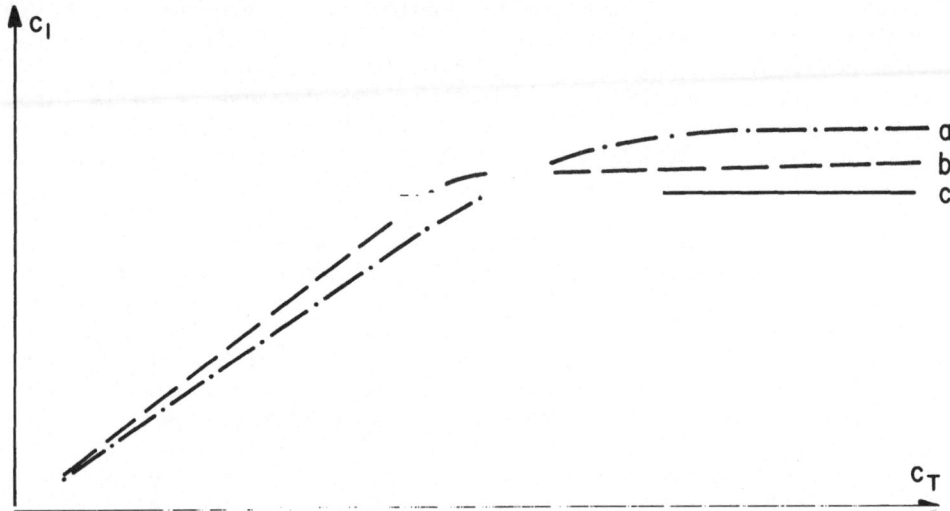

Figure 1. Monomer concentration versus surfactant inventory, single surfactant in a single phase. a: mass-action model (polydisperse micelles), b: Bury-Hartley limit (monodisperse micelles), c: phase-separation model (Brobdingnagian micelles).

Now if surfactant monomers are positively adsorbed while complete micelles are negatively adsorbed, $\Gamma_1 = H_1 c_1$ and $\Gamma_N = -H_N c_N$ (note minus sign). The interfacial excess inventory $\Gamma_1 + \Gamma_N$ vanishes at

$$c_1 = (\frac{H_1}{H_N NK})^{1/(N-1)} \tag{23}$$

which is therefore the monomer concentration at a surface tension minimum (stm). The ratio of surfactant present at this condition to that at the cmc is

$$\frac{(c_T)_{stm}}{(c_T)_{cmc}} = (\frac{H_1}{H_N})^{1/(N-1)} \tag{24}$$

This ratio is greater than one if exclusion of micelles from the interface is weaker than incorporation of monomers, for then $H_N < H_1$. Moreover, at the minimum

$$(\frac{\partial c_1}{\partial c_T})_{stm} = \frac{1}{1 + NH_1/H_N} \tag{25}$$

Thus if $H_N < H_1$, a local minimum in tension can only occur at a total surfactant concentration in excess of the cmc. If either the aggregation number N is large or micellar exclusion is weak ($H_N/H_1 \ll 1$), the minimum is necessarily broad and shallow, because it falls into a concentration range where monomer chemical potential is insensitive to surfactant concentration and $\partial\sigma/\partial c_T \ll 1$, according to Equations (13) and (25).

Thus the Bury-Hartley limit indicates that even a micellizable, *pure* surfactant can in principle display a weak s.t. minimum at some concentration higher — probably much higher — than the cmc range, *if* micelles are to some extent excluded from the interfacial region. No experimental data confirming this possibility of a shallow, post-cmc minimum have come to the authors' attention, and it may be that other effects intervene even where the possibility exists.

This is the *only* possibility of a s.t. or i.t. minimum with a single surfactant, under the premises listed. The conclusion is unaltered by a second bulk phase in contact with the first, but has to be reconsidered when the solvent component of the second phase can be solubilized into micelles in the first phase.[31]

TWO NON-INTERACTING SURFACTANTS

By non-interacting surfactants is meant those which obey the quasi-ideal solution Equation (7) and do not enter into mixed micelles. When they are added in fixed ratio to the system, Equations (8), (10) and (12) give

$$c_{1A} + N_{1A}K_{1A}c_{1A}^{N_{1A}} + \phi(P_1c_{1A} + N_{1B}K_{1B}P_1^{N_{1B}}c_{1A}^{N_{1B}}) = \alpha\, c_T$$

(26)

$$c_{2A} + N_{2A}K_{2A}c_{2A}^{N_{2A}} + \phi(P_2c_{2A} + N_{2B}K_{2B}P_2^{N_{2B}}c_{2A}^{N_{2B}}) = (1-\alpha)c_T$$

where α is the mole fraction of 1 in the total surfactant. All monomer concentrations are obviously monotonic in c_T. Therefore $\partial c_{1A}/\partial c_T > 0$ and $\partial c_{2A}/\partial c_T > 0$ and, according to Equation (6), the only possibility for a local minimum is for one or both net interfacial excess inventories to turn negative, in the way illustrated in the preceding section. This conclusion applies whether each surfactant is present in both phases or is restricted to one.

TWO SURFACTANTS WITH MIXED MICELLES IN ONE PHASE

Mixed Micelles of Fixed Composition

In this case the monotonicity arguments of the preceding sections fail. The possibility of a maximum in monomer concentration is best illustrated by the special case in which pure micelles can be neglected and the mixed micelles can be regarded as having a fixed size N and a definite mole fraction, x, of surfactant 1. Equations (11) and (12) give

$$c_1 + xNK_x c_1^{xN} c_2^{(1-x)N} = \alpha\, c_T$$

$$(27)$$

$$c_2 + (1-x)NK_x c_1^{xN} c_2^{(1-x)N} = (1-\alpha) c_T$$

By the method of Lagrange multipliers it is readily shown that the condition for an extremum in c_1 versus c_T with x fixed is

$$NK_x c_1^{xN} c_2^{(1-x)N-1} = \frac{\alpha}{N(1-x)(x-\alpha)} \qquad (28)$$

The necessary and sufficient condition for a solution $\{c_1, c_2, c_T\}$ of Equations (27)-(28) is that $x > \alpha$. If the micelle were a stoichiometric association complex, x would truly be independent of c_1 and c_2, and K_x would not be simply related to the micellization equilibrium constants of the pure surfactants, viz. K_1 and K_2.

Mixed Micelles of Variable Composition

In reality, however, there are systems which behave as though the composition of micelles varies with monomer concentrations of both surfactants, and does so virtually continuously from pure micelles of one surfactant to pure micelles of the other.[19,20] Therefore it is appropriate to put Equation (28) aside and allow the micelle composition x to depend on c_1 and c_2. The micelle composition which actually obtains is that which minimizes the free energy of the system, and this can be derived as follows.

For a concentration of micelles c_m the material balance Equations (12) are (N is taken to be independent of x):

$$c_1 + xNc_m = \alpha c_T$$

$$(29)$$

$$c_2 + (1-x)Nc_m = (1-\alpha) c_T$$

If the molecular weight of the solvent is M_s, its own molality is $c_s \equiv 1000/M_s$. Then the mole fraction of solvent is

$$x_s = \frac{c_s}{c_s + c_T - Nc_m + c_m} \tag{30}$$

The total free energy of the system per 1000g of solvent is $G = c_s\mu_s + c_1\mu_1 + c_2\mu_2 + c_m\mu_m$; from Equations (7), (29)–(30) this is

$$G(c_m,x) = c_s\{\mu_s^o + RT\ln[c_s/(c_s + c_T - Nc_m + c_m)]\}$$

$$+ [\alpha c_T - xNc_m]\{\mu_2^o + RT\ln[\alpha c_T - xNx_m]\}$$

$$+ [(1-\alpha)c_T - (1-x)Nc_m]\{\mu_2^o + RT\ln[(1-\alpha)c_T - (1-x)Nc_m]\}$$

$$+ c_m\{\mu_m^o + RT\ln c_m\} \tag{31}$$

The first condition of equilibrium is $\partial G/\partial c_m = 0$. This yields

$$\mu_m = xN\mu_1 + (1-x)N\mu_2 + (N-1)RT\left(\frac{c_T + c_m - Nc_m}{c_s + c_T + c_m - Nc_m}\right) \tag{32}$$

Because $c_T + c_m - Nc_m \ll c_s$, the last term can be neglected, which leaves an expression equivalent to Equation (4):

$$\mu_m = xN\mu_1 + (1-x)N\mu_2 \tag{33}$$

The second condition of equilibrium is $\partial G/\partial x = 0$. This yields

$$\frac{1}{N}\frac{\partial \mu_m^o}{\partial x} = \mu_1 - \mu_2 \tag{34}$$

To proceed it is necessary to relate the standard free energy of mixed micelles at infinite dilution to the standard free energies of pure micelles of surfactants 1 and 2, from which the mixed micelles could be formed by a mixing process:

$$\frac{\mu_m^o}{N} = \frac{x\mu_{1N}^o}{N} + \frac{(1-x)\mu_{2N}^o}{N} + RT[x\ln x + (1-x)\ln(1-x)] + x(1-x)w(x) \tag{35}$$

On the assumption of *ideal* micellar mixing, $w(x) = 0$ and the last term vanishes. Equation (33) then gives

$$\frac{c_m}{c_1^{xN} c_2^{(1-x)N}} = \exp\left(-\frac{\mu_m^o - xN\mu_1^o - (1-x)N\mu_2^o}{RT}\right) \equiv K_x \tag{36}$$

which is an example of Equation (11). From Equations (10) and (35) with $w(x) = 0$, it follows that

$$K_x = K_1^x K_2^{1-x} x^{-xN} (1-x)^{-(1-x)N} \tag{37}$$

Equation (34), together with Equation (10) and (35) with $w(x) = 0$, leads to an equation for equilibrium micelle composition in terms of monomer concentrations of both surfactants:

$$\frac{x}{1-x} = \frac{c_1}{c_2} \left(\frac{K_1}{K_2}\right)^{1/N} \tag{38}$$

The micellization equilibrium constants can be replaced by cmc's, c_i^*, in virtue of Equation (19); thus

$$\frac{x}{1-x} = \frac{c_1}{c_2} \left(\frac{c_2^*}{c_1^*}\right)^{(N-1)/N} \tag{39}$$

The monomer concentrations are also related to x by the material balance Equations (19), which with Equations (36)-(38) can be expressed in terms of the cmc's:

$$c_1 \left[1 + \left(\frac{c_1}{xc_1^*}\right)^{N-1}\right] = \alpha c_T \tag{40}$$

$$c_2 \left[1 + \left(\frac{c_2}{(1-x)c_2^*}\right)^{N-1}\right] = (1-\alpha) c_T \tag{41}$$

These three Equations (39)-(41) together determine the monomer concentrations c_1 and c_2 and the micelle composition x, given the surfactant amount c_T and make-up α, the critical micelle concentrations of the pure surfactants c_1^* and c_2^*, and the aggregation number N. In general they must be solved numerically.

A convenient choice for the cmc of mixed surfactant is one that is compatible with Equation (18):

$$c^* \equiv c_1 + c_2 \quad \text{when} \quad c_1 + c_2 = c_T/2 \tag{42}$$

Equations (39)-(42) must be solved simultaneously for c^* as a function of c_1^*, c_2^* and α (for given N).

Local minima in i.t. are identified with Equation (6). This depends not only on $\partial c_1/\partial c_T$ and $\partial c_2/\partial c_T$, which can be evaluated from Equations (39)-(41), but also on other factors which must be taken into account in a complete analysis. As pointed out at Equation (6), one circumstance that can lead to a minimum is a change in sign of $\partial c_1/\partial c_T$ (or, equivalently, of $\partial c_2/\partial c_T$). If $c_1 = f(c_T)$ is a smooth function, this circumstance is signaled by a minimum or maximum which can be located by the method of Lagrange multipliers. The result of doing so is

$$(\frac{1-\alpha}{\alpha}) (\frac{c_1^*}{c_2^*}) (\frac{c_1}{xc_1^*})^N + (\frac{c_2}{(1-x)c_2^*})^N$$

$$= [1 + N(\frac{c_2}{(1-x)c_2^*})^{N-1}] (\frac{c_2}{(1-x)c_2^*}) \frac{1}{x(N-1)} \tag{43}$$

Equations (39)-(41) and (43) must be solved simultaneously for c_T, the amount of surfactant at which $\partial c_1/\partial c_T$ changes sign (it is necessary to confirm that the extremum actually is a maximum or minimum).

Pseudo-phase-separation Limit

Because these equations are complicated — reflecting the complex interactions of solution behavior and mixed micellization in the two-surfactant system — it is instructive to examine a limiting case. This limiting case turns out to demonstrate that mixed micellization can indeed account for a s.t. minimum.

The limit is that as aggregation number grows large while the amount of surfactant in micelles remains constant. From Equations (7) and (35) with $w(x) = 0$,

$$\lim_{N \to \infty} \frac{\mu_m}{N} = \lim_{N \to \infty} \frac{\mu_m^o}{N} = x(\lim_{N \to \infty} \frac{\mu_{1N}^o}{N}) + (1-x)(\lim_{N \to \infty} \frac{\mu_{2N}^o}{N})$$

$$+ RT[x\ln x + (1-x)\ln(1-x)] \tag{44}$$

Substituting Equations (7),(21),(22) and (33) in (44) yields

$$x \ln\left(\frac{c_1}{xc_1^*}\right) + (1-x) \ln\left(\frac{c_2}{(1-x)c_2^*}\right) = 0 \tag{45}$$

On the other hand, as $N \to \infty$ the limit of Equation (39) is

$$\frac{c_1}{xc_1^*} = \frac{c_2}{(1-x)c_2^*} \tag{46}$$

Together the last two equations imply that at and above the cmc,

$$c_1 = xc_1^* \quad \text{and} \quad c_2 = (1-x)c_2^* \tag{47}$$

These are Clint's[20] Equations (7) and (8), derived from the pseudo-phase-separation model on the assumption that the mixed micellar 'phase' behaves as an ideal solution of pure micellar 'phases' of surfactants 1 and 2. Moreover, as $N \to \infty$, Equations (39)–(41), with $c_2^* > c_1^*$ (this condition entails no loss of generality), lead to

$$x = \frac{1}{2} + \frac{-c_T + \sqrt{(c_T - c_2^* + c_2^*)^2 + 4\alpha(c_2^* - c_1^*)}}{2(c_2^* - c_1^*)} \tag{48}$$

and, for the amount of micellar 'phase' per 1000g of solvent,

$$c_P = \frac{c_T - c_1^* - c_2^* + \sqrt{(c_T - c_1^* - c_2^*)^2 + 4c_T[\alpha c_2^* + (1-\alpha)c_1^*] - 4c_1^* c_2^*}}{2} \tag{49}$$

At the cmc,

$$x^* = \frac{\alpha}{\alpha + (1-\alpha)c_1^*/c_2^*} \tag{50}$$

$$\frac{1}{c^*} = \frac{\alpha}{c_1^*} + \frac{1-\alpha}{c_2^*} \tag{51}$$

The last is Clint's[20] Equation (13). So in the case of two surfactants with mixed micelles, the phase-separation model is again the limit, as aggregation number grows very large, of a Bury-Hartley limit.

Differentiation of Equations (47) and (48) reveals that the monomer concentration of the surfactant with the lower cmc necessarily decreases monotonically with increasing total surfactant content beyond the cmc of the mixture:

$$\frac{\partial c_1}{\partial c_T} = \frac{c_1^*}{2(c_2^* - c_1^*)} \left[-1 + \frac{c_T - c_2^* + c_1^*}{\sqrt{(c_T - c_2^* + c_1^*)^2 + 4\alpha(c_2^* - c_1^*)}} \right] < 0 \qquad (52)$$

Interchanging 1 and 2 and the roots of the underlying quadratic equation for x yields $\partial c_2/\partial c_T$, which is also related to $\partial c_1/\partial c_T$ by Equations (47); thus

$$\frac{\partial c_2}{\partial c_T} = - \frac{c_2^*}{c_1^*} \frac{\partial c_1}{\partial c_T} > 0 \qquad (53)$$

and the possibility of a s.t. minimum exists. Indeed, if only monomers are adsorbed, Equations (6), (46) and (53) establish that at and above the cmc

$$\frac{\partial \sigma}{\partial c_T} = - [\Gamma_1 - (\frac{x}{1-x})\Gamma_2] \frac{RT}{c_1} \frac{\partial c_1}{\partial c_T} \qquad (54)$$

where x is given by Equation (48), whereas below the cmc,

$$\frac{\partial \sigma}{\partial c_T} = - [\Gamma_1 + \Gamma_2] \frac{RT}{c_T} \qquad (55)$$

Since $\Gamma_1 > 0$ and $\Gamma_2 > 0$ for surfactants, a necessary condition for a sign change, and thus a minimum, is

$$\frac{\Gamma_1}{\Gamma_2} \geq \frac{x}{1-x} \qquad (56)$$

The inequality corresponds to a cusp-minimum at the cmc, the equality to a smooth minimum above the cmc. To proceed it is necessary to know the equation of interfacial state. Invoking Szyszkowski's, Clint[20] shows that if 1 gives a lower post-cmc s.t. than 2 and $\partial c_1/\partial c_T$ is negative beyond the cmc, there is a cusp-minimum of s.t. at the cmc: see Figure 2. The minimum cannot be less than the smaller post-cmc s.t. of the two surfactants (this conclusion holds only for ideally mixed micelles). Probably the cusp-minimum is a peculiarity of the phase-separation limit. Calculations for finite N are not yet available.

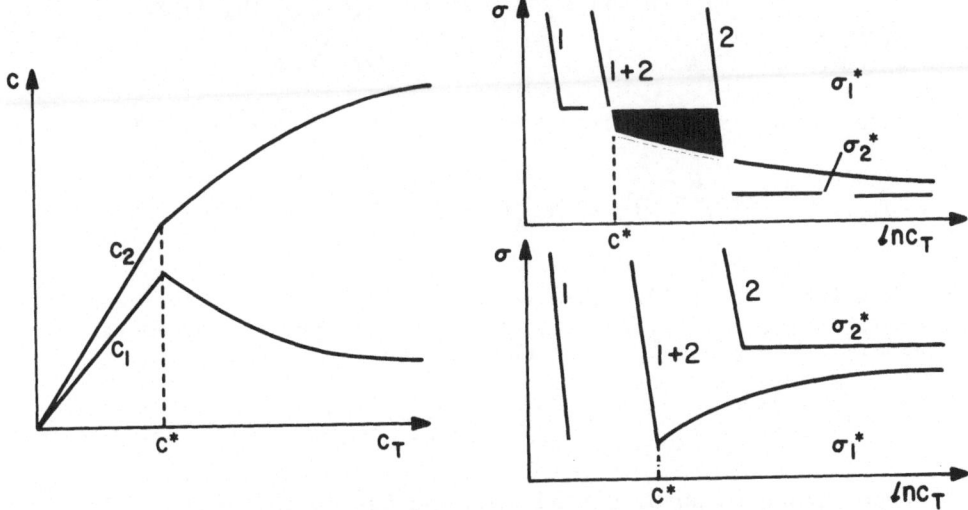

Figure 2.   Pseudo-phase-separation limit, ideally mixed micelles in a single phase. (Left) Surfactant monomer concentrations versus total surfactant inventory. (Right above) Surface tension when $(c_1^* - c_2^*)(\sigma_1^* - \sigma_2^*) < 0$, with Szyszkowski equation of surface state. (Right below) Surface tension when $(c_1^* - c_2^*)(\sigma_1^* - \sigma_2^*) > 0$: note the cusp-minimum at the mixed-micelle cmc, c^*.

TWO SURFACTANTS IN TWO PHASES

Bury-Hartley Limit

The special case of mixed micelles of mole fraction x_A in phase A and x_B in phase B, suffices to illustrate the approach. Overall equilibrium can be represented as a superposition of homogeneous equilibria in each phase and partitioning equilibria between the phases. Equation (8) gives for the latter

$$c_{1B} = P_1 c_{1A} , \quad c_{2B} = P_2 c_{2A} \tag{57}$$

From Equation (39) it follows that for mixed micelles of size N_A in phase A and N_B in phase B,

$$\frac{x_A}{1-x_A} = \frac{c_{1A}}{c_{2A}} \left(\frac{c_{2A}^*}{c_{1A}^*}\right)^{(N_A-1)/N_A} , \qquad \frac{x_B}{1-x_B} = \frac{c_{1B}}{c_{2B}} \left(\frac{c_{2B}^*}{c_{1B}^*}\right)^{(N_B-1)/N_B} \qquad (58)$$

The material balance equation for 1 is (compare Equation (40))

$$c_{1A}\left[1 + \left(\frac{c_{1A}}{x_A c_{1A}^*}\right)^{N_A-1}\right] + \phi P_1 c_{1A}\left[1 + \left(\frac{P_1 c_{1A}}{x_B c_{1B}^*}\right)^{N_B-1}\right] = \alpha \, c_T \qquad (59)$$

and that for 2 is similar (see Equation (41)). These four equations determine c_{1A}, c_{2A}, x_A and x_B, given c_T, α and the parameters. In general they too must be solved numerically.

Pseudo-phase-separation Limit

Some insight can be gained from the pseudo-phase-separation limit. At and above the cmc in phase A (see Equation 47))

$$c_{1A} = x_A c_{1A}^* , \qquad c_{2A} = (1-x_A) c_{2A}^* \qquad (60)$$

and at and above the cmc in phase B

$$c_{1B} = x_B c_{1B}^* , \qquad c_{2B} = (1-x_B) c_{2B}^* \qquad (61)$$

If both micellar 'phases' are present, Equations (57),(60) and (61) require that

$$\frac{P_1 c_{1A}}{c_{1B}^*} + \frac{P_2 c_{2A}}{c_{2B}^*} = 1 \quad \text{and} \quad \frac{c_{1A}}{c_{1A}^*} + \frac{c_{2A}}{c_{2A}^*} = 1 \qquad (62)$$

If both micellar 'phases' were to form at the same total surfactant content, the material balance equations up to infinitesimally greater content would require that

$$c_{1A} + \phi c_{1B} = \alpha c_T , \qquad c_{2A} + \phi c_{2B} = (1-\alpha) c_T \qquad (63)$$

and c_T would be the critical overall concentration. But since the compositions would already be set by Equations (57) and (62), (63) could only be satisfied by a particular phase ratio ϕ at a given α (and vice versa). Therefore, the cmc in one phase or the other is reached first, in general.

If the cmc (in the pseudo-phase-separation limit) is reached first in phase A, the *total* surfactant content at that cmc is given by Equation (51) modified for partitioning:

$$\frac{1}{c_T^*} = \frac{\alpha}{c_{1A}^* (1 + P_1\phi)} + \frac{1-\alpha}{c_{2A}^* (1 + P_2\phi)} \tag{64}$$

The results of the preceding section show that there could be an i.t. minimum at this cmc, depending on the equation of interfacial state: cf. Equation (56) and see Figure 3.

As more surfactant is added, the micellar mole fraction x_A approaches α, starting from the value given by Equation (50) suitably modified, viz.

$$x_A^* = \frac{\alpha}{\alpha + (1-\alpha)c_{1A}^* (1 + P_1\phi)/c_{2A}^*(1 + P_2\phi)} \tag{65}$$

If this range includes the x_A dictated by Equations (60) and (62), then the second micellar pseudophase forms in phase B at a value of c_T which is the *second* critical overall concentration. This is readily found from Equations (60) and (62) and the material balances:

$$c_{1A}(1 + P_1\phi) + x_A c_{PA} = \alpha c_T^{**} , \quad c_{2A}(1 + P_2\phi) + (1-x_A)c_{PA} = (1-\alpha)c_T^{**} \tag{66}$$

Two cases are summarized in Figure 3, which is based on Szyszkowski's equation of interfacial state. The qualitative features of the curves are independent of partitioning behavior, so long as partitioning is monotonic. The presence of micellar pseudo-phases in both phases above the second cmc would imply invariant compositions if the pseudo-phase-separation model were accurate, which it is not, in this regard.

DISCUSSION

Solution behavior, partitioning, micellization, adsorption and mass conservation interact with enough complexity to demand systematic mathematical modeling of two-phase micellar systems, as this paper shows. In reality there are distributions of premicellar and micellar aggregates of surfactant molecules. These distributions can be modeled mathematically by means of multiple association equilibria, with appropriate sets of association equilibrium constants K_{in} and K_{xn}.[31-33] The mathematics, however, is so convoluted that it is expedient to sacrifice all detail about the distributions in order to model the behavior of surfactant monomers and micelles more simply. This is the justification for the Bury-Hartley limit employed here, and for the more extreme pseudo-phase-separation limit, in which all information about the effects of micelle size is sacrificed. These limits fall into a hierarchy of mathematical models, and for investigations of mixed micellization it appears that the pseudo-phase-separation limit may often be a

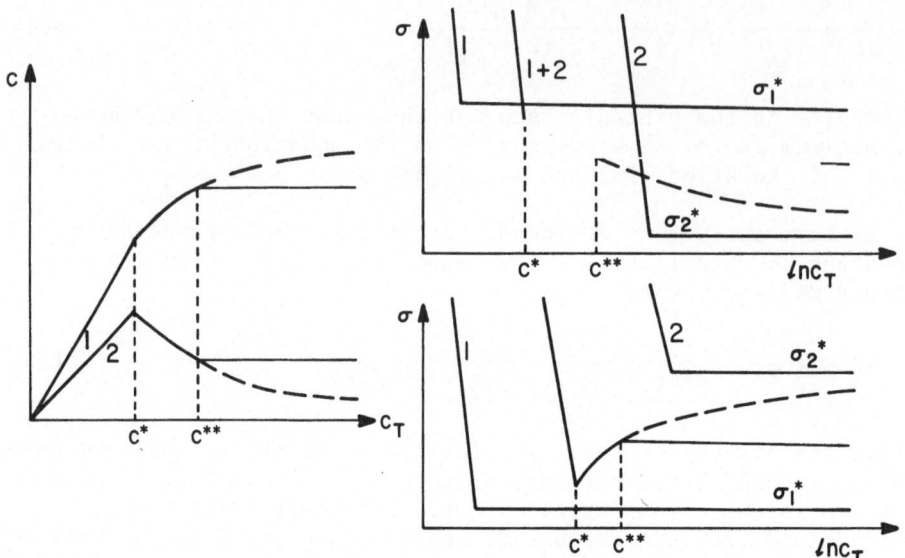

Figure 3. Pseudo-phase-separation limit, ideally mixed micelles in two phases. (Left) Surfactant monomer concentrations versus total surfactant inventory. Broken curves apply if micelles form in only one phase. (Right above) Interfacial tension, based on Szyszkowski equation, when $(c_1^* - c_2^*)(\sigma_1^* - \sigma_2^*) < 0$. (Right below) Interfacial tension when $(c_1^* - c_2^*)(\sigma_1^* - \sigma_2^*) > 0$: note the cusp-minimum at the first cmc and constant value beyond the second cmc.

useful first approximation, as it is in the preceding two sections.

When only one nonionic surfactant is present and it forms pure micelles in a single solvent or in both of a pair of immiscible solvents in contact, its chemical potential increases monotonically with total surfactant content. This is true in the Bury-Hartley limit at least. Though this conclusion is established here for quasi-ideal solution behavior, it holds for any stable solution because $\partial\mu_j/\partial c_j$ is necessarily positive. The conclusion extends to nonlinear partitioning and adsorption provided these are monotonic in every related concentration. To decide whether it is true in other circumstances, further mathematical modeling will be necessary. Of high interest are cases of solubilization, in which micelles in one solvent can incorporate the opposing, immiscible solvent.[31] The somewhat more complicated cases of ionic surfactants are of great interest too.

In order to have a maximum in chemical potential it appears that mixed micelles are necessary. Variable micelle composition, which is observed,[19] can be modeled by mass-action equilibria with variable stoichiometry, as shown here. This approach is more general than the pseudo-phase-separation limit and requires merely a revaluation of the free energy minimum at equilibrium. The consequences of non-ideal mixing within micelles remain to be treated, as do those of a variable micelle size N(x) and of solubilization of an immiscible solvent.

ACKNOWLEDGEMENT

This research was supported by the National Science Foundation and the University of Minnesota. The Graduate School provided a fellowship for one of the authors (M.S.B.). The helpful suggestions of H. T. Davis are gratefully acknowledged.

REFERENCES

1. W. R. Foster, J. Pet. Tech. 25, 205 (1973).
2. P. M. Wilson (Dunlap) and C. F. Brandner, (1973), paper presented at A.C.S. Meeting, Dallas, Texas, April 1973; J. Colloid Interface Sci., in press.
3. P. M. Wilson, C. L. Murphy and W. R. Foster, (1976), SPE paper 5812, presented at Soc. Pet. Eng. Symposium on Improved Oil Recovery, Tulsa, OK, March 1976.
4. J. L. Cayias, R. S. Schechter and W. H. Wade, in "Adsorption at Interfaces," K. L. Mittal, Editor, ACS Symposium Series No. 8, pp. 234-247, American Chemical Society, Washington, D.C., 1975.
5. J. L. Cayias, R. S. Schechter and W. H. Wade, (1976), J. Colloid Interface Sci., submitted for publication.
6. P. H. Doe and W. H. Wade, (1976), preprint for A.C.S. Meeting, New York City, April 1976.
7. J. C. Morgan, R. S. Schechter and W. H. Wade, (1976), manuscript to be published.
8. D. Anderson, M. S. Bidner, H. T. Davis, C. D. Manning and L. E. Scriven, (1976), SPE paper 5811, presented at Soc. Pet. Eng. Symposium on Improved Oil Recovery, Tulsa, OK, March 1976; final version submitted to Soc. Pet. Eng. J.
9. H. Saito and K. Shinoda, J. Colloid Interface Sci. 32, 647 (1970).
10. R. N. Healy, R. L. Reed and D. G. Stenmark, Soc. Pet. Eng. J. 16, 147 (1976).
11. K. J. Mysels and A. T. Florence, in "Clean Surfaces," G. Goldfinger, Editor, pp. 227-268, M. Dekker, N.Y., 1970.
12. G. D. Miles, J. Phys. Chem. 49, 71 (1945).

13. L. Shedlovsky, J. Ross and C. W. Jacob, J. Colloid Sci. $\underline{4}$, 25 (1949).
14. D. Reichenberg, Trans. Faraday Soc. $\underline{43}$, 467 (1947).
15. D. G. Hall and B. A. Pethica, in "Nonionic Surfactants," M. J. Schick, Editor, Ch. 16, pp. 516-557, M. Dekker, N.Y. 1967.
16. E. Hutchinson, J. Colloid Sci. $\underline{3}$, 413 (1948).
17. J. L. Moilliet, B. Collie and W. Black, "Surface Activity" 2nd ed., pp. 73-81, Van Nostrand, London, 1961.
18. P. Mukerjee and A. Anavil, in "Adsorption at Interfaces," K. L. Mittal, Editor, ACS Symposium Series No. 8, pp. 107-128, American Chemical Society, Washington, D. C., 1975.
19. K. J. Mysels and R. J. Otter, J. Colloid Sci. $\underline{16}$, 462 (1961).
20. J. H. Clint, J. Chem. Soc. Farad. Trans. I $\underline{71}$, 1327 (1975).
21. G. Stainsby and A. E. Alexander, Trans. Faraday Soc. $\underline{46}$, 587 (1950).
22. K. Shinoda, in "Colloidal Surfactants. Some Physicochemical Properties," K. Shinoda, T. Nakagawa, B.-I. Tamamushi and T. Isemura, pp. 25-29, Academic Press, N.Y. 1963.
23. R. Defay, I. Prigogine, A. Bellemans and D. H. Everett, "Surface Tension and Adsorption," pp. 86,89,112, Wiley, N.Y. 1966.
24. D. J. Crisp, Trans. Faraday Soc. $\underline{43}$, 815, 816 (1947).
25. K. Denbigh, "The Principles of Chemical Equilibrium," 3rd ed., p. 276, Cambridge University Press, Cambridge 1971.
26. E. R. Jones and C. R. Bury, Phil. Mag., Ser. 7, $\underline{4}$ 841 (1927).
27. R. C. Murray and G. S. Hartley, Trans. Faraday Soc. $\underline{31}$, 183 (1935).
28. M. S. Bidner, R. G. Larson and L. E. Scriven, Lat. Am. J. Chem. Eng. Appl. Chem. $\underline{6}$, 1 (1976).
29. G. S. Hartley, "Aqueous Solutions of Paraffin-Chain Salts," pp. 24-25, Hermann, Paris, 1936.
30. J. M. Corkill, J. F. Goodman and S. P. Harrold, Trans. Faraday Soc. $\underline{60}$, 202 (1964).
31. M. S. Bidner, E. Franses and L. E. Scriven, (1976), in preparation.
32. C. Tanford, J. Phys. Chem. $\underline{78}$, 2469 (1974).
33. N. Muller, J. Phys. Chem. $\underline{79}$, 287 (1975).

EQUILIBRIUM BICONTINUOUS STRUCTURES

L. E. Scriven

Department of Chemical Engineering & Materials Science

University of Minnesota, Minneapolis, MN 55455

For certain ranges of phase-volume ratio, there
are two-phase structures in which both phases are continuous and interfacial area is less than in dispersions of globular units having the same volume ratio
and average repeat distance. Included are bicontinuous structures defined by multiply connected minimal
surfaces of very high genus and everywhere saddle-
shaped.

The topology and thermodynamics are examined of
systems containing subdivided or multiply connected
interface. A new idea is developed: submicroscopic
bicontinuous structures may exist as equilibrium
states in microemulsions and mesomorphous phases of
amphiphile systems. Effects of thermal agitation on
such structures can account for the gradual transition
of microemulsion from water-continuous to oil-continuous. Related hypotheses are also put forward for
experimental investigation.

INTRODUCTION

What are the ways of filling space with two material phases,
or with two compositions of matter? Conventional thinking about
fluid phases is that one or the other forms a continuum in which
the second is dispersed as discrete globules or as less symmetric
blobs. Water/oil emulsions are supposed to be either water-continuous or oil-continuous. For microemulsions, micellar solutions,
mesomorphous phases and certain lyotropic liquid crystalline phases

877

the possibilities are held to be blobs of one composition dispersed
in another, or tubules of one threading the other, or lamellae of
one alternating with the other.[1-5] Among common solid/fluid interspersions are examples of another possibility: *both* phases continuous, as in sandstone, fritted glass, sponge and many other
porous materials. This possibility is uniquely three-dimensional,
an apt subject of stereology,[6] for to represent it on the page
requires serial sections or a perspective likeness (cf. Figure 3).

Submicroscopic, bicontinuous interspersions of two compositions
are well-known in glass-forming systems and multicomponent polymeric
systems.[7,8] Of special note are those created by spinodal decomposition. The resulting bicontinuous structure, though to some
extent random, has an average phase repeat length which is controlled directly by intermolecular forces.[7-10] Similar structures
are reported in phase inversion membranes.[11] At the molecular
level these systems are amorphous; they are subcooled liquids in
essence. Ordered networks of one composition embedded in another
probably do occur in crystals, however, and can be regarded as
bicontinuous or virtually so. Luzzati, Tardieu *et al.* adduce
evidence of network arrays of rod-like micelles in certain cubic
phases of water/lipid systems.[12,13,3,4]

Many aqueous lipid and other water/amphiphile systems have a
temperature range in which they are fluid microemulsions, or what
Winsor named amorphous solution phases and denoted by S.[1,3] They
yield continuously under applied shear stress, at moderate shear
levels, at least; they may scatter light appreciably but are not
birefringent; by small-angle x-ray analysis they appear to be
amorphous; and by thermodynamic standards they are equilibrium
states. At lower temperatures they transform into more highly
ordered mesomorphous or liquid crystalline states: middle phases
(M) consisting of long tubular micelles arrayed in parallel on
hexagonal centers (and hence nematic); gel phases (G) consisting of
large, multilayered lamellar micelles (and hence smectic); viscous
isotropic phases (V) which show cubic symmetry and are variously
thought to arise in globular or rodlet microstructure; and uncommon
viscous isotropic phases (S_c) which are believed to involve
spherical micelles constrained in cubic array. The designations
are Winsor's. In Figure 1 his idealized phase schemes have been
clarified and extended into a thermodynamic phase diagram. It is
hypothetical but summarizes the problem of understanding the inter-
relationships of the phases (not all of which occur in any one system). As drawn, the intermediate phase transitions are all between
states that differ in symmetry. The S-states (excepting S_c) are
amorphous. The S_1-state consists of globular amphiphile micelles
randomly dispersed in water. The S_2-state consists of globular
inverted micelles with water solubilized in molten amphiphile.
Between these two states is a continuous progression of S-states
which Winsor argued are thermally roiled solutions of fragments of

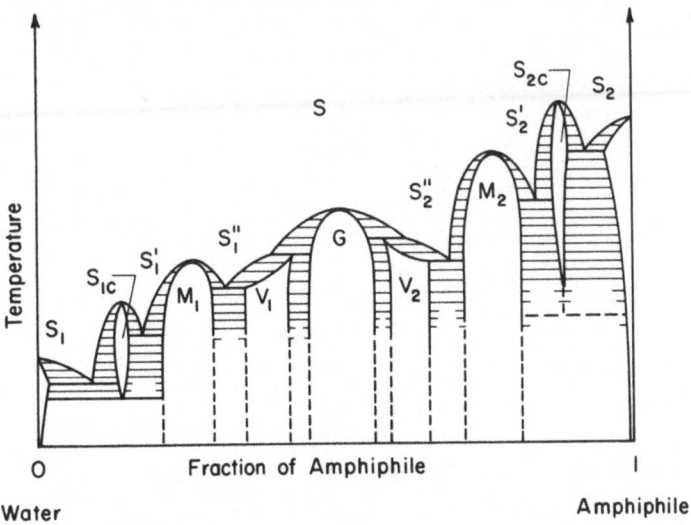

Figure 1. Hypothetical water-amphiphile binary phase diagram,
generated by clarifying and extending Winsor's phase schemes.

globular, tubular and lamellar micellar forms.

    Phase behavior of water and amphiphile becomes even more
interesting scientifically, and directly relevant to detergency and
other fields of application, when hydrocarbon oil or other nonpolar
liquid is also present.

    At moderate concentrations of amphiphile some water/oil/
amphiphile systems display a temperature range in which an amorphous solution phase extends all the way across the thermodynamic
ternary phase diagram. The situation is illustrated by Figure 2, a
hypothetical diagram generated from Winsor's phase schemes for
"Type III systems" and from other sources.[1-4,14-16] The S_1-side of
the S-corridor corresponds to globular amphiphile micelles, more or
less swollen with solubilized oil, but randomly dispersed in water.
On the S_2-side are globular inverted micelles which contain more or
less of solubilized water, and are randomly dispersed in oil.
Between is a continuous progression of states which though seemingly
amorphous are probably related to the distinctive mesomorphous or
liquid crystalline states which lie above them in amphiphile concentration. As oil/water ratio increases, electrical resistivity
increases smoothly and viscosity may show two mild maxima, according
to the limited data that are available[1,3,15] (the viscosity is not
Newtonian and in fact these systems show complicated rheology including elastic recoil in some cases). Along the binodal curve at
the bottom of the corridor are microemulsions S_2^*, which exist in

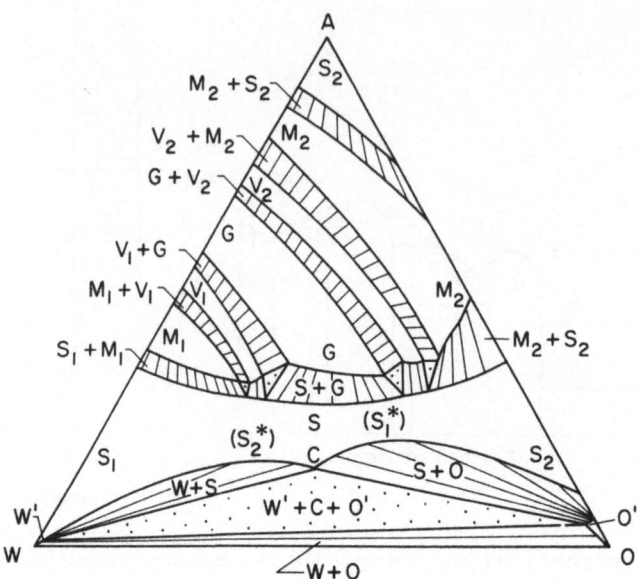

Figure 2. Hypothetical water-oil amphiphile ternary phase diagram, generated from Winsor's phase schemes and other sources.

equilibrium with nearly pure water; because they are immiscible with the water phase Winsor and others presumed them to be oil-continuous. Likewise, microemulsions S_1^* exist in equilibrium with nearly pure oil and are presumed to be water-continuous. Most remarkable is the doubly saturated microemulsion, C, which coexists as a third phase in contact with both an aqueous phase and an oil phase. Healy *et al*. conclude that in it the continuous zone is predominantly neither water nor hydrocarbon, and if it contains discrete structures they are not globular, tubular or lamellar on the average.[15] Indeed C appears to be a very confused state which can well be called "crazy mixed up stuff;" it consists of comparable amounts of water and oil and only a modest amount of amphiphile, yet does not mix with water or oil. Its interfacial tension with both is low — extremely low in some cases — which is of considerable scientific interest and great practical significance.[14,15]

If the presumptions were correct, a traverse of the S-corridor would carry one from water-continuous to *oil*-continuous to *neither*-continuous to *water*-continuous to oil-continuous states. One wonders if there are other ways to explain the continuous progression of states, and the state C in particular.

Equilibrium bicontinuous structures are a possibility. They
may bear on other parts of the phase diagram. They may occur in
other types of dispersed systems. They have not been considered
before, because the idea of an *equilibrium* bicontinuous structure
is new.[17] The idea is basically simple yet it has many ramifications.

GEOMETRY AND TOPOLOGY

With a suitable surface a volume can be divided into two
interpenetrating, labyrinthine subvolumes, each of them physically
continuous (which means mathematically connected). The two subvolumes can then be filled with distinct material phases, or compositions of matter, to create a bicontinuous structure.[17] A
specimen is said to be bicontinuous only if both subvolumes intersect its outer boundary in more than one place; practically, both
should be connected from each side of the specimen to the opposite
side. A specimen which is not bicontinuous may nevertheless contain
domains which are, both subvolumes being multiply connected in the
domain. Disjoint tube-like surfaces are excluded by requiring the
surface to be continuous and to have positive genus (which means
multiply connected). Genus is the topological property of holeyness: the genus of a sphere is zero; of a torus, one; of an eight,
two; and so on. The genus of any infinite network or framework is
infinity (distinctions can be made by means of various Betti numbers of the network). Surfaces that intersect themselves do not
accord with reality, nor do Klein bottles. Suitable surfaces are
continuous, orientable, of positive genus and without self-intersection. Inscribed in a volume, such a surface produces a bicontinuous partitioning of space from which a bicontinuous structure
can be created. Figures 3 and 4 show exceptionally regular
examples of bicontinuous structures. Obviously bicontinuous
structures need not have any geometrical regularity at all, however.
The reader can easily picture examples that are totally devoid of
crystal-like order yet are bicontinuous over unlimited distances.

It is obvious that *multi*continuous partitionings of space and
*multi*continuous structures can exist. An example of particular
interest is a formerly bicontinuous structure the partitioning
surface of which has been everywhere thickened to yield a tricontinuous structure. For an example one can imagine the surface in
Figure 3 or 4 thickened into a uniform layer. Conversely, a
physical layer can be idealized as a mathematical surface.

The structures in Figures 3 and 4 are periodic. They are
described by their symmetries and repeat distances or lattice
parameters (three in general). They happen to be self-conjugate
in the sense that both subvolumes are the same. Their topologies
are characterized by the connection or coordination numbers of a

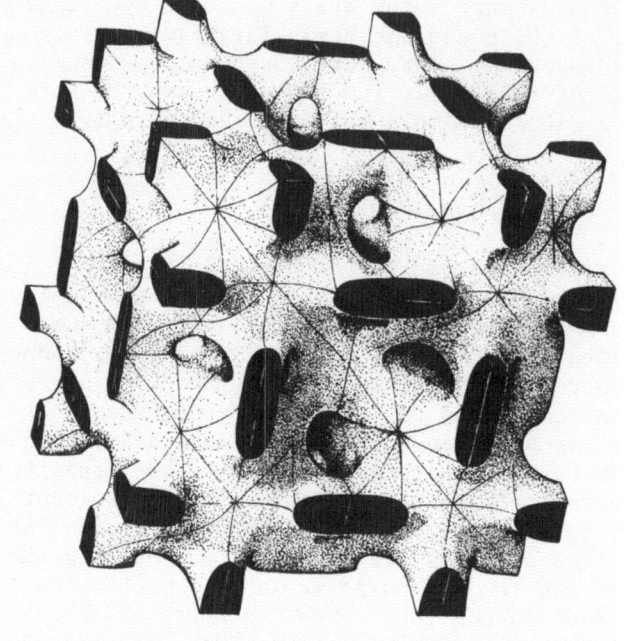

Figure 4. Neovius's periodic minimal surface, with simple cubic symmetry and topological indices {4,6,4} . Almost every point is a saddle point. This surface partitions space into two equal, infinitely connected, interpenetrating subvolumes: an exceptionally regular bicontinuous structure.

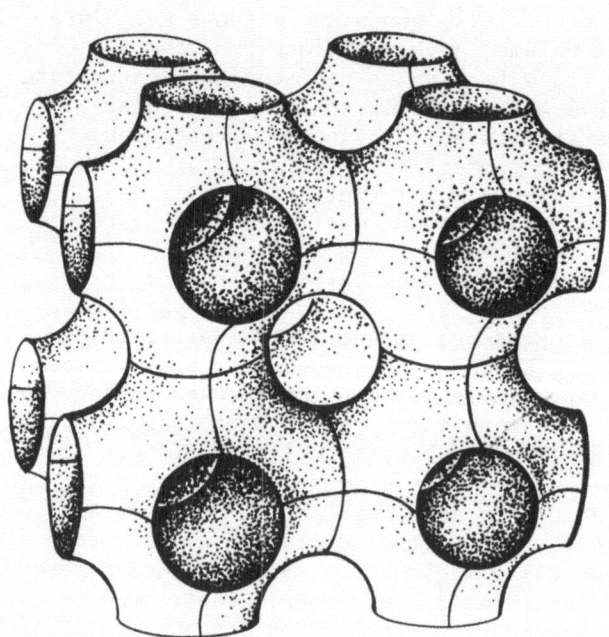

Figure 3. Schwarz's primitive periodic minimal surface, with simple cubic symmetry and topological indices {1,6,2} . Except for flat points on the three-fold axes, every point is a saddle point. This surface partitions space into two equal, infinitely connected, interpenetrating subvolumes: an exceptionally regular bicontinuous structure.

unit cell. A convenient standard unit cell is one which encompasses
the largest structural segment, called an *atrium* (the space in
which the largest inscribed sphere will fit). Between atria there
are vestibula. With Schwarz's surface there is just *one* path between two nearest-neighbor atria; each atrium connects through
vestibula to *six* others; and each vestibulum opens onto *two* atria.
Thus its topological indices are {1,6,2}. The nearest intersection
of two distinct paths leading out of an atrium is in a next-nearest
neighbor. Altogether there are twelve distinct paths that cross
(by fours) in next-nearest atria and so the genus of the unit cell
is twelve. With Neovius's surface the indices are {4,6,4}. The
nearest intersections of distinct paths are in nearest atria and
the genus of the unit cell turns out to be eighteen (thus Neovius
discovered a minimal surface which has 50% greater holeyness than
the one found by his mentor Schwarz).

Both Schwarz's[18] and Neovius's[19] surfaces are periodic minimal
surfaces, which means that for given symmetry and lattice parameter
(and equal subvolumes) the surface area is the least possible.[20]
Schwarz's is topologically equivalent to a simple cubic lattice of
fused, truncated octahedra, and the latter is known to have less
area than discrete spheres on the same lattice spacing, provided
the two fractional volumes fall in the range 0.40-0.60.[17] Thus
Schwarz's surface, which separates equal subvolumes, and others
like it have less area per unit volume than a simple cubic array of
any kind of globular units with the same lattice parameter, for any
fractional volumes falling in a range which certainly spans 0.40 to
0.60.

Five periodic minimal surfaces without self-intersection were
discovered by Schwarz and his student Neovius. A number of new
ones have been reported but not confirmed.[21] Besides cubic
symmetry, orthorhombic, tetragonal, hexagonal and trigonal symmetries are represented. Neither the areas of the surfaces, nor the
volume fractions of the subvolumes, nor the stability of the forms
appear to have been studied systematically. Incidentally, Kelvin's
famous problem on the division of space with minimum partitional
area refers not to a bicontinuous partitioning but to forming a
space-filling set of unconnected cells all of the same volume and
zero genus.[22]

In bicontinuous structures the partitioning surface is curved
(polyhedral forms are simply limits as all of the curvature is
concentrated in edges and vertices). Because surface curvature is
not fully appreciated in even the latest publications on micelles,
microemulsions and mesomorphous phases, certain facts bear listing.
Locally the principal curvatures are $\kappa_1 \equiv 1/r_1$ and $\kappa_2 \equiv 1/r_2$,
where r_1 and r_2 are the principal radii of curvature. The
local mean curvature is $H \equiv (\kappa_1 + \kappa_2)/2$. The local Gaussian
curvature is $K \equiv \kappa_1\kappa_2$. These two measures can be used

to distinguish the sorts of points that make up different kinds of surface, as indicated in Table I. Saddle-shaped, or anticlastic, surfaces have been completely overlooked by previous writers.

Table I.   Local Curvatures

	Plane	Cylinder	Inverted Cylinder	Sphere	Inverted Sphere	Concave	Convex	Saddle-shaped
H	0	>0	<0	>0	<0	>0	<0	any
K	0	0	0	>0	>0	>0	>0	<0

Minimal surfaces have no mean curvature.[20]   The two principal curvatures κ_1 and κ_2 either are equal and opposite or else vanish together.   If it is not a plane, a minimal surface is everywhere saddle-shaped, as are the surfaces in Figures 3 and 4. Minimal surfaces are also solutions of the Young-Laplace equation of capillarity under the condition that normal stress be continuous across the surface.   Thus a minimal surface can separate two bulk fluid phases at rest at the same hydrostatic pressure.   Almost periodic minimal surfaces, if they exist, have not been studied mathematically, nor have otherwise less regular minimal surfaces of high genus (cf. Nitsche[20]).

Though partitioning surfaces that are not minimal have not been studied either, it is certain that they, like minimal ones, have substantial fractions that are saddle-shaped if they produce bicontinuous partitionings.

A bicontinuous structure that is not periodic may be topologically equivalent to one that is; i.e. the two may be related by a continuous deformation that destroys symmetry but preserves genus. The original structure though lacking in symmetry has a high degree of topological order and can be said to be completely ordered topologically.   If the departures from periodicity are random, there is some length scale over which the structure is approximately periodic, within whatever tolerance is chosen.   Such a structure is said to be almost periodic, and average "phase" repeat distances or average lattice parameters can be defined statistically for it. Periodicity may fail locally only, in lattice imperfections — vacancies, interstitials, dislocations, domain boundaries, interfaces, and so on — just as in the discrete arrays known to crystallography.   There must be not one but two classes of defects: those in which the partitioning surface is continuous, and those in which it is torn and so has edges.   Some types of imperfections would be expected to exist in equilibrium numbers per unit volume, others to arise in deformation processes. Many types would be expected to control one or another aspect of the rheology of bicontinuous structures.

In more highly disordered bicontinuous structures average distances over which each composition or "phase" repeats can be defined, but the distinction between atria and vestibula may be untenable geometrically. What remains is the identification with nodes and connectors in some kind of three-dimensional network which is topologically equivalent. Beyond its network equivalent a bicontinuous structure has the topological order of *insideness vs. outsideness,* or *two-sidedness* more generally. Thus a domainally bicontinuous specimen has long-range topological order on the scale of the domains. Topological order plainly is distinct from the geometrical order of crystalline materials.

Bicontinuous structures can be topologically chaotic, with distributions of topological indices best described in terms of averages and moments; an example would be a suitably defined average genus per unit volume. The appropriate mathematics seems not to have been developed yet. A possibility appears in the next section.

ENERGY AND ENTROPY

If bicontinuous structures exist in thermodynamic equilibrium states they are brought about by intermolecular forces and thermal motions of the molecules. One factor is the short-ranged repulsive forces between molecules, which in effect confer shapes on them and thereby influence their aggregation into structures. Though shape and packing are emphasized in literature on mesomorphous and microemulsion phases, they are by no means alone in controlling the association of amphiphile molecules into partitioning layers. Nor do long-ranged intermolecular attractions and repulsions dominate the scene. An instructive example of the way all the forces share together with thermal motions the responsibility for a supramolecular structure is an ordinary fluid interface, the thin zone of steep density or composition gradients between liquid and vapor or between two liquids.[23] Even spherical molecules form the layer-like, fluid structure. Recent theoretical work shows that spherical molecules, over certain ranges of the chemical potential-temperature-pressure combination, can be organized into periodic, space-filling, layered structures which represent constrained equilibria at the least.[24] Whether such structures, with their definite, molecularly controlled "phase" repeat length, are thermodynamically stable in the absence or presence of amphiphile is another matter although the ordinary fluid interface, which is a limiting form, certainly can be stable.

Given that intermolecular forces can induce layer-like aggregates of amphiphile and solvent molecules at equilibrium, state variables that characterize internal configuration of the material are needed. If one idealizes the layer-like aggregate as a mathematical surface, one identifies in micellar forms and bicontinuous structures the following descriptors: total volume fractions V_A and V_B of compositions A and B respectively; total area of surface between A and B per unit volume, namely S; mean separations of portions of S as measured through A and through B, respectively λ_A and λ_B; the symmetry group G and lattice parameters α_1, α_2 and α_3 of surface if it is periodic; the total mean curvature of surface, H_T; the total of the square of the mean curvature, J_T; the total Gaussian curvature of surface, K_T; and higher order curvature terms (all per unit volume). If defects involving torn edges or if recognizable fragments of micellar forms are present in appreciable numbers, then the list should be augmented with the total length L_T of edge of surface (and, in principle, higher order terms in edge curvature). Small aggregates are accounted for in the solution thermodynamics of materials A and B. Except for the volume fractions V_A and V_B, all of these descriptors are of effects of *inhomogeneity* on the total potential energy of the intermolecular forces. In theories of interfaces these effects are represented by gradient energy, the difference from the energy density of a homogeneous bulk phase having the same mean density and composition.[23] Gradient energy is responsible for the spontaneous development of inhomogeneity, through diffusion without nucleation, in spinodal decomposition.[7-10]

The shorter are the "phase" repeat distances λ_A and λ_B, the greater the gradient contributions to energy densities, E_A and E_B, of the two compositions; the greater too the potential for sensitivity to symmetry properties. Only in the limit as λ_B grows large does E_A approach the bulk-phase value (and similarly E_B versus λ_A). The surface energy E_S is that of a unit area of planar surface, a surface of which both principal curvatures, κ_1 and κ_2, vanish everywhere.

The total curvatures are related to κ_1 and κ_2:

$$H_T \equiv \sum_j \int (\kappa_1 + \kappa_2) dS_j \; , \; J_T \equiv \sum_j \int (\kappa_1 + \kappa_2)^2 dS_j \; , \; K_T \equiv \sum_j \int \kappa_1 \kappa_2 dS_j$$

The integrals are over the surfaces of individual, unconnected structures or domains; the summations are of all of these in a unit volume. What is often called *the* integral curvature, $\int \kappa_1 \kappa_2 dS_j$, has a remarkable property: if S_j is a closed surface of genus p, its integral curvature equals $4\pi(1-p)$, according to the Gauss-Bonnet theorem.[25] If p is large, as in an extensive bicontinuous structure, this formula is still a good approximation when the surface is not entirely closed, i.e., when it has a small proportion

of torn edges. Thus K_T is positive and proportional to the number N of globular units when one "phase" is dispersed and the
other is continuous. When both are continuous, K_T is negative and
very nearly proportional to the surface genus of the bicontinuous
structure. K_T can also be influenced by the presence of appreciable amounts of surface fragments (through a contribution
$\Sigma \int \kappa_g dL$; cf. Kreyszig[25]). Another view of the sign of K_T is that
it is negative when on the average the partitioning surface is
anticlastic, or saddle-shaped; and it is positive when the surface
is synclastic, or convex, toward either A or B. The average side
of convexity is given by the sign of H_T, provided of course that
K_T is positive. The other total curvature, J_T, is always positive.
Various cases — all of them conceivable in water/oil/amphiphile
systems — are distinguished in Table II. When the surface is
curved, the added gradient energy contributions corresponding to
these curvature measures are E_H, E_J, and E_K.

Table II.   Total Curvatures and Edge Length

	Lamallae	Tubules	Inverted Tubules	Globules	Inverted Globules	Bicontinuous Small Domains	Bicontinuous Large Domains
H_T	~0	>0	<0	>0	<0	any	any
J_T	~0	>0	>0	>0	>0	≥ 0	≥ 0
$K_T/4\pi$	0	$\leq N$	$\leq N$	N	N	$-N(\bar{p}-1)$	<<0
L_T	>0	≥ 0	≥ 0	0	0	0	0

In terms of the descriptors and associated energy densities,
the total potential energy per unit volume is

$$E = V_A E_A + V_B E_B + S E_S + H_T E_H + J_T E_J + K_T E_K + L_T E_L$$

($V_A + V_B = 1$ in the same approximation that denies S volume). It
is instructive to consider the case when intermolecular forces
conspire to elevate the first three contributions way over the
curvature energies and edge energy. Then if (1) E_A and E_B are
greatly reduced by segregating water and oil molecules, say, on
either side of a predominantly amphiphile layer for which (2) E_S
is positive, and if (3) $V_A E_A + V_B E_B$ has a sharp minimum with
respect to simple cubic symmetry and a specific lattice parameter
α (hence also λ_A AND λ_B), then a bicontinuous structure like

that in Figure 3 minimizes the total potential energy over a certain
midrange of volume fractions. It minimizes the energy because it
minimizes the surface area S under the imposed constraints.
Schwarz's surface, which is shown, gives the minimum when $V_A/V_B = 1$.
Under the alternate constraint of body-centered or face-centered
cubic symmetry, bicontinuous structures may not be favored around
$V_A/V_B = 1$ but certainly would be over a higher and a lower range of
volume ratios.[17]

In another case E_H is overriding and has a deep, steep
minimum at a particular value of local mean curvature $(\kappa_1 + \kappa_2)/2$.
Then the contending structures must have constant mean curvature
and the energy minimization problem is more difficult. Evidently
the possibility of periodic solutions has not been investigated:
indeed, this area of mathematics is relatively undeveloped. In
more general cases the dependence of E_A and E_B on repeat distances λ_A and λ_B is significant and curvature contributions are
important. The problems are to find surface configurations,
periodic and otherwise, which minimize energy under integral curvature constraints. (In the course of a computer simulation of
spinodal decomposition, Cahn noticed a multiply connected, chaoti-
cally irregular, isoconcentration surface for which H is zero,
and area is probably nearly minimal.[9])

Equilibrium states are minima in free energy, not potential
energy alone. There is of course a kinetic energy contribution
but the chief difference is entropic. The more liquid the system
the more appreciable the composition fluctuations and diffusion
caused by thermal motions of molecules. In structured liquids with
internal gradient zones, these motions produce wave-like fluctuations in zone shape and thickness as well as composition. Molecular motions cause the system almost continually to migrate among
states of nearly the same energy and to make wider excursions with
less frequency. Energy fluctuates and so do other state properties,
which here include length scales, curvature measures, genus and
other topological indices. In a periodic bicontinuous structure,
thermal fluctuations can certainly affect lattice distance and
symmetry and sustain equilibrium populations of certain lattice
imperfections. With rising temperature the ever-greater, wave-
like shape excursions and the composition fluctuations of the partitioning layer can destroy periodicity and symmetry, connectivity
and genus. Necking off and rewelding of narrow vestibula in the
structure cause connectivity fluctuations. As composition or temperature changes and the free energy minimum shifts, these fluctuations can gradually yet completely destroy the connectivity, or
physical continuity, of one of the subvolumes. At first it may be
left as large domains of high genus, but as the trend continues its
mean connected length is reduced until it is a dispersion of
globular, tubular, or lamellar units of zero genus in the other

subvolume. This is the limit when the coherence length of the partitioning layer is many molecular dimensions, and its molecular
aggregation number is comparatively high. Which shape emerges is
a matter of curvature and edge effects. Another limit, which
obviously should be approached as temperature increases, is a
molecular solution in which coherence lengths and aggregation numbers are small — as in ordinary associated and solvated solutions.

Measurements are limited time averages. In statistical thermodynamics the fluctuation time is presumed to be infinitesimal compared to measurement time, and so the measurements are postulated
to equal averages in an ensemble of all the states accessible to
the system. Both types of averages in practice are supposed to be
dominated by the states of highest probability. To understand
measurements then requires accurate information about the potential
energies of all of the more probable states. There is a presumption
that all of these are readily accessible from a given initial state.

When the potential energy E has a minimum deep and steep
compared to NkT, the most probable of accessible states around the
minimum very likely is indistinguishable from the minimum energy
state. When the minimum is shallow, or when there are multiple
local minima, and especially when they are all shallow, the picture
becomes fuzzy and perhaps cannot be understood by applying established concepts in standard ways, even when potential energy is
replaced by the appropriate free energy. Equilibrium states are
free energy minima — with respect to changes that can proceed
appreciably during the interval of observation. Whether and which
equilibrium is observed depends on initial state, kinetic rates,
and time waited. Polymorphism — the existence of metastable
equilibrium states at the same external intensive variables as an
absolutely stable state — corresponds to multiple local minima of
free energy, deep enough to have distinct domains of attraction
among the full range of nonequilibrium states. Geometrical polymorphism is well known in crystalline systems and is impossible in
ordinary, unstructured liquids. Bicontinuous structures bring the
possibility of topological polymorphism and combinations of geometrical polymorphism with the topological type.

When micellar forms and bicontinuous structures of different
genus are involved, not only may there be multiple free energy
minima but also transformations between them may require collective
molecular motions that are relatively improbable, or translational
and rotational diffusion that is slow on the time-scale of observation. The former is illustrated by ordinary nucleation, which
generates a new phase of genus zero by cluster formation. The
latter is exemplified by spinodal decomposition, in which diffusion
slowly evolves a pair of phases of high genus.

Long-range topological ordering into a bicontinuous structure
would be expected to take longer to establish than long-range
crystalline order in an array of separate objects on the same
scale, the objects themselves being given and all else being the
same: compare the structure in Figure 3 and the corresponding
simple cubic crystal of spherical units. Once any sort of long-
range topological ordering is established in a system, spontaneous
changes in the ordering would be expected to be slow. For example,
after monomers are combined into polymers their topology is
enforced by the covalent bonding of chains and crosslinks. In
silicate melts it is the ionic bonding of chains, rings, sheets,
and networks. In both cases the consequences include the glass-
transition phenomenon and other evidence of hindered approach to
equilibrium.

Thus where bicontinuous structures are involved one would not
be surprised by experimental observations of entropic effects:
slow approach to equilibrium, hysteresis in phase transition, and
metastable equilibrium states, or polymorphism. Equilibrium is
always relative to rate processes. Laboratory experience indicates
that the problems are much greater with amphiphile-containing
liquids and semisolids than with more common solutions in the same
solvents.[1,3,14,15,26,27]

APPLICATION AND CONCLUSION

These concepts can be used to construct hypotheses about the
molecular behavior responsible for the phase behavior of the
amphiphile systems which were introduced above and are schematized
in Figures 1 and 2. The hypotheses can be tested experimentally,
as the Minnesota group, for one, is attempting to do.

The first issue is the continuity of states in the S-corridor
of the water/oil/amphiphile diagram, Figure 2. In the middle of
the corridor, where water and oil proportions are comparable, the
microemulsion S could be a thermally disordered bicontinuous
structure. This structure would consist of predominantly water
regions (A) and predominantly oil regions (B), both of submicroscopic dimensions and separated by a connected, amphiphile-rich
partitioning layer of up to comparable thickness. This would also
be the structure of the "crazy mixed up stuff," C, which is
immiscible with both aqueous solution W' and oily solution O' and
contains, in a strictly ternary system, invariant concentrations of
water, oil and amphiphile. As water-to-oil ratio is increased the
structure could, owing to curvature effects, lose its water continuity and be merely oil-continuous in the S_2^*-range. Then with
further increase in water it would first revert to bicontinuous
states, then develop large bicontinuous domains in water, and at

last break down into oil-swollen, globular micelles dispersed in water, in the S_1-range. Or bicontinuity could persist through the S_2^*-range, the rest of the sequence being the same. Either way would account for the leftward progression, and totally analogous hypotheses would account for the rightward progression when water-to-oil ratio is decreased.

The alternative hypothesis is Winsor's,[1,3] that the micro-emulsion S is a solution of small fragments from lamellar, tubular, and globular structures and that these fragments make water and oil miscible. At sufficiently high temperature this may be so, but then the phase diagram is probably different. To generate the lower part of the diagram in Figure 2, a free energy function apparently requires two of the features of partitioning layers: the screening of water-oil interaction, and the configurational entropy of large, deformable aggregates of amphiphile.[27] This argues against the alternative hypothesis. Winsor argues in its favor, in terms of a balanced ratio of tendencies of the partitioning layer to become convex toward, respectively, the oil-rich and water-rich sides, but the argument is incomplete. The reason is that the layer may be convex toward *neither* side yet may be quite curved, as explained above. Indeed, if H and J were both precisely zero, the layer could only be a minimal surface of high genus and that implies a microemulsion that is a bicontinuous structure. Experiments by electron microscopy, transport measurements and other means are underway to resolve this issue. The continuity of states in the S-roof of the water-amphiphile diagram, Figure 1, is a similar issue, for which experimental evidence is needed also.

The concepts of equilibrium bicontinuous structure, topological ordering, equilibrium between coexisting phases that are differently ordered, and gradient zones of transition between coexisting phases lead directly to a spectrum of models for interfaces of structured liquids, for example the low tension interfaces between "crazy mixed up stuff" and dilute solutions of amphiphile in water or oil. Models of C/W' and C/O' interfaces will be discussed elsewhere.

The states M_1, V_1, G, V_2 and M_2 all exhibit internal symmetry and a different range of microemulsion compositions can exist in two-phase equilibrium with each. Insofar as S is bicontinuous there is the possibility that in its disorder it becomes topologically biased, in the vicinity of a given segment of the binodal curve, toward the conjugate phase there. If nematic M_1 contains ordered, tubular micelles, S in equilibrium with M_1 might contain bent and twisted tubular structures. Whether they were connected or separated, the deformed tubules would represent topological ordering in the face of geometrical disorder, and the geometrical disorder-order transition would be the phase transition.

Likewise S in equilibrium with smectic G might contain bent and
warped pancake structures, connected or not. And so on along the
binodal curve.

The optically isotropic phases V_1 and V_2, which exhibit
cubic symmetry and noticeably higher viscosity than adjacent phases,
might be periodic bicontinuous structures, rather than the cubic
packings of globular micelles, or the almost tricontinuous structures with rodlet micelles which have been deduced.[1-4,12,13] The
latter are close to being two distinct, three-connected, three-
dimensional, interpenetrating "phases" of the same composition,
surrounded by a multiply connected "phase" of a different make-up.
One wonders if this sort of model would fit crystalline forms for
which rodlet models are contemplated. The rodlet models are
probably more convenient for volume and area accounting in the
interpretation of x-ray scattering data. Nevertheless the prospect
of fully bicontinuous structures and of bicontinuous domains should
be entertained until turned out by weight of data. This should be
as true of crystalline states as of amorphous ones in amphiphile
systems.[3,4,28]

One wonders if endoplasmic reticulum and other membranous
parts of organelles in living cells are partitioning layers of
bicontinuous structures at or near stable or metastable equilibrium.

Bicontinuous and multicontinuous structures are ways of filling
space with more than one phase or composition of material. On the
submicroscopic scale they are related to ordinary condensed phases
as porous media are to homogeneous solids on a larger scale. As
possible equilibrium states of matter they are relatively unexplored,
mathematically, thermodynamically, and experimentally.

ACKNOWLEDGEMENT

This research was supported by the National Science Foundation
and the University of Minnesota. Discussions with W. R. Schowalter,
J. C. C. Nitsche, W. J. Swiatecki, H. T. Davis and W. G. Miller
were very helpful. It must be recorded that the referee objects
strongly to the appearance of the term blob and the expression
crazy mixed up stuff in a technical article.

REFERENCES

1.   P. A. Winsor, Chem. Rev. <u>68</u>, 1 (1968).
2.   P. Ekwall, L. Mandell and K. Fontell, in "Proc. Fifth Internat.
      Congr. on Surface Active Substances," 443-453, Ediciones
      Unidas, Barcelona, 1968.

3. P. A. Winsor, in "Liquid Crystals and Plastic Crystals," Vol. 1, G. W. Gray and P. A. Winsor, Editors, Ch. 5, pp. 199-287, Ellis Horwood Ltd., Chichester, 1974.
4. P. Ekwall, Adv. Liq. Cryst. 1, 1 (1975).
5. I. F. Efremov, in "Surface and Colloid Science," E. Matijević, Editor, Vol. 8, Ch. 2, pp. 85-192, Wiley-Interscience, New York, 1976.
6. E. E. Underwood, "Quantitative Stereology," Addison-Wesley, Reading, Mass. (1970).
7. P. F. James, J. Materials Sci. 10, 1802 (1975).
8. L. P. McMaster, in "Copolymers, Polyblends and Composites," N.A.J. Platzer, Editor, Ch. 5, pp. 43-65, Adv. Chem. Ser. No. 142, American Chemical Society, Washington, D. C., 1975.
9. J. W. Cahn, J. Chem. Phys. 42, 93 (1965).
10. J. S. Langer, in "Fluctuations, Instabilities and Phase Transitions," T. Riste, Editor, pp. 19-42, Plenum Press, New York, 1975.
11. R. E. Kesting, "Preprints. Amer. Chem. Soc. Div. Org. Coatings," 34 (No. 1), 553 (1974).
12. A. Tardieu and V. Luzzati, Biochim. Biophys. Acta 219, 11 (1970).
13. K. Fontell, J. Colloid Interface Sci. 43, 156 (1973).
14. K. Shinoda and S. Friberg, Adv. Colloid Interface Sci. 4, 281 (1975).
15. R. N. Healy, R. L. Reed and D. G. Stenmark, Soc. Pet. Eng. Journ. 16, 147 (1976).
16. M. L. Robbins, Soc. Pet. Engrs., Paper No. 5839, preprinted and presented at Tulsa, OK, 24 March 1976.
17. L. E. Scriven, Nature 263, 123 (1976).
18. H. A. Schwarz, "Gesammelte Mathematische Abhandlung," Vol. 1, pp. 6-125, Springer, Berlin, 1890.
19. E. R. Neovius, "Bestimmung zweier speziellen periodischen Minimalflächen," J. C. Frenkel, Helsingfors, 1883.
20. J. C. C. Nitsche, "Vorlesungen über Minimalflächen," Springer-Verlag, Berlin, 1975.
21. A. H. Schoen, Nat. Aero. Space Adm. Tech. Note D5541 (1970).
22. W. Thomson (Kelvin), Acta Math. 11, 121 (1887-1888).
23. V. Bongiorno, L. E. Scriven and H. T. Davis, J. Colloid Interface Sci., in press (1976).
24. V. Bongiorno, L. E. Scriven and H. T. Davis, (1976) J. Colloid Interface Sci., to be submitted.
25. E. Kreyszig, "Differential Geometry," Univ. of Toronto Press, 1959.
26. K. Fontell, J. Colloid Interface Sci. 44, 318 (1973).
27. D. R. Anderson, M. S. Bidner, H. T. Davis, C. D. Manning and L. E. Scriven, Soc. Pet. Engrs., Paper No. 5811, preprinted and presented at Tulsa, OK, 24 March 1976.
28. Fontell, K., in "Colloidal Dispersions and Micellar Behavior," K. L. Mittal, Editor, pp. 270-277, ACS Symp. Ser. No. 9, American Chemical Society, Washington, D. C., 1975.

INTRAMACROMOLECULAR MICELLES

Ulrich P. Strauss

Wright and Rieman Laboratories
Rutgers, the State University
of New Jersey
New Brunswick, New Jersey   08903

Polyelectrolytes with hydrophobic side chains may under appropriate conditions exhibit behavior typical of soap micelles. They may solubilize organic molecules normally insoluble in water and assume conformations distinctly more compact than those attainable by ordinary polymer molecules in solution. Studies on the physical chemical properties of such hydrophobic polyelectrolytes will be reviewed with special emphasis on the relevance of the findings to analogous properties of ordinary micelles. Topics to be treated include the effects of electrical charge and hydrophobic group size of the polyelectrolytes as well as general and specific effects of added small ions and solubilizates on the conformation and intramolecular interactions of the macromolecules.

INTRODUCTION

One of the essential ingredients in the chemical and physical processes occurring in natural biological systems is the hydro-phobic-hydrophilic balance of the macromolecules and membranes in or near which the processes take place.

To understand these effects in biological macromolecules it is desirable to study them in properly designed synthetic macro-molecules. Synthetic macromolecules have an advantage over biolo-gical macromolecules in that they allow one to vary systematically the relevant structural features and thus to get an insight into the underlying physical and chemical mechanisms.

In achieving the synthesis of appropriate macromolecules we may take as a starting point our knowledge that soap molecules form micelles in aqueous solution. Thus, one would anticipate that by chemically attaching soap molecules to a polymer chain micelles might be formed inside a polymer molecule. This was, in fact, found to be the case with "polysoaps" consisting of vinylpyridinium polymers with dodecyl side chains.[1,2]

Later it was also shown that by shortening the size of the alkyl group and using for the polymer chain a weak polyacid, micelles existing at low degrees of ionization could be broken up by increas-ing the ionization through appropriate changes in the pH.[3] The resulting conformational transition mimics the denaturation of globular proteins both in cause and in effect. What follows will be a brief review of some of the highlights encountered in the studies carried out in our laboratory with these synthetic intra-molecular micelle-forming polymers.

EVIDENCE FOR INTRAMOLECULAR MICELLE FORMATION

The first polysoap studied by us was prepared by the reaction of a poly-2-vinylpyridine sample with n-dodecyl bromide so that 33.7% of the pyridine groups were quaternized.[1] As might be expected from intramolecular micelle formation, this macromolecule showed a relatively compact structure in aqueous solution as evi-denced from viscosity measurements. The micelle formation was con-firmed by demonstrating that the polysoap solubilized isoöctane. The solubilization efficiency of the polysoap was found to be larger than that of its monomeric relative, N-n-dodecyl-pyridinium bromide. In contrast to the monosoap, the polysoap needed no criti-cal micelle concentration before it was able to solubilize the hydrocarbon. For polysoap concentrations up to 3% the solubiliza-tion of isoöctane per unit amount of polysoap was constant. These results demonstrated that the micelles were intramolecular.[1]

TRANSITION FROM POLYELECTROLYTE TO POLYSOAP

The effects of varying the number of soap molecules attached to the polymer chain were studied with derivatives of poly-4-vinyl-pyridine obtained by quaternizing samples of the parent polymer with n-dodecyl bromide to various extents and then, in order to keep the ionic character of the polymer backbone invariant, to quaternize the remaining free pyridine groups of each sample with ethyl bromide.[2,4] By including in such a series one sample quaternized only with ethyl bromide, the transition from polyelectrolyte to polysoap could be investigated. Viscosity and light scattering studies on such series in aqueous solutions showed the macromolecules to become increasingly compact with increasing dodecyl group content.[4-6] However, the effects were not linear; instead there was a rather sudden increase in compactness at about 10 mole percent dodecyl group content and relatively little change in the high compactness achieved beyond that composition. The results pointed to the existence of a critical dodecyl group content which might be considered the intramolecular equivalent of the critical micelle concentration observed in solutions of ordinary colloidal electrolytes.[5] No solubilization of decane, benzene or 1-heptanol could be detected at compositions below the critical dodecyl group content.[4] Beyond this point the amount of hydrocarbon solubilized per unit amount of polysoap increased much faster than linearly with increasing dodecyl group content while at the same time remaining constant as the concentration of a particular polysoap was varied. These results indicate that it is the synergistic effect of a large number of dodecyl groups in the same polysoap molecule which controls the solubilization of hydrocarbons. In contrast, the amount of heptanol solubilized appeared to be approximately proportional to the dodecyl group content, indicating a different solubilization mechanism.[4]

The polysoap molecules showed a tendency to form soluble aggregates.[4-6] This tendency increased with increasing dodecyl group content, with increasing polysoap concentration, with increasing concentration of added simple electrolytes, and with decreasing temperature.[4,5] The aggregation and disaggregation rates were followed by viscosity measurements, and the results indicated that a substantial activation energy stabilized the aggregated states.[5]

The existence of a critical dodecyl content in a similar series of poly-4-vinylpyridine derivatives was confirmed by Woerman and Wall[7] who found, furthermore, a specific counterion effect. The critical dodecyl content was observed to be lower in the presence of bromide ion than in the presence of chloride ion. Inoue[8] found critical micelle compositions both with series of dodecyl and with series of octyl quaternization products of poly-2-vinylpyridine. The critical mole percent of dodecyl groups was considerably smaller than that of octyl groups. These effects of counterions and of alkyl group size are similar to the analogous effects on the critical micelle concentration of ordinary colloidal electrolytes.

EFFECTS OF SOLUBILIZATION ON MACROMOLECULAR
DIMENSIONS AND INTERACTIONS

The solubilization of hydrocarbons and other compounds of low water solubility affected the reduced viscosity of polysoap solutions in rather characteristic ways.[9-11] Aliphatic hydrocarbons caused a small decrease independent of polysoap concentration, pointing to small effects on the macromolecular size and shape.[9] Increasing solubilization of benzene and other aromatic hydrocarbons produced viscosity maxima whose sharply increasing magnitude with increasing polysoap concentration indicated that the aromatic hydrocarbons strongly affect the interactions between polysoap molecules.[10] This effect was attributed to an increase in the ratio of hydrophobic to hydrophilic groups on the macromolecular surface brought about by a contraction of the hydrophilic portion of the surface by solubilized benzene located close to the polar groups.[10] Similar though still more complex effects on the viscosity were observed with amphiphilic solubilizates such as long chain aliphatic alcohols and nitroaromatics.[4] It is noteworthy that viscosity maxima as a result of benzene solubilization have also been observed in 10% sodium oleate solutions.[12]

SURFACE ACTIVITY OF POLYSOAPS

Surface activity studies of polysoaps derived from poly-4-vinylpyridine by quaternization with n-dodecyl bromide and ethyl bromide at air-water and air-hydrocarbon interfaces showed that in the absence of simple electrolyte the surface activity was very small, suggesting that under similar conditions micelles of ordinary colloidal electrolytes are not significantly adsorbed to air-water interfaces. However, there were sizable depressions of the interfacial tensions between water and heptane and between water and benzene. In all cases the surface activities were significantly enhanced by the addition of simple electrolytes.[13]

CONFORMATIONAL TRANSITIONS

The chemical nature of the groups which lend water solubility to polysoaps is not expected to be critical. These groups may be cationic, anionic or non-ionic. A useful and promising class of anionic macromolecules are the alternative 1-1 copolymers of alkyl vinyl ethers and maleic anhydride. If the alkyl group has about ten or more carbon atoms, these copolymers upon hydrolysis of the anhydride groups, behave as typical polysoaps.[14-16] If the alkyl group has one or two carbon atoms, the copolymers show typical polyelectrolyte behavior.[17,18] The most interesting members of this series lie between these extremes. Specifically, the butyl copolymer has been investigated extensively and found to exhibit upon neutralization with a base, a well-defined transition from a hypercoiled conformation typical of a polysoap to an extended conformation typical of a polyelectrolyte.[3,17] The phenomenon has been investigated by viscosimetry,[17] potentiometric titration,[3,17] solubilization of a water-insoluble dye,[16] and by following the changes in fluorescence of chemically attached optical probes sensitive to the polarity of the effective local solvent environment.[19] The hexyl copolymer undergoes a similar transition which, however, occurs at higher values of the degree of neutralization than that required for the butyl copolymer.[17] These conformational transitions resemble the denaturation of globular proteins whose native state is also stabilized by hydrophobic forces. A typical denaturant for such proteins is urea. Urea has also been shown to destabilize the hypercoiled state of the butyl and hexyl copolymers. The denaturant effect was shown to be caused by an enhancement of the solvent affinities of both hydrophobic and hydrophilic groups upon the addition of urea.[20]

CONCLUSION

The results reviewed here indicate that polymers and hydrophobic polyacids mimic in a number of ways the behavior of both ordinary soaps and proteins. Because of the greater control over their chemical structure that these synthetic macromolecules afford the investigator, they will continue to form a fertile field for future exploration.

REFERENCES

1. U.P. Strauss and E.G. Jackson, J. Polymer Sci., 6, 649 (1951).
2. U.P. Strauss, S.J. Assony, E.G. Jackson and J.H. Layton, J. Polymer Sci., 9, 509 (1952).
3. P. Dubin and U.P. Strauss, J. Phys. Chem., 71, 2757 (1967).
4. U.P. Strauss and N.L. Gershfeld, J. Phys. Chem., 58, 747 (1954).
5. U.P. Strauss, N.L. Gershfeld and E.H. Crook, J. Phys. Chem., 60, 577 (1956).
6. U.P. Strauss and B.L. Williams, J. Phys. Chem., 65, 1390 (1961).
7. D. Woerman and F.T. Wall, J. Phys. Chem., 64, 581 (1960).
8. H. Inoue, Kolloid Z., 195, 102 (1964).
9. E.G. Jackson and U.P. Strauss, J. Polymer Sci., 7, 473 (1951).
10. L.H. Layton, E.G. Jackson and U.P. Strauss, J. Polymer Sci., 9, 295 (1952).
11. U.P. Strauss and L.H. Layton, J. Phys. Chem., 57, 352 (1953).
12. A. Hahne, Z. dtsch. Öl u. Fettind., 45, 245 (1925).

13. H.E. Jorgensen and U.P. Strauss, J. Phys. Chem., $\underline{65}$, 1873 (1961).

14. K. Ito, H. Ono and Y. Yamashita, J. Colloid Sci., $\underline{19}$, 28 (1964).

15. R. Varoqui and U.P. Strauss, J. Phys. Chem., $\underline{72}$, 2507 (1968).

16. P.L. Dubin and U.P. Strauss, "Hypercoiling in Hydrophobic Polyacids" in "Polyelectrolytes and Their Applications", A. Rembaum and E. Sélegny, Editors, pp. 3-13, D. Reidel Publishing Co., Dordrecht, Holland.

17. P.L. Dubin and U.P. Strauss, J. Phys. Chem., $\underline{74}$, 2842 (1970).

18. A.J. Begala and U.P. Strauss, J. Phys. Chem., $\underline{76}$, 254 (1972).

19. U.P. Strauss and G. Vesnaver, J. Phys. Chem., $\underline{79}$, 2426 (1975).

20. P. Dubin and U.P. Strauss, J. Phys. Chem., $\underline{77}$, 1427 (1973).

SOLUBILIZATION BY NONIONIC SURFACTANTS IN THE HLB-TEMPERATURE

RANGE

Stig Friberg

Department of Chemistry, UMR

Rolla, MO 65401

Irena Buraczewska

The Swedish Institute for Surface Chemistry

Stockholm, Sweden

Jean Claude Ravey

Universite de Nancy

Nancy, France

A general pattern is presented of the solubilization of hydrocarbons in nonionic surfactants and water in HLB-temperature range. The separation of a third phase, the surfactant phase, with increasing temperatures gives a rational explanation to the ultra-low interfacial tensions between the water and the hydrocarbon solutions, which has been observed in these systems. Light scattering determinations gave evidence of a new kind of micellar aggregation in the liquid emulsifiers with added water. In contrast to the regular inverse micelles these aggregations show marked anisotropic properties.

INTRODUCTION

The solubilization of hydrocarbon and water by nonionic surfactants is characterized by a high sensitivity towards temperature and towards the kind of hydrocarbon involved. Figures 1 and 2 are good illustrations of the pronounced changes of solubilization brought about by altering the nature of the hydrocarbon.

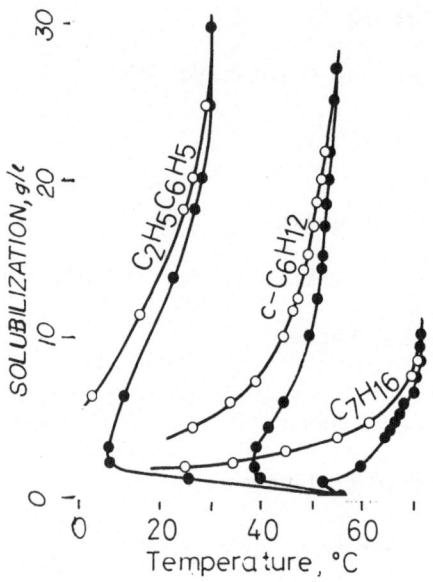

Figure 1. The solubilization of hydrocarbon by nonionic surfactants is sensitive to temperature and nature of the hydrocarbon. (C-C_6H_{12} = cyclohexane)

According to Shinoda, however, the complex behaviour may be systematized using temperature as the variable. Each combination of surfactant and hydrocarbon is characterized by a narrow temperature range, the HLB-temperature[1-8]. This temperature was originally called the Phase Inversion Temperature (PIT) and is operationally defined as follows. Below the HLB-temperature range a nonionic surfactant of the polyoxyethylene glycol alkyl ether type is preferentially soluble in water; above it in hydrocarbon. In the HLB-temperature range a new phase is often experienced. It is called the surfactant phase, and may concurrently be in equilibrium with both an aqueous and an oil phase. The conditions may be described according to Figure 3 which is a schematic summary of Shinoda's pioneering contributions to the field. The figure is essentially self-explanatory.

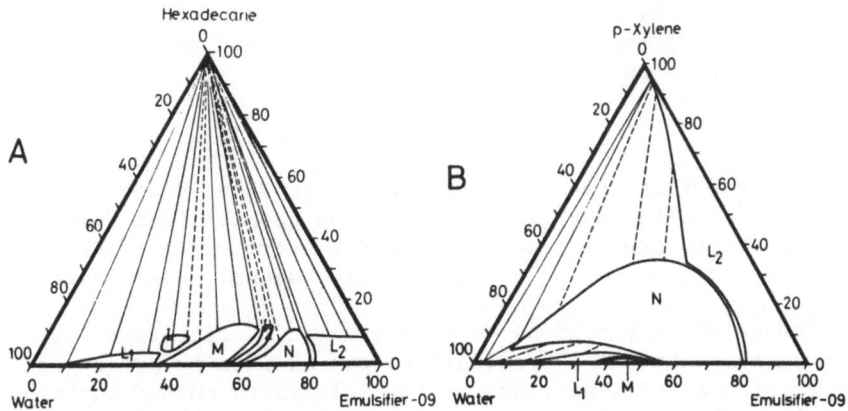

Figure 2.   The phase regions in a system water, nonionic surfactant and hydrocarbon may drastically change when an aliphatic
hydrocarbon (A) is exchanged with an aromatic one (B).

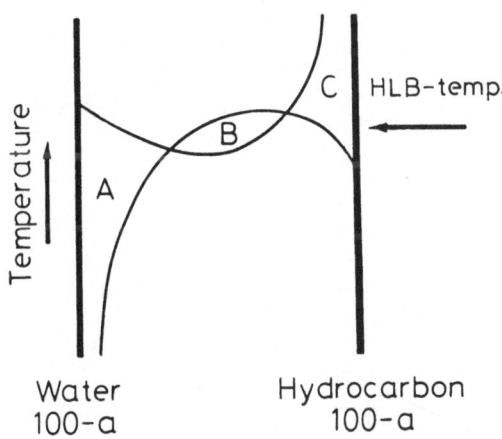

Figure 3.   At temperatures below the HLB-temperature hydrocarbon
is solubilized in water; above it water is solubilized into the
hydrocarbon.

Since the nonionic surfactants are of pronounced importance for the tertiary oil recovery, due to extremely low interfacial tension between the surfactant phase and the oil or the aqueous phase respectively we have considered it of value to contribute to the clarification of the different phases and their compositional variation in the HLB-temperature range.

EXPERIMENTAL

The materials used were pure polyoxyethylene dodecyl ethers, $C_{12}(EO)_n$, (n is the number of oxyethylene groups) from Nikkon Co., Japan (gas-chromatographically checked), hydrocarbon of the highest purity obtainable and twice distilled water.

The solubility regions were determined by titration with one of the components and the solubility was checked by prolonged storage of compositions close to the solubility border on both sides.

RESULTS

At temperatures not far below the HLB-temperature range the conditions are illustrated by Figure 4. The surfactant was

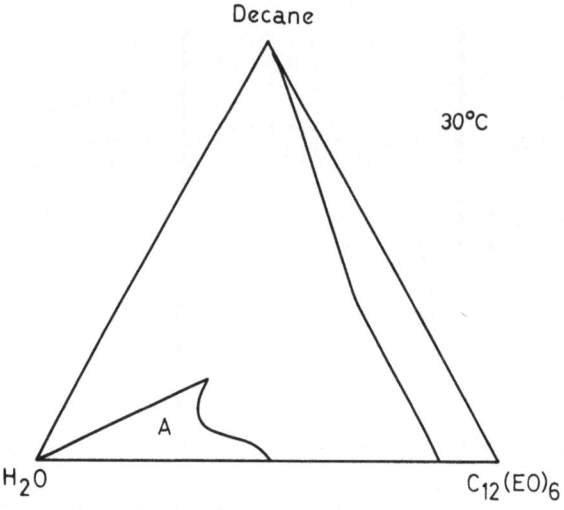

Figure 4. At temperatures below the HLB-temperature the surfactant ($C_{12}(EO)_6$=hexaoxyethylene dodecyl ether) is soluble in water forming normal micelles which solubilize hydrocarbon (A). Water is soluble in the decane/surfactant solution approximately in proportion to the amount of emulsifier. t = 30°C

soluble in water and it solubilized hydrocarbon. It is reasonable
to expect the presence of normal micelles in the aqueous area and
the contributions on the thermodynamics of micellization of
nonionics refer to these conditions.

As the temperature is increased a new solubility area (A) is
developed parallel to the aqueous micellar solution (B) (Figure
5A). The solubility area (A) covered a sectorial region from the
aqueous corner; this means that this kind of solution is stable
only for a certain range of hydrocarbon/emulsifier ratios and that
a minimum hydrocarbon/emulsifier ratio was necessary for its
stability.

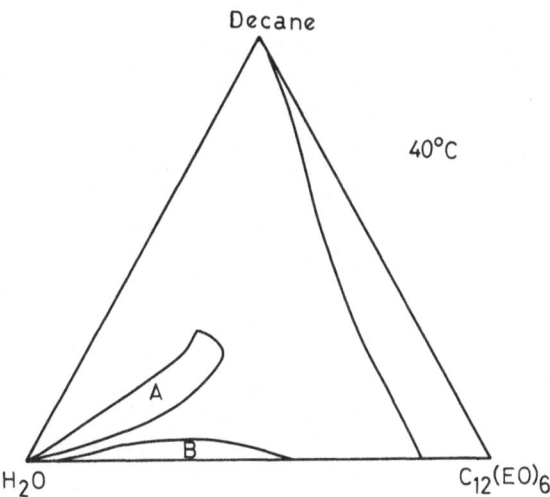

Figure 5. When the temperature approached the HLB-region
separate solubility areas (A) and (B) were formed from the aqueous
micellar solution. t = 40°C.

At higher temperatures (Figure 6) the solubility of the
surfactant in water disappeared entirely and an isolated area, a
surfactant phase (7), was formed. The minimum content of water
in water/emulsifier solution (B) was further reduced. Corresponding phenomena but at lower temperature were found in the
system water tetra-oxyethylene dodecyl ether ($C_{12}(EO)_4$) and
hexadecane (Figure 7). At higher temperatures the surfactant
phase solubility region moved towards higher emulsifier and
hydrocarbon content (Figure 8 A,B) and a coalescence took place
with the hydrocarbon/surfactant solution. A narrow solubility

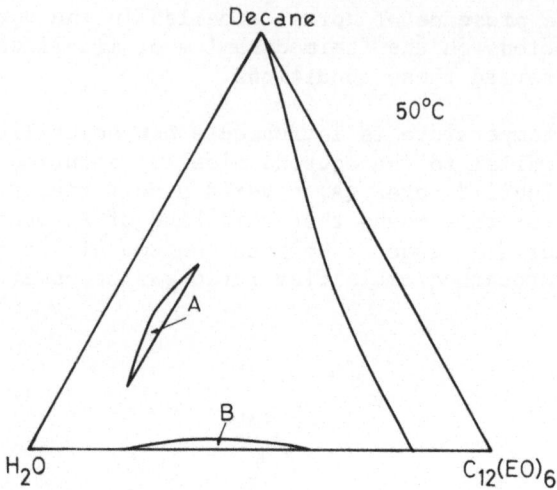

Figure 6.   At the HLB-temperature the amount of emulsifier needed
to form the surfactant phase (A) was a minimum.   t=50°C.

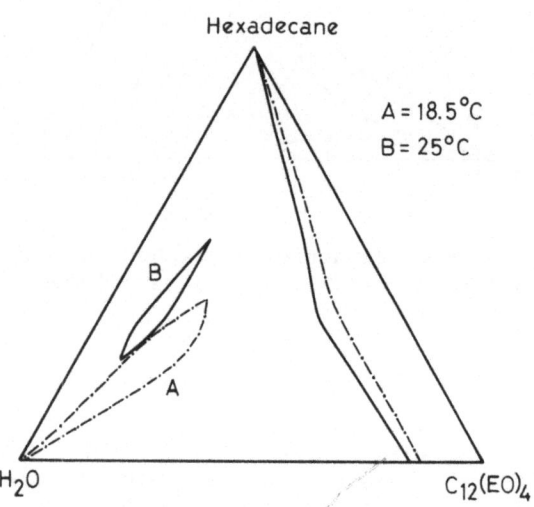

Figure 7.   A corresponding change with temperature was found for
the system water, tetra-oxyethylene dodecyl ether ($C_{12}(EO)_4$) and
hexadecane.
—.—.—.—   t=18.5°C   ————————   t=25°C.

channel reaching in the direction of the water corner was found
(Figure 8,C). With increasing temperature this channel required
higher emulsifier/hydrocarbon ratios. In some cases the solubility area was formed according to Figure 9 leaving a solubility
gap for certain water/emulsifier ratios. At high temperatures
the solubility region of hydrocarbon always assumed the shape as
in Figure 10.

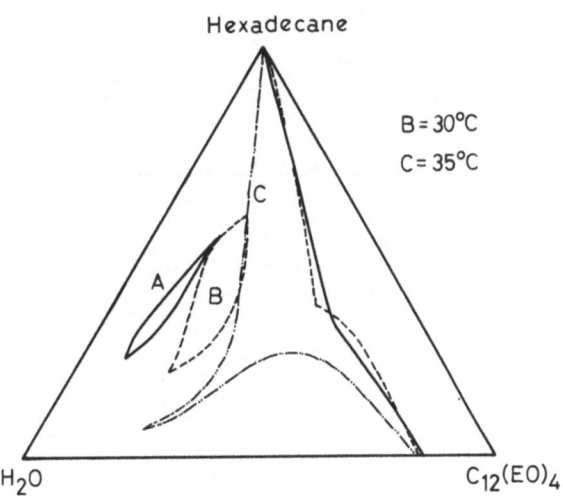

Figure 8. At temperatures above the HLB-temperature the
surfactant phase (A) was at first transferred towards higher
emulsifier concentrations and finally a coalescence with the
emulsifier/hydrocarbon took place (C). ———— t=25°C
----- t=30°C   .—.—.— t=35°C.

DISCUSSION

The results show the early interpretations by Shinoda to be
correct; at low temperatures the surfactant is water-soluble and
at high temperatures it is soluble in hydrocarbon. Figures 2 and
10 illustrate this behaviour.

However, the behaviour in the HLB-temperature range is
interesting and deserves further comments both from the point of
view of solubility and of interfacial energies.

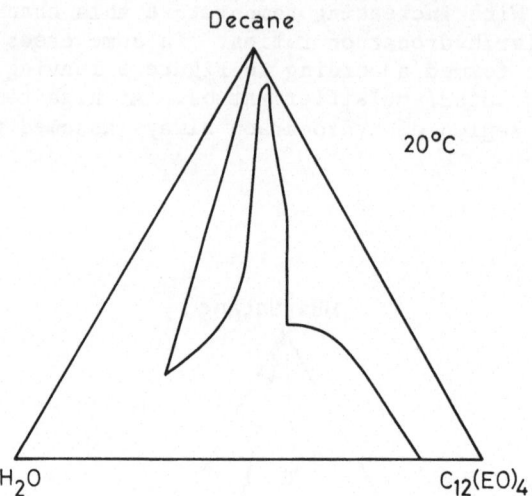

Figure 9.   For some systems in excess of but close to the HLB-temperature the water solubilization took place in a sector from the hydrocarbon corner.   (Cf. Figure 5 and 7).

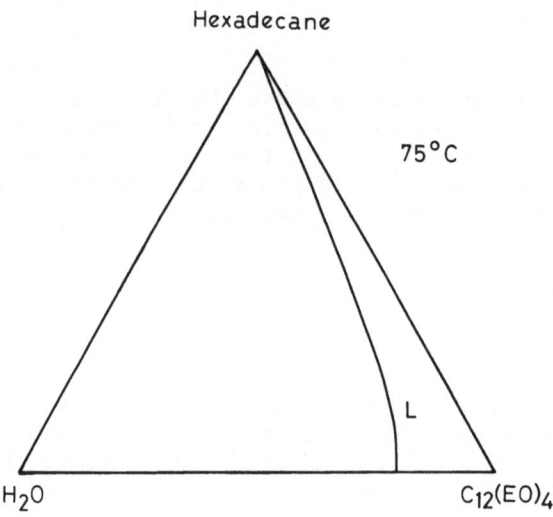

Figure 10.   At temperatures far in excess of the HLB-temperature the hydrocarbon/surfactant solution is the only one present.

Solubilization

At temperatures close to but below the HLB-temperature the surfactant was not soluble in water (Figure 5B) meaning temperatures in excess of the cloud-point. However, certain ratios of hydrocarbon to emulsifier gave solubility (Figures 5,A and 7,A). A corresponding solubility sector was detected from the hydrocarbon corner at higher temperatures (Figure 9). This latter sector has earlier been recognized and investigated as the "second solubilization"[9] although the connection with the surfactant phase has not been realized.

It is tempting to interpret these solubilization regions using the bending or splay energy concept developed by Shinoda[10], Prince[11] and Robbins[12] and rigorously treated by Murphy[13]. The splay energy concept essentially[13] implies the negative of the change in interfacial energy per unit interfacial area with change in the mean curvature at constant area and gaussion curvature. It is important for small interfacial energies and small particles, e.g. micellar systems.

Using the bipartite model of Shinoda and Prince the solubility sectors from the water and hydrocarbon corner may be interpreted as arising from the minimization of splay energy. Too small hydrocarbon emulsifier ratios (Figures 5,A and 6,A) should give too large an angle between the emulsifier molecules, an increase of the interfacial splay energy and instability. The solubility conditions in Figure 9 may be similarly explained exchanging the role of hydrocarbon and water.

It would be tempting to continue the analogy to the maximum water solubility in the normal micellar solution (Figures 5,A, 6,B) discussing autosolubilization of the surfactant. An assumption of the surfactant serving both as solubilizing agens and being solubilized leads to difficulties for the explanation of the solubility deficit for certain hydrocarbon/emulsifier (Figure 5) and water/emulsifier (Figure 9) ratios.

The form of the emulsifier/hydrocarbon solution to the right (Figures 4, 5, 6, 7) does not give reason to expect the presence of inverse micelles; the solubility of water approximately being proportional to the amount of emulsifier. The studies of Kitahara[14] showed the nonionic surfactants not to form micelles in various solvents.

Preliminary results of light-scattering[15] indicated association structures to be present. The investigations included separate determinations of the two components H_v and V_v and Figure 10 shows the variations of these components with water

content for three hydrocarbon/emulsifier ratios. The sudden
increase of the intensity of both components at a certain concentration of water and the reduction of the H_v/V_v ratio with higher
amounts of hydrocarbon are evident. A reasonable interpretation
of the results is the formation of anisotropic aggregates in the
solution of water in the emulsifier.

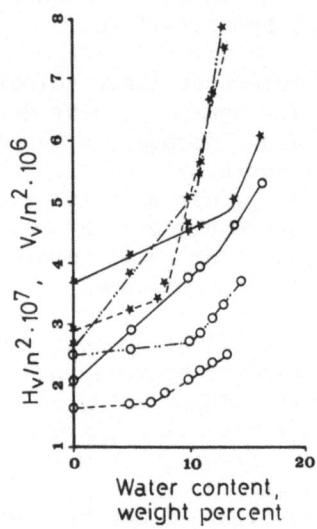

Figure 11. The two components H_v and V_v of the polarized light
indicated the formation of anisotropic aggregates. O V_v, ★ H_v;
- - - - - $C_{10}/C_{12}(EO)_4 = 0$; ---..--- $C_{q0}/C_{12}(EO)_4 = 3/7$;
————— $C_{10}/C_{12}(EO)_4 = 1.5$ (W/W).

A detailed interpretation will be given elsewhere[15], but one
essential conclusion may be drawn. The association structures in
this solution are different from the inverse micelles found at
higher temperatures in the narrow solubilization region (Figure
8,C). These (1) are characterized by an optimal hydrocarbon/
emulsifier ratio; the associations presented in this article are
most pronounced when no hydrocarbon is present and are of the
nature of water/emulsifier associations. It is tempting to
describe the aggregates as preassociation structures in the
isotropic liquid prior to the formation of a lamellar liquid
crystal.

Interfacial tension

No interfacial tension was determined but the present results may contribute to the understanding of the tertiary oil recovery systems[16].

Figures 5,A-B and 6,A-B show the surfactant phase to separate from the aqueous corner and the initial two-phase system is changed to a three-phase system, the third phase being the surfactant phase. The interfacial tension is drastically reduced with temperatures reaching extremely small values when the surfactant phase is formed[10]. This fact is of great importance for tertiary oil recovery where interfacial tensions of the magnitude 10^{-4} mN/m are required. In practice such low interfacial tensions have been found when the third phase, the surfactant phase, is formed within the system. A quantitative theory has recently been worked out[17] to relate low interfacial tension to phase separation and micellar size.

REFERENCES

1. K. Shinoda and T. Ogawa, J. Colloid Interface Sci. 24, 56 (1967).
2. H. Saito and K. Shinoda, J. Colloid Interface Sci. 24, 10 (1967).
3. H. Saito and K. Shinoda, J. Colloid Interface Sci. 32, 649 (1970).
4. K. Shinoda, J. Colloid Interface Sci. 34, 278 (1970).
5. K. Shinoda, J. Colloid Interface Sci. 24, 4 (1967).
6. H. Saito and K. Shinoda, J. Colloid Interface Sci. 35, 359 (1971).
7. S. Friberg and I. Lapczynska, Progr. Colloid & Polymer Sci. 56, 16 (1975).
8. J.B. Brown, I. Lapczynska and S. Friberg, in Proceedings, Intern. Conf. Colloid Surface Sci., Budapest 1975, Vol. 1.
9. S.G. Frank and G.J. Zogrofi, J. Colloid Interface Sci. 29, 27 (1969).
10. K. Shinoda and S. Friberg, Adv. Colloid Interface Sci. 4, 281 (1975).
11. L.M. Prince, J. Colloid Interface Sci. 23, 165 (1967).
12. M. Robbins, "The Theory of Microemulsions", paper presented at the Nat. A.I.Ch.E. Meeting, Tulsa, Oklahoma, March 1974.
13. C. Murphy, Thesis, University of Minnesota, 1966.
14. K. Kou-no and A. Kitahara, J. Colloid Interface Sci. 35, 636 (1970).
15. J.-C. Ravey (1976), to be published.
16. R.N. Healy and R.L. Reed, Soc. Pet. Eng. J. 14, 491 (1974).
17. C. Miller (1976), personal communication.

THE EFFECT OF LYSOPLASMALOGEN ON SOME PHYSICAL PROPERTIES OF

DIPALMITOYLLECITHIN BILAYERS: A FLUORESCENT PROBE STUDY

D.A.N. Morris and J.K. Thomas

Department of Chemistry and Radiation Laboratory*

University of Notre Dame, Notre Dame, Indiana 46556

The effect of lysoplasmalogen on some physical
properties of dipalmitoyllecithin bilayers has been
determined by fluorescence probe techniques. The
solvent dependencies of the fluorescence spectra
(pyrene and pyrene-3-aldehyde) in simple solvents have
been established. By using the established properties
of the probes, information is obtained on the nature of
the mixed lipid systems. A dramatic increase in the
polarity of the probe environment was observed on
increasing the lysoplasmalogen up to 40 mole percent.
Above 50 mole percent lysoplasmalogen, the probe environment is relatively non-polar and insensitive to lysoplasmalogen content. Kinetic features in these mixed
systems also depend on the composition of the mixture.
It is suggested that the lyso-compounds disrupt the
lecithin bilayer and cause entry of water into the
paraffin or hydrocarbon-like core of the vesicle bilayers.
Above 50 mole percent of lysoplasmalogen the system
breaks up into small mixed micelles rather than the
mixed lysoplasmalogen/ lipid bilayer system. Water is
again excluded from the micelles and the kinetic and
fluorescent properties of the system tend to return to
those noted in the pure lecithin bilayer, or lysoplasmalogen micelle.

INTRODUCTION

It is generally accepted that the "lyso compound" content of membranes has implications for the stability of the cell interface[1] Increased lysolecithin concentrations favor conditions for cell fusion[2,3] and excess of exogenously added lysolecithin causes lysis. The lytic process has been described in terms of subtle changes in the bilayer orientation causing increase in cation permeability. The disturbance of the Donnan equilibrium then causes entry of water into the cell which bursts by osmotic force.[4,5] Another explanation of the effect of lyso compounds proposes that the thickness of the paraffin core is critical in order to act as an effective permeability barrier. X-ray studies have demonstrated a decrease in bilayer thickness of egg lecithin with increasing lysolecithin content. It has been suggested that decreasing thickness of the paraffin core causes a decrease in interaction between the lipid molecules. If the bilayer thickness decreases below a critical value a stable association of the lipid molecules is no longer warranted.[2,6,7]

Fluorescence probe techniques have been used to study lipid-protein interactions in black lipid membranes[8] and features of factors controlling the permeability of simple micelles[9,10] and phospholipid dispersions.[11] The fluorescence probes pyrene and pyrene-3-aldehyde have been shown to be sensitive to solvent polarity.[12,13]

In the present work the effect of lysoplasmalogen on dipalmitoyl lecithin bilayers is investigated by the aforementioned fluorescence probe techniques. In particular, these techniques have been used to study the effect of added lyso compounds on the permeability and polarity of the paraffin core environment. Fluorescence probe techniques also permit an accurate determination of the critical micelle concentration of lysoplasmalogen.

EXPERIMENTAL SECTION

Materials Dipalmitoyl lecithin, dilauryl lecithin, and lyso-plasmalogen were purchased from Calbiochem. Pyrene was obtained from Fluka and was purified on a silica column. Pyrene-3-aldehyde was obtained from Aldrich and recrystallized from 95% ethanol. Nitromethane was obtained from Eastman organic chemicals.

Sample Preparations The lecithins (2mM) were dispersed in water by ultrasonication of the deoxygenated samples with a Branson sonifier for five minutes at power setting 3. Lysoplasmalogen (2mM) was dispersed in water by mixing. Pyrene (or pyrene-3-aldehyde) was deposited on the vessel wall by evaporation of pyrene/chloroform

solution, and sonicated with added lecithin or lysoplasmalogen
dispersions by means of a Bendix ultrasonic bath at 50°C for 20
minutes. Measured amounts of dipalmitoyl lecithin and lysoplasmalogen, each containing the fluorescent probe, were mixed and
sonicated with the Bendix sonifier for 15 minutes at 50°C to form
mixed lipid systems. Solutions were bubbled with N_2 before soni-
cation and prior to fluorescence measurements.

<u>Apparatus</u> Laser photolysis experiments were carried out with a
Korad K1Q ruby laser with a frequency doubler. The wavelength of

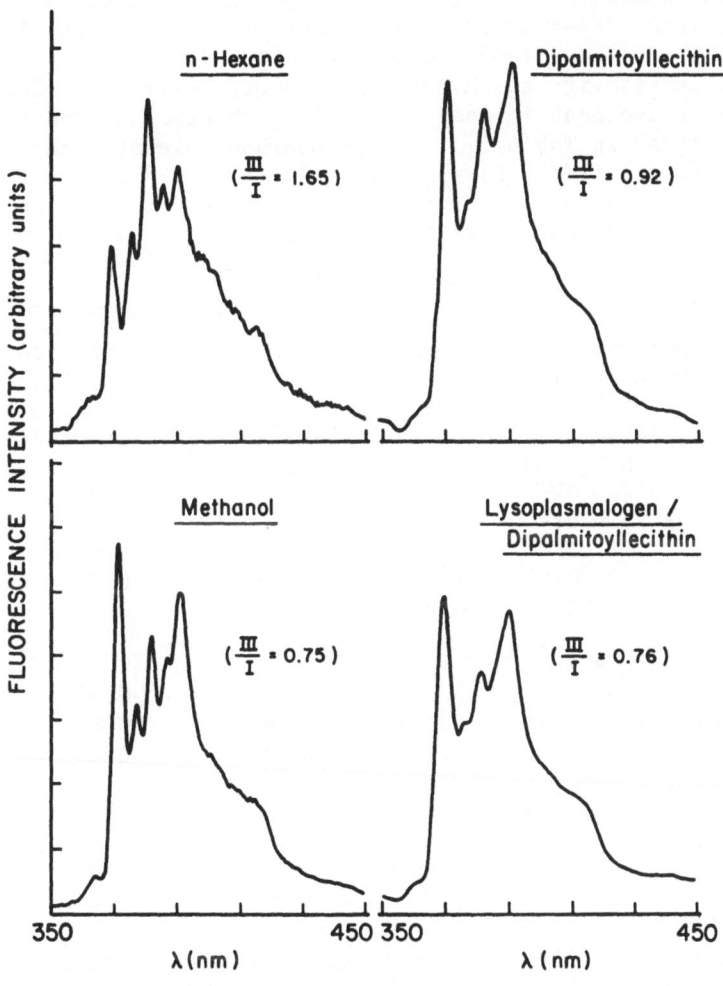

Figure 1. Fluorescence spectrum of pyrene in solvents and lipid
dispersions.

the light was 347.1 nm, and the pulse width was 20 ns, with an
energy of 0.1J per pulse. Steady state fluorescence measurements
were performed on an Aminco-Bowman spectrophotofluorometer with
excitation wavelengths of 330 and 365 for pyrene and pyrene-3-
aldehyde, respectively.

RESULTS

Fluorescence Emission Spectrum of Pyrene

The fluorescence spectra of pyrene in n-hexane and methanol
solvents and in lecithin and lecithin/lysoplasmalogen dispersions
are shown in Figure 1. The ratios shown in the figure are the
relative intensities of the middle vibronic band to the highest
energy vibronic band (0-0 transition). The III/I ratio showed the
greatest sensitivity to changes in solvent. The solvent dependence
of the relative peak intensities of the vibrational fine structures
are summarized in Table I. In hydrocarbon solvents the relative
peak intensity, III/I, is approximately 1.7, in aromatic solvents
the value is 0.8 to 0.9 and in polar solvents about 0.5 to 0.8. The
III/I value of pyrene in water is 0.62.

Table I. Solvent Dependencies of Vibrational Fine Structures
of Pyrene Monomer Fluorescence.*

Solvent	Solvent D.M.	ϵ	Relative Peak Intensities I	III
Simple Polar Solvents				
Acetone	2.88	20.70	1.00	0.68
Methanol	1.70	32.70	1.00	0.75
Chloroform	1.01	4.80	1.00	0.78
Aromatic Solvents				
Benzene	0	2.28	1.00	0.88
Benzyl alcohol	1.71	13.10	1.00	0.82
Toluene	0.36	2.37	1.00	0.90
Hydrocarbon Solvents				
Iso-Octane			1.00	1.68
n-hexane		1.89	1.00	1.65
Dodecane		2.01	1.00	1.67

*Partial list of solvents from K. Kalyanasundaram study.[12]

An example of the association of pyrene with lysoplasmalogen micelles but not with the monomers may be inferred from Figure 2, which shows the dependence of the III/I ratio on lyso plasmalogen concentration. In water the value is 0.62, and in $2x10^{-5}$M lyso the value is only slightly greater at 0.63. At concentrations of $4x10^{-5}$ and $6x10^{-5}$M lyso, aggregates of the lyso-monomers may occur and association of pyrene with these aggregates results in an increase in the III/I ratio. At lyso concentrations of $8x10^{-5}$M and greater the III/I ratio is maximal and independent of concentration. If one assumes that this invariance of III/I ratio with concentration results from the presence of uniform micelles whose size is independent of concentration, then the concentration of $8x10^{-5}$M lyso represents the critical micelle concentration of lysoplasmalogen in water. This value compares favorably with the reported value of $<12x10^{-5}$ obtained by the light scattering technique for lysphosphatidyl choline.[14,15]

The III/I ratio of pyrene is 0.86 in lysoplasmalogen and about 0.92 in both dilauryl lecithin and dipalmitoyl lecithin vesicles as shown in Table II. The results indicate that the probe environment in the lipid aggregates is less polar than that in water, but not quite as non-polar as that in pure hydrocarbon solvents. The reported differences in bilayer thickness in dilauryl- and dipalmitoyl lecithin[2] vesicles do not appear to affect the observed polarity of the probe environment.

Mixing of lecithin and lysolecithin leads to the formation of mixed lipid aggregates. For mole ratios of lysolecithin to lecithin up to 0.5 these mixed aggregates are reported to be of the bilayer

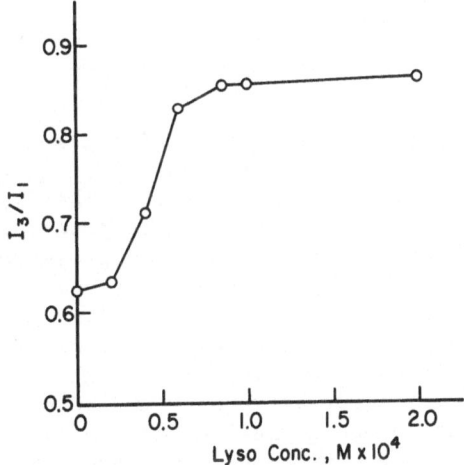

Figure 2. Dependence of pyrene vibrational structure band intensities on lysoplasmalogen concentration.

type. The bilayer thickness shows a decrease with increasing lyso
content, and a corresponding increase in potassium ion permeability
up to 25 mole % lyso.[2] At lysolecithin/lecithin mole ratios
greater than 0.5 smaller micelles are formed.[2]   An examination of
the III/I ratio of the pyrene fluorescence emission in lysoplas-
malogen/lecithin mixtures indicates a sharp increase in polarity of
the probe environment over the range 0-30% mole percent lyso. The
results are shown in Figure 3. The III/I ratios for pyrene in
dilauryl lecithin and dipalmitoyl lecithin are approximately
equal. Bilayers formed by these lipids are significantly different
in thickness, suggesting that the increase in polarity shown in
Figure 3 is not simply due to a change in bilayer thickness with
increasing lyso- content.

Table II. Vibrational Fine Structures of Pyrene* Monomer
Fluorescence in Lipids.

Lipid	Relative Peak Intensities	
	I	III
Lysolecithin	1.00	0.86
Lysoplasmalogen	1.00	0.86
Dilauryl lecithin	1.00	0.91
Dipalmitoyl lecithin	1.00	0.92
**H_2O	1.00	0.62

*10 μm pyrene in 2mM lipid in water.
**1 μM pyrene added to H_2O.

The results also suggest that pyrene remains associated with the
lipid aggregate as the III/I values remain much greater than the
value of 0.62 observed for pyrene in water. The lysoplasmalogen
disruption of lecithin bilayers may lead to penetration of water
into the disrupted hydrocarbon region. The presence of water in
the bilayer may result in an increase in the polarity of the probe
environment. The discontinuity at 30 mol % lyso may be due to a
sudden change in the properties of the paraffin core prior to
breakdown of the vesicles.   The vesicle breakdown into small
micelles is expected to occur between 40 and 50 mol % lyso.[2] In
the concentration region where small mixed micelles are expected,
the polarity of the probe environment decreases gradually with
increasing lyso content to a III/I ratio of 0.86 in pure lyso
solution.

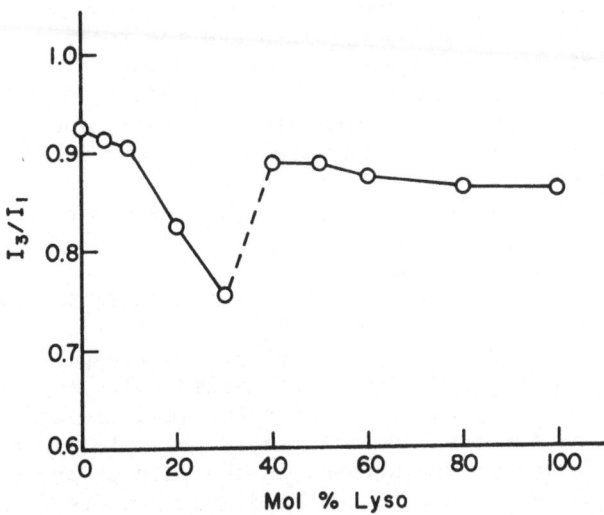

Figure 3. Relative peak intensities (III/I) of pyrene fluorescence vibrational bands _versus_ mol % of lysoplasmalogen and dipalmitoyl lecithin. 5×10^{-6}M pyrene, 2×10^{-3}M total lipid.

Fluorescence Lifetime of Pyrene

The critical micelle concentration (CMC) of lysoplasmalogen in water may be also determined by measuring the pyrene fluorescence lifetime as a function of lipid concentration. The first order decay rate of pyrene monomer fluorescence is shown as a function of lysoplasmalogen concentration in Figure 4a. The increase in decay rate with increasing lyso- concentration below the CMC may be due to quenching of the excited state of pyrene by the choline moiety of lysoplasmalogen. At lysoplasmalogen concentrations of 1×10^{-4}M and greater, there is a dramatic decrease in the decay rate to a value which becomes independent of lipid concentration. The dramatic decrease in decay rate may be attributed to micellization of lyso-plasmalogen with incorporation of pyrene into the micelles. Pyrene solubilized within the micelles is protected from the quenching effects of the hydrophilic choline head groups. At lyso concentrations below the CMC, pockets of loosely associated aggregates of lyso-plasmalogen may be attractive solubilization sites for pyrene. Encounters of the excited state of "unprotected" pyrene with choline groups of the lysoplasmalogen monomers, or un-micellized aggregates would lead to fluorescence quenching. The results of Figure 4a indicate that the onset of micellization occurs at 1×10^{-4}M lysoplasmalogen.

    The dependence of 1/k versus lysoplasmalogen concentration is
shown in Figure 4b. The ordinate values are the reciprocals of the
values shown in Figure 4a and have the units of seconds. The data
represented in Figure 4b, therefore, show "lifetime" of pyrene
fluorescence decay versus lysoplasmalogen concentration. The
fluorescence lifetime of pyrene in water, not shown in Figure 4b,
is about 140 nanoseconds. The fluorescence lifetime of pyrene in
the lysoplasmalogen micelles is about 280 ns, as shown in Figure
4b.

    The fluorescence lifetime of pyrene in mixed lipid systems is
shown in Figure 5. In the region previously noted for dramatic
increases in probe polarity [see Figure 3], a corresponding dramatic
decrease in the fluorescence lifetime is noted which may be related
to an increase in water penetration into the vesicles and concomittant
quenching of pyrene fluorescence. A second possible explanation of
the decreased lifetime is an increase in the rate of pyrene migration
through the disrupted bilayer to the head groups where fluorescence
quenching may occur. However it appears unlikely that pyrene which
is soluble in the hydrocarbon region of pure lecithin bilayers and
lysoplasmalogen micelles should prefer the head-group choline
regions in the mixed lipid system.

    In the region of small mixed micelles viz. lyso greater than
50 mol %, the pyrene fluorescence lifetime is independent of lyso
concentration. This is analogous to the constancy of the probe
polarity environment shown earlier in Figure 3.

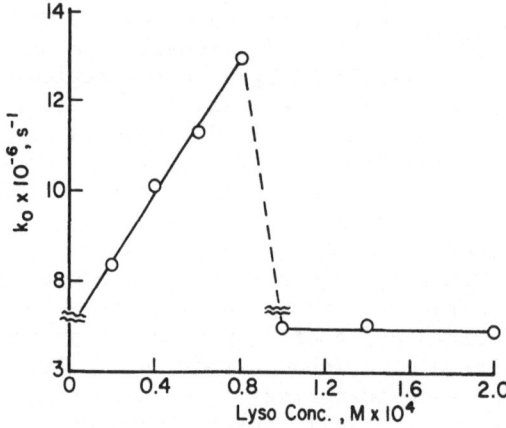

Figure 4a. Dependence of pyrene monomer fluorescence decay rate
on lysoplasmalogen concentration.

In summary, the pyrene fluorescence lifetime initially decreases dramatically with increasing lyso content. The observed effect can be attributed to the entry of water into the bilayer to quench the excited probe. The fluorescence lifetime of pyrene, and the polarity of the probe environment, is relatively insensitive to lyso content greater than 50 mol % lyso.

Pyrene Fluorescence Quenching by Nitromethane

The quenching kinetics of pyrene fluorescence in micelles by nitromethane has been discussed elsewhere.[16] The quenching rate constant is a measure of the rate of entry of the quencher molecule into the micelle where the fluorescence probe resides. The quenching rate constant is thus a measure of the permeability of the hydro-carbon region to the quencher molecule. The quenching rate constants k_q for nitromethane reacting with pyrene in dipalmitoyl lecithin (C_{16}) and dilauryl lecithin (C_{12}) vesicles are $5 \times 10^7 M^{-1} s^{-1}$ and $3.6 \times 10^8 M^{-1} s^{-1}$, respectively. The difference in the values suggests that the permeability of nitromethane is influenced by the thick-ness of the bilayer. The quenching rate constants in lysopalmitoyl choline (C_{16}) and lysoplasmalogen are identical, and equal to

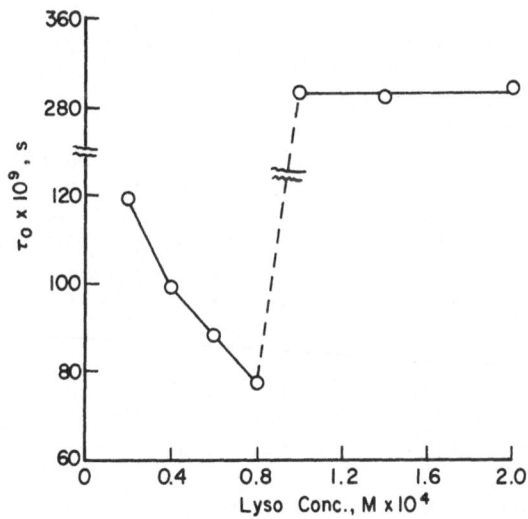

Figure 4b. Dependence of pyrene monomer fluorescence radiative lifetime on lysoplasmalogen concentration.

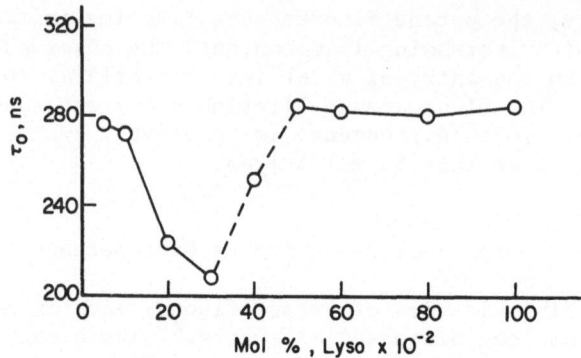

Figure 5.   Pyrene fluorescence lifetime versus mol % of lysoplasmalogen and dipalmitoyl lecithin. $5x10^{-6}$M pyrene, $2x10^{-3}$M
total lipid.

$5x10^8 M^{-1}s^{-1}$.

   The effect of disruption of the dipalmitoyl lecithin bilayer
by lysoplasmalogen is shown in Figure 6.   The permeability of
nitromethane increases sharply with lyso content up to 40 mol %
lyso.   This increase may be attributed to changes in permeability
properties of the paraffin core due to increased water content.
However, the reported decrease in bilayer thickness may also be
considered a factor,   A discontinuity in quencher penetration is
observed at 50 mol %, and an increase in quencher penetration
efficiency with increasing lyso content appears above 50 mol %
lyso.   This dependence on lyso content at higher concentrations is
in contrast to the insensitivity of the polarity and pyrene fluorescence lifetime shown in Figures 3 and 5. The results may be
reconciled by considering that the quenching results of Figure 6
depict the ability of the quencher to diffuse into the vesicles or
micelles and react with the probe, whereas the results of Figures 3
and 5 depict the static condition of the environment at the site of
the probe.   The discontinuity of Figure 6 may depict the change of
lipid structure from vesicle to micelle.

   In summary, the increased permeability of the mixed lipid
vesicles to nitromethane may be due to the reported decrease in
bilayer thickness.   The reported thickness of dilauryl lecithin is
equal to the thickness of a mixed lipid vesicle containing 37 mol %
lyso in dipalmitoyl lecithin.[2]   The quenching rate constants of
nitromethane with excited pyrene in these two systems are $3.6x10^8$
and $3.8x10^8 M^{-1}s^{-1}$, respectively.   Mixed micelles formed at lyso
concentrations greater than 50 mol % lyso may be expected to

decrease in size with increasing lyso content. Such a decrease in
size would account for the increase in efficiency of nitromethane
penetration shown in Figure 6.

Fluorescence of Pyrene-3-Aldehyde

Pyrene-3-aldehyde is only weakly fluorescent in hydrocarbons
at room temperature but fluoresces strongly in ethanol or acetic
acid.[13] Generally in solvents of increasing dielectric constant the
quantum yield of monomer fluorescence increases and the fluorescence
maximum is shifted to longer wavelengths. It has been shown that
the fluorescence quantum yield ability of an aromatic aldehyde is
determined by the difference between the energies of the n→π* and
π→π* singlet states.[13] The magnitude of the energy difference
between these weakly fluorescent, n→π* and fluorescent π→π* states
may be changed by solvent interaction. It has been suggested that
the lowering of the fluorescent π→π* energy results from solvent
relaxation by reorientation of solvent dipoles around the excited
state of the probe molecule.[13] The π→π* energy of pyrene-3-aldehyde
is lowered in solvents of increasing polarity. Concurrent work in
this laboratory has demonstrated a direct relationship between the
lowering of π→π* energy and solvent dielectric constant.[12] The
π→π* energy represented as fluorescence maximum is shown in Table

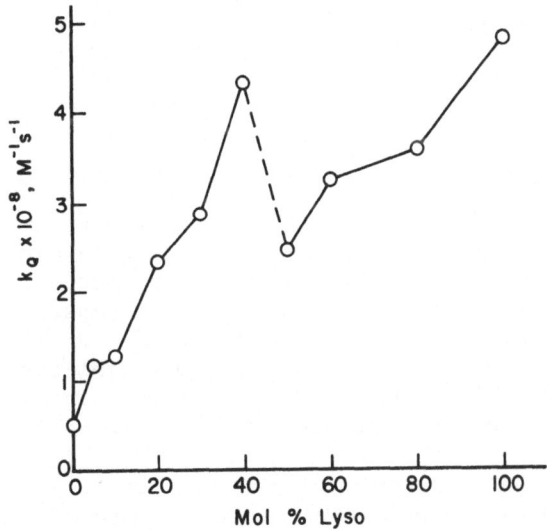

Figure 6. Quenching efficiency of pyrene by added nitromethane
versus mol % lysoplasmalogen and diplamitoyl lecithin.
1×10^{-5}M pyrene, 2×10^{-3}M total lipid.

III for a number of solvents and solvent mixtures of known dielectric constant. The fluorescence maxima of the probe dissolved in dipalmitoyl lecithin and lysoplasmalogen aqueous dispersions are indicated and extrapolated values of the dielectric constant are shown.

Table III. Variation of Pyrene-3-aldehyde Fluorescence Maximum with Solvent Polarity.

Solvent System	ε	Observed λ_{max}, nm
2-propanol	18	438
1-propanol	20.3	442
ethanol	24.6	444
90% ethanol-H_2O	29	448
methanol	33	450
90%methanol-H_2O	36	451
ethylene glycol	38	454
40% dioxane-H_2O	43	456
glycerol	43	456
60% methanol-H_2O	52	457
20% dioxane-H_2O	61	464
20% ethanol-H_2O	69	467
10% ethanol-H_2O	74	471
water	79	474
Dipalmitoyl letichin (DPL)	29*	447
Lysoplasmalogen (LP)	20*	442.5
50 mol % LP/DPL	49*	457

*Estimated from best fit ε vs. λ_{max} relationship.

In dipalmitoyl lecithin, the probe environment is the ordered crystalline array of the hydrocarbon chains which have been shown by NMR to be relatively immobile at room temperature. The single hydrocarbon chains of lysoplasmalogen are by comparison relatively mobile at room temperature. The dependence of fluorescence maxima on dielectric constant indicates an apparent dielectric constant of 29 in DPL and 20 in LP. The greater value of ε in DPL may be due to a greater influence of bound water on the probe.

The behavior of pyrene-3-aldehyde fluorescence maximum in mixed lysoplasmalogen/dipalmitoyl lecithin systems is shown in Figure 7. The values of dielectric constant shown at the right ordinate were extrapolated from the relationship between fluorescence maximum and dielectric constant shown in Table III. The apparent dielectric constant of the probe environment is dramatically

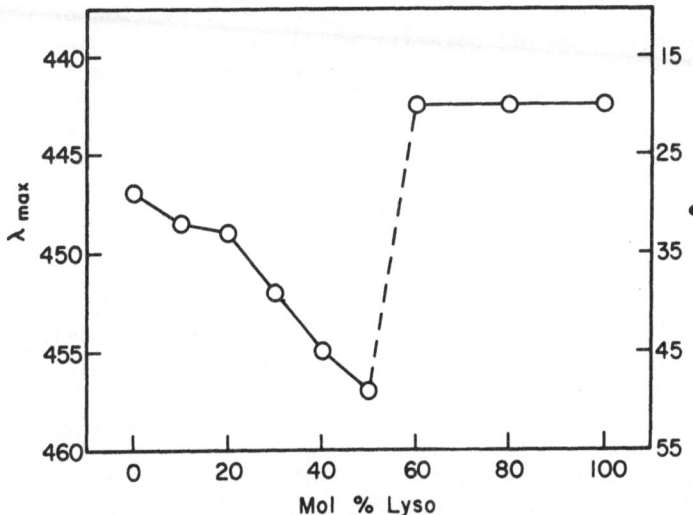

Figure 7. Fluorescence maximum of pyrene-3-aldehyde versus mol %
lysoplasmalogen and dipalmitoyllecithin. Values of dielectric
constant extrapolated from Table III. 1×10^{-5} M pyrene, 2×10^{-3} M total
lipid.

increased to values much greater than that experienced in either
pure lecithin or lysoplasmalogen dispersions. Above the discontinuity
the apparent dielectric constant is low and independent of lyso
content. The data are best interpreted in terms of an enhanced
access of water to the lipid system on addition of lysoplasmalogen.
This explanation is also successful in explaining the earlier
kinetics and fluorescence data.

CONCLUSIONS

The contention that disturbance of the Donnan equilibrium by
addition of lysoplasmalogen causes entry of water into lipid
vesicles is supported by observed increases in polarity or dielectric
constant in these mixtures.   The bursting of the vesicle to form
micelles is indicated by the discontinuities at 40-50% lyso.   The
decrease in bilayer thickness with increasing lyso content is
described by the increased permeability of nitromethane into the
bilayer.

These results may relate to the haemolytic process if it is
assumed that the results may be extrapolated to erythrocyte membranes.
The observations support the hypothesis that an important stage in
the lysis by lysolecithin is a change in cation permeability,
inducing swelling of the cell, and leading to a progressive osmotic

breakdown of the membrane. The specific contribution of this work
is evidence of increased polarity or dielectric constant of the
paraffin core of the bilayer in the presence of a lyso lipid. This
increased polarity may be caused by the .entry of water into the
bilayer.

REFERENCES

*The Radiation Laboratory of the University of Notre Dame is
operated under contract with the U.S. Energy Research and Develop _
ment Administration. This is ERDA Document No. COO-38-1050.

1.   D.A. Haydon and J. Taylor, J. Theoret. Biol. $\underline{4}$,
     281 (1963).

2.   J.G. Mandersloot et. al., Biochim. Biophys. Acta $\underline{382}$,
     22 (1975).

3.   J.A. Lucy, Nature $\underline{227}$, 8157 (1970).

4.   E. Rideal and F.H. Taylor, Proc. Roy. Soc. B. $\underline{148}$,
     450 (1958).

5.   F.C. Reman, Thesis, Utrecht, The Netherlands (1971).

6.   H.Hauser and M.D. Barratt, Biochim. Biophys. Res.
     Commun. $\underline{53}$, 399 (1973).

7.   M.C. Phillips, R.M. Williams and D. Chapman, Chem.
     Phys. Lipids $\underline{3}$, 234 (1969).

8.   (a) E. Smekal, H.P. Ting, L.G. Augenstein and H.T. Tien,
     Science $\underline{168}$, 1108 (1970).
     (b) G. Strauss, Photochem. Photobiol. $\underline{24}$, 141 (1976).

9.   S.C. Wallace and J.K. Thomas, Radiat. Res. $\underline{54}$, 49 (1973).

10. M. Gratzel and J.K. Thomas, J. Amer. Chem. Soc. $\underline{95}$, 6885
     (1973).

11. S. Cheng and J.K. Thomas, Radiat. Res. $\underline{60}$, 268 (1973).

12. K. Kalyanasundaram and J.K. Thomas, to be published.

13. K. Bredereck, Th. Forster and H.-G. Oesterlin, in
     "Luminescence of Organic and Inorganic Materials," H.P.
     Kallmann and G.M. Spruch, editors. Wiley-Interscience, New
     York, 1962.

14. (a) M.S. Lewis and M.H. Gottlieb, Fed. Proc. $\underline{30}$, 1303
     Abs.
     (b) L. Saunders, Biochim. Biophys. Acta $\underline{125}$, 70 (1966).

15. C. Tanford, "The Hydrophobic Effect: Formation of
     Micelles and Biological Membranes," Wiley-Interscience, New
     York, 1973.

16. P.P. Infelta, M. Gratzel and J.K.Thomas, J. Phys. Chem. $\underline{78}$,
     190 (1974).

CONCLUDING REMARKS

G. S. Hartley

57 Aurora Terrace, Hillcrest

Hamilton, New Zealand

This has been an extremely interesting and lively symposium and I am sure you would first wish me to express our great thanks to Dr. Kashmiri Mittal for the hard work and time he has put into organizing it. Most of us know how difficult it is to get contributions in and questions replied to on time but we still need a driving force. Thank you, Dr. Mittal.

As a winding-up speaker, I have a wonderful opportunity to be provocative without incurring riposte. I will try not to take advantage but confine myself to non-controversial points. It is useful, I think, to remind ourselves what a very small fraction of the whole diagram - water, amphiphile, hydrophobe - has been examined with precision and interpreted with clarity. These extreme corners are easy - the aqueous corner in which I spent some time, and the oil corner where Dr. Kertes finds no evidence for a critical concentration. He does well to remind us that the behavior dictated by water aggregation when this solvent is the continuum can be very different when it is absent or contained only in the discrete aggregates.

The most important part of the whole field for future development, particularly for its physiological importance, is the labyrinthine complex of separate phases, some anisotropic, occupying the large central area. Like the impressive and elegant SUNY campus in which we have been privileged to meet, the correlations are often obscure and the direction confusing. I found Dr. Scriven's description of interpenetrating continuous phases in sponge-type structures most stimulating.

It is obviously desirable to understand the simple before proceeding to the complex but I feel that I was very unadventurous in retreating into my simple corner as soon as the complexities appeared.

927

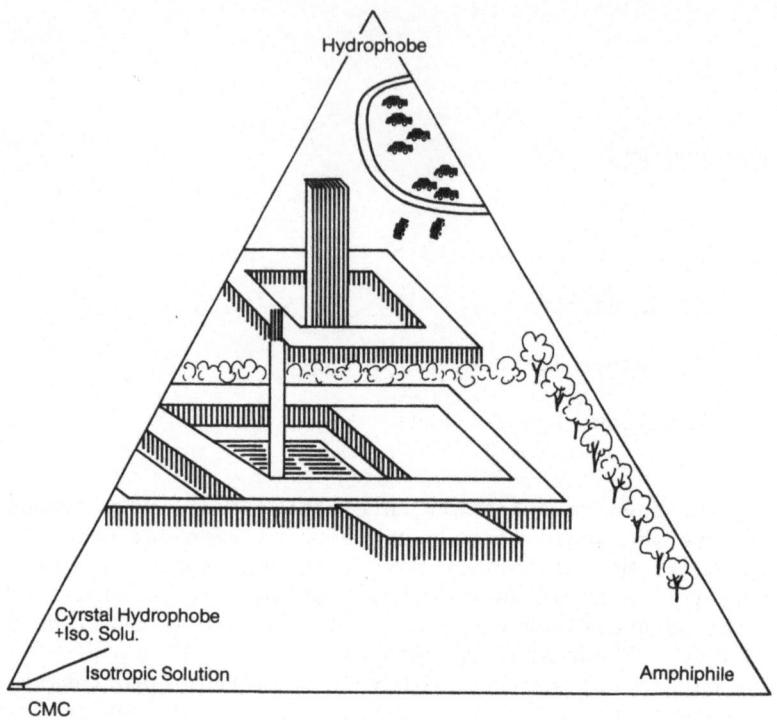

Diagram – Water, Amphiphile, Hydrophobe

I pay my respects to those who are actively investigating the spontaneous formation of laminae or rods which are, so to speak, equilibrium systems in one or two dimensions but not in all. I hope
the work in the corners has been of some help to them.

They have had the work in the corners to build on. They also
have sophisticated tools at their disposal which were not dreamed
of at the time of my own contribution - the use of photochemical
probes and relaxation methods, discussed in this symposium by Dr.
Turro and Dr. Aniansson. Light scattering technique has become
much more advanced but Dr. Mazer draws conclusions from its use
which Dr. Tanford finds at variance with those from a combination
of less sophisticated measurements.

Unquestionably, these new tools have great power, but I wonder
whether as much effort is put into the interpretation of the results
they provide as is put into the design of the tools themselves. I
raise this question with some hesitation because I know that an old
man can become bewildered with innovations and feel jealous of those
who can understand them, but I recall the errors of the first
interpretation of X-ray diffraction data in this field and could

illustrate my point with historical examples from other fields. I
would have liked to have heard more argument over the analysis of
just how certain and unique is the interpretation of some of the
new data.

The micelle state makes possible some interesting possibilities
in chemical reaction through promoting molecular contacts which would
otherwise occur relatively infrequently. This has importance both
in industry and in vital chemistry. We have had some illustrations
of these reactions. They are necessarily tied up with the complexities of structure and make the unravelling of these complexities
difficult, though it is all the more worthwhile. This symposium
has helped considerably the mutual understanding of the many interests in micellization, solubilization and micro-emulsions.

ABOUT THE CONTRIBUTORS

Here are included biodata of only those authors who have contributed to this volume. Biodata of contributors to Volume I are included in that volume.

W. G. M. Agterof is preparing his Ph.D. thesis at the Van't Hoff Laboratory for Physical and Colloid Chemistry, Utrecht. He graduated in 1971 from the State University of Utrecht.

Reuben Balagot is Department Chairman and Professor of Anesthesiology, Chicago Medical School. He obtained his M.D. degree from the University of the Philippines. His research interests include the application of physical chemical concepts to problems in the clinical sciences.

David Z. Becher is employed at the PPG Research Center in Pittsburgh. He received his Ph.D. degree in Chemistry from Lehigh University in 1976.

I. V. Berezin is the Chairman of the Chemical Enzymology Department at Lomonosov State University, Moscow. He is an Associated Member of the U.S.S.R. Academy of Sciences. He has co-authored (with Dr. Martinek) the book "Physicochemical Fundamentals of Enzyme Catalysis" published in Moscow, 1975.

M. S. Bidner was a Visiting Professor at the University of Minnesota from 1974 to early 1976 from La Plata, Argentina. Her current research interests are interfacial phenomena, flow in porous media, and applications to enhanced petroleum recovery.

Irena Buraczewska is at the Swedish Institute for Surface Chemistry, Stockholm. She received her M. Sci. in Chemical Engineering from the Technical University in SZCZECIN in 1973. She has several publications in the area of nonionic emulsifiers and their solubilization.

A. A. Calje graduated in chemistry at the State University of Utrecht in 1974. From September 1975 he worked for one year on w/o microemulsions at the Van't Hoff Laboratory for Physical and Colloid Chemistry.

Chien-Chung Chao is employed with the National Carbide and Carbon at their Tarrytown Research Center. He received his Ph.D. degree from the University of Minnesota.

Bernard Ecanow is Professor of Pharmaceutical Sciences, University of Illinois. He received his Ph.D. degree from the University of Minnesota. His research interests include application of physical chemical concepts to biological and health science problems.

Janos H. Fendler is Professor of Chemistry, Texas A&M University, College Station. He obtained his Ph.D. degree in 1964 from the University of London, England. He has published more than 90 papers in the areas of micellar catalysis, physical organic, bioorganic and radiation chemistry. He is the co-author (with A. M. Fendler) of the book "Micellar Catalysis" published in 1975.

Frederick M. Fowkes is Professor and Chairman, Department of Chemistry, Lehigh University. He received his Ph.D. degree from the University of Chicago. He has had 26 years of industrial experience before joining Lehigh University in 1968.

Arthur J. Frank has been a Visiting Scientist for the past year and a half at the Hahn-Meitner Institute in Berlin. He obtained his Ph.D. degree in 1975 from the University of Florida. His current research concerns pulse radiolysis, flash photolysis, and heterogeneous electron-transfer reactions in solution.

Thomas C. Franklin is Professor of Chemistry, Baylor University, Waco, Texas. He received his Ph.D. from Ohio State University in 1951. His research interests include electrodeposition and electro-organic synthesis. He is a member of the Editorial Board of Surface Technology.

Elias Franses is a doctoral candidate in Chemical Engineering, University of Minnesota, working on the properties of micellar systems pertaining to enhanced oil recovery.

Stig Friberg is Chairman, Department of Chemistry, University of Missouri-Rolla. He was the Director of the Swedish Institute for Surface Chemistry from 1969-1976. He obtained his Ph.D. degree in 1966 from Stockholm University. He has published two books and more than 100 publications on surfactants and their association structure. He is a Fellow of the Royal Swedish Academy for Engineering Sciences.

E. D. Goddard is Research Associate with the Union Carbide in Tarrytown, New York, where he leads a skill center concerned with various aspects of surface chemical research. He obtained his Ph.D. in Physical Chemistry from Cambridge University, England, in 1951.

Bernard H. Gold is Professor in the Graduate College and the Departments of Pharmacy Practice and Psychiatry, University of Illinois at the Medical Center, Chicago. He obtained his Ph.D. degree in 1953 from the University of Chicago. His research interest is in the area of physical chemistry of stress related disease.

Michael Grätzel is Professor of Physical Chemistry, ETH Lausanne, Switzerland. He carried out his Ph.D. thesis research (1968-1971) at the Hahn-Meitner Institute. His research interests include laser and pulse radiolysis investigations with micellar systems. He has about 50 publications to his credit.

R. B. Hannan is employed as a Chemist by Union Carbide Corporation, Tarrytown, New York.

Shelley A. Harkaway is working in the Eastman Kodak Research Laboratories, Rochester, New York.

Tadatoshi Honda obtained an M.S. in Chemistry from Baylor University while on leave from Mitsui Toatsu Chemicals, Japan, from 1973 to 1975.

Charles E. Jones is a graduate student working toward his Ph.D. degree in the Chemistry Department at Drexel University, Philadelphia.

K. Kalyanasundaram is a graduate student working toward his Ph.D. degree at the University of Notre Dame.'s Radiation Laboratory. His graduate research concerns physical studies on the dynamical structures of micellar assemblies.

Oh-Kil Kim is a Research Chemist in the Organic Chemistry Branch, Chemistry Division, Naval Research Laboratory. He received his Ph.D. degree in 1966 from the University of Tokyo. He has published about 20 papers in the area of polymer chemistry and related areas.

Ayao Kitahara is Professor at the Science University of Tokyo. He obtained his Dr. Sc. degree in 1959 from the University of Tokyo. He has many publications in the area of solution behavior of surfactants.

Kijiro Kon-no received his Dr. Sc. degree in 1976 from the Science University of Tokyo.

Kathy Letts is presently teaching at Cecilian Academy, Mt. Airy, Pennsylvania. She joined the order of the Sisters of St. Joseph in 1975. She received her M.S. in Chemistry from Drexel University, 1974.

A. V. Levashov is associated with the Lomonosov State University, Moscow.

R. Saul Levinson is Assistant Professor, Pharmaceutical Sciences, University of Oklahoma. He received his Ph.D. from the University of Illinois. His research interests include the colloid sciences and their application in the health sciences.

Raymond A. Mackay is a faculty member in the Chemistry Department at Drexel University. He received his Ph.D. degree in Chemistry in 1966 from the State University of New York at Stony Brook. His current research interests include reactions in microemulsions.

Michael Marmo is employed by Eastman Kodak Company in Rochester, New York. He received his M.S. in Chemical Engineering in 1975 from Lehigh University.

Karel Martinek is head of the team at Lomonosov State University studying catalytic mechanisms of enzyme action and other related systems. He is author of more than 100 scientific publications and co-author (with I. V. Berezin) of the book "Physico-Chemical Fundamentals of Enzyme Catalysis" published in Moscow in 1975.

Kashmiri Lal Mittal* is presently at the IBM Corporation, Poughkeepsie, New York. He received his Ph.D. degree in Physical Chemistry (Surface and Colloid) from the University of Southern California in 1970. He has organized and chaired a number of international symposia. In addition to these volumes, he is also the editor of "Adsorption at Interfaces" and "Colloidal Dispersions and Micellar Behavior" (both published by the ACS), and of "Adhesion Measurement of Thin Films, Thick Films, and Bulk Coatings" (to be published by the ASTM). His research interests include: Colloid science, adhesion science and technology, corrosion and passivation of metals, and surface properties of polymers. He has published about 35 scientific or technical papers, and has given many invited talks on the various facets of adhesion on the invitation of various societies or organizations.

David A. N. Morris is a graduate student (on a leave of absence from Miles Laboratories) working toward his Ph.D. degree at the University of Notre Dame. He is investigating physical properties of lipid dispersions via various techniques.

---

*As the editor of this volume. His contribution as an author is included in Volume 1.

Robert A. Moss is Professor of Chemistry at Rutgers University. He received his Ph.D. in 1963 from the University of Chicago. His research concerns reactive organic intermediates, and bio-organic chemistry, especially the development of micellar catalysts as analogs of enzymic catalysts. He is co-editor of the Wiley Series "Carbenes".

Robert C. Nahas is a graduate Research Assistant at Rutgers University. At Rutgers, he has done graduate research in micellar organic chemistry.

Larry K. Patterson is Associate Professional Specialist, Radiation Laboratories, University of Notre Dame. He received his Ph.D. degree in 1967 from Kansas State University. His research interests include free radical processes in micellar and model membrane systems, micellar effects on photophysical and photochemical processes.

Suryakumari Ramaswami is a Postdoctoral Fellow at Rutgers University. She received her Ph.D. degree in organic chemistry in 1963 from Poona University in India. Her present research concerns micellar organic chemistry.

Jean-Claude Ravey is "Charge de recherches" in the Centre National de la Recherche Scientifique in the Laboratory of Biophysics in Nancy. He received his thesis d'Etat (physical science) from the University of Nancy (France) in 1973. He has about twenty publications.

J. L. Redpath is a Radiobiologist in the Department of Medical Physics, Michael Reese Hospital, Chicago. He received his Ph.D. degree in 1968 from the University of Newcastle-upon-Tyne, England. His research interests are in radiobiology and radiation chemistry.

Richard L. Reeves is a Research Associate, Eastman Kodak Research Laboratories, Rochester, New York. He obtained his Ph.D. degree in 1955 from the University of Wisconsin. His research interests are dye-dye, dye-polymer, and dye-surfactant interactions.

Max L. Robbins is Research Associate, Corporate Research Laboratories, Exxon Research and Engineering Company, Linden, New Jersey. He received his Ph.D. in 1961 from Columbia University. His research interests include inorganic colloids in hydrocarbons, emulsions, monolayers, and microemulsions and micellar solutions for enhanced oil recovery.

Lawrence S. Romsted is a Postdoctoral Research Associate, University of California, Santa Barbara. He did his graduate work at Indiana University. His research interests include using

kinetics as a probe of micelle structure, introducing selectivity
into reactions with micelles, and oscillating chemical reactions.

Eli Ruckenstein is Faculty Professor of Engineering and Applied
Sciences, State University of New York, Buffalo. He was Professor
at the Polytechnic Institute, Bucharest, Romania, 1949-1969. He
received the Gheorghe Spacu Award of the Romanian Academy of Sciences
(1963) for research in surface phenomena, and two awards for research of the Romanian Department of Education (1958, 1964). He
has published extensively dealing with many subjects. His research
interests are catalysis, colloids and interfaces, and biophysics.

L. E. Scriven is Professor and Associate Head, Department of
Chemical Engineering and Materials Science, University of Minnesota. He received his Ph.D. in 1956 from the University of Dela-
ware. Dr. Scriven has served in editorial capacities for the J.
Colloid Interface Sci., and J. Fluid Mechanics. His researches
have included thermodynamics of low tension and highly curved
interfaces, fluid mechanics of interfaces, meniscus configurations,
and the theory of imperfections in surface crystals.

Cesare Silebi is a graduate student working toward his Ph.D.
degree at Lehigh University.

Ulrich P. Strauss is Professor and Chairman of the Chemistry
Department, Rutgers University. He received his Ph.D. degree in
1944 from Cornell University. Dr. Strauss has many publications
in synthetic and biological polyelectrolytes, with particular
emphasis on the role of hydrophobic groups and interaction with
metal ions.

John Kerry Thomas is Professor of Chemistry, University of
Notre Dame. He obtained his Ph.D. degree in 1957 from the University of Manchester, England. While working at the Argonne National
Laboratory (1960-1970) he developed the short-pulsed nanosecond
laser and pulse radiolysis technique. He is the author of over one
hundred research publications. In 1974 he was given the research
award of the Radiation Research Society. Dr. Thomas was a Gäst
Professor at the Hahn Meitner Institute in Berlin in 1975.

A. Vrij became Professor of Physical Chemistry in 1968 at the
University of Utrecht. He obtained his Ph.D. degree in Physical
Chemistry at the University of Utrecht in 1959. In 1972/73 he was
a Visiting Professor in the Department of Chemical Engineering at
M.I.T. His research interests are in the field of colloid and sur-
face chemistry.

Maureen Mulien Wong joined the University of Notre Dame in 1972 as a graduate student to work toward her Ph.D. degree. Her graduate research is in the field of kinetics in micelles and membranes.

A. K. Yatsimirski is associated with the Lomonosov State University, Moscow, U.S.S.R.

SUBJECT INDEX

Pages 1-488 appear in Volume 1
Pages 489-930 appear in Volume 2